The International Space Station

Building for the Future

John E. Catchpole

The International Space Station

Building for the Future

Published in association with
Praxis Publishing
Chichester, UK

John E. Catchpole
Gosport
Hampshire
UK

SPRINGER–PRAXIS BOOKS IN SPACE EXPLORATION
SUBJECT *ADVISORY EDITOR*: John Mason B.Sc., M.Sc., Ph.D.

ISBN 978-0-387-78144-0 Springer Berlin Heidelberg New York

Springer is part of Springer-Science + Business Media (springer.com)

Library of Congress Control Number: 2008923104

Printed in Germany

Cover design: Jim Wilkie
Project management: Originator Publishing Services, Gt Yarmouth, Norfolk, UK

Printed on acid-free paper

Contents

APPENDICES

Preface

INTRODUCTION TO VOLUME 2

Creating the International Space Station, written by David Harland and John Catchpole, was published by Springer-Praxis in 2002 (ISBN 1-85233-202-6). It described the American and Soviet/Russian national space station programmes, as well as the long, convoluted history of the International Space Station (ISS), from its conception through to the safe recovery of the Expedition-3 crew in December 2001.

The International Space Station: Building for the Future continues the coverage of the construction and occupation of ISS, but first there is a brief résumé of the hardware that is already in orbit. These early flights and the politics of constructing ISS during this period are covered in full in the original volume, which ended with the delivery of the Expedition-4 crew to ISS onboard the STS-108 Shuttle flight.

The original flight coverage in this new volume returns to the launch of STS-108, and the beginning of the Expedition-4 crew's occupation of ISS. It ends with a review of how the modules developed by the European and Japanese partners will be added to the station, enhancing its research capability, and, finally, there is a brief look at the early designs for the Orion spacecraft and its Ares-1 launch vehicle. Plans for Project Constellation to carry humans back to the Moon and on to Mars are not covered as they have no bearing on the ISS programme as presently defined.

Appendices include a Flight Log and an Extravehicular Activity (EVA) Log for the period covered in this volume. Both of these logs continue from those included in the original volume. There is also a List of abbreviations and descriptions of the major ISS hardware.

As this volume begins, all was well with the station, with the exception of the Russian budget. Although many scientists were sceptical about the quantity and quality of science being performed on the station, at least science was being performed daily. Russian experience on their Salyut and Mir space stations had suggested that on average 2.5 crew members were required simply to keep up with the

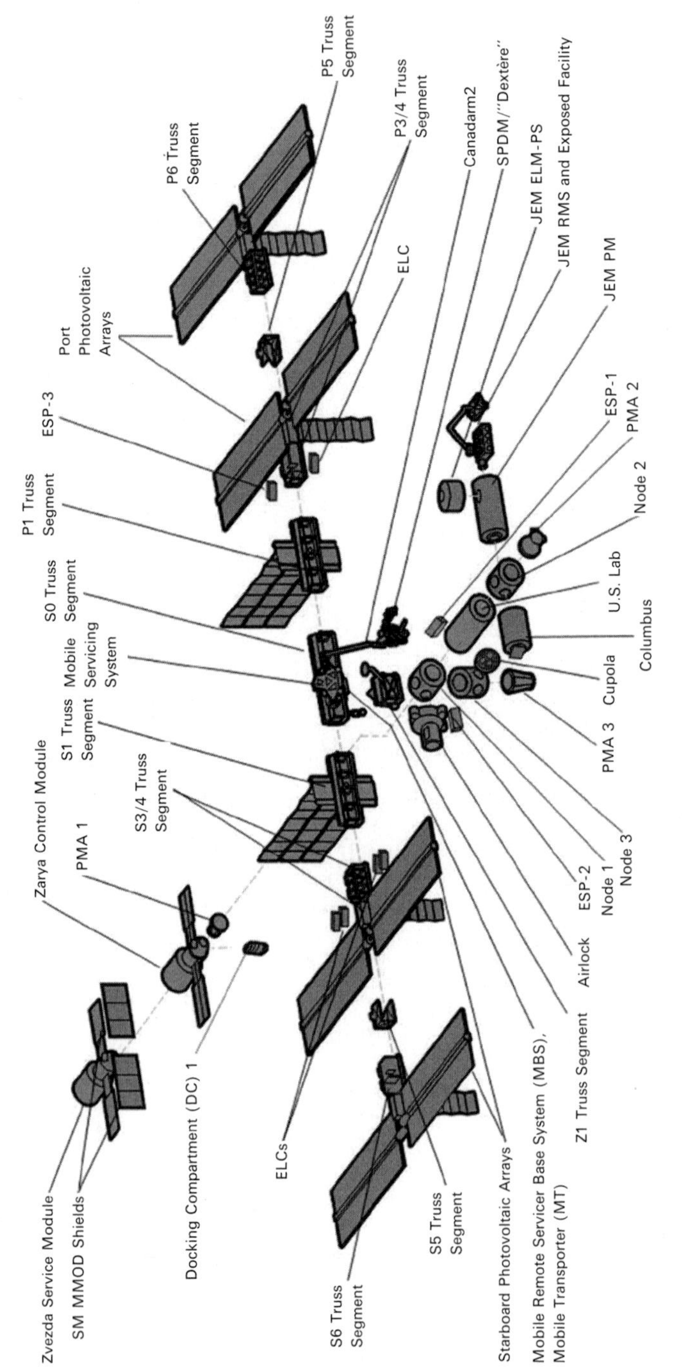

Figure I.1. Exploded view of the International Space Station.

ever-present requirement to maintain the station's systems, while the third crew member spent part of their time performing scientific experiments. While Russia struggled even to fund the contracted number of Soyuz and Progress vehicles, America prepared to move forward with the construction of the Integrated Truss Structure (ITS), the huge cross beam that would house the station's eight Solar Array Wings (SAWs), associated storage batteries, and cooling radiators. Construction of the ITS would allow the station's primary power and cooling systems to be configured and take over from the temporary systems put in place when the Port-6 ITS was temporarily located on the Z-1 truss. The ITS also had to be at least partially constructed, with its power and cooling systems functional, before Node-2 could be launched and docked in place to serve as a mount for the European and Japanese modules. Following the cancellation of the American Habitation Module, the European Space Agency had been paid to provide additional living quarters inside Node-3, which would now be the final pressurised module delivered to ISS, some time in 2008. This would allow the Expedition crew to be expanded to six people.

It was not to be.

The tragic loss of STS-107, Columbia, on February 1, 2003, grounded the Space Shuttle fleet and threw the ISS construction schedule in the rubbish bin. Plans to have the station fully constructed by 2008, and maximise its potential through permanent occupation until 2016 were no longer realistic. The period that followed the loss of Columbia stretched the goodwill of the partners involved in the ISS programme to the full.

Ever short of money, the Russians claimed that they could not afford to produce the extra Soyuz and Progress spacecraft required to keep ISS occupied. They suggested that it be abandoned until Shuttle flights were resumed. NASA stated that they would only abandon the station as a last resort. Russia was therefore forced to find the additional funding, and permanent occupation continued, with two-man "caretaker crews" flying to and from ISS in Russian Soyuz spacecraft and being resupplied by Russian Progress cargo ships. Despite major differences of opinion on Earth over the American-led invasion of Iraq and its aftermath, the Russians continued to work amicably with the Americans on ISS.

When the Columbia Accident Investigation was over, and the Shuttle was preparing to Return to Flight, the Russians felt that they had paid the debt incurred when their first module, Zvezda, was only fitted out with American financial assistance and then launched two years late. They insisted on the return of the ISS experiment time that the Americans had negotiated away from them at that time. The relationship between the two major partners had changed significantly.

Dedicated to the memory of the crew of Soyuz-11

Georgi Dobrovolsky
Vladislav Volkov
Viktor Patsayev

They were called forward at short notice to occupy the world's first space station, Salyut-1, and perished during their return to Earth

Figures

"I think, historically ... when we look back fifty years to this time, we won't remember the experiments that were performed, we won't remember the assembly that was done, we may barely remember any individuals. What we will know was that countries came together to do the first joint international project, and we will know that that was the seed that started us off to the Moon and Mars. Because then, I know, when we're looking back from Mars, for example, it won't be just the United States, or it won't just be China or Russia: it will be an international mission. And it will have come out of the very fact that we're doing the International Space Station today."

Michael Foale, NASA astronaut

Resources

All of the information included in this manuscript has come from official sources within the space agencies of the countries involved in the International Space Station programme, unless otherwise stated in the manuscript.

LIST OF ISS NATIONAL SPACE AGENCY WEBSITES

National Aeronautics and Space Administration	*www.nasa.gov*
Russian Federation	*www.roscosmos.ru*
European Space Agency	*www.esa.int*
Japan	*www.jaxa.jp/index_e.html*
Canada	*www.space.gc.ca/asc/eng/default.asp*
Brazil	*www.aeb.gov.br*

PHOTOGRAPHS

All of the photographs used in this manuscript are from the NASA Human Spaceflight Gallery on the NASA website quoted above. In illustrating the manuscript, I have attempted to include an image of each individual who has visited the International Space Station in the period covered in this volume. Due to the large number of people involved, Shuttle crews are represented by their official crew portrait. Shuttle mission in-flight images generally show external views. Expedition crews and Soyuz "taxi" crews are generally shown in images of each individual at work inside the station. Occasionally, group photographs are used as these show an individual, usually a commercial spaceflight participant, who is not available in an individual view.

ACKNOWLEDGEMENTS

Many people have helped me to find the information in this manuscript. The Public Relations personnel at the space agencies listed above have, as always, been tireless in their assistance. David Harland, my co-aurthor on the first volume in this series, has had an editorial input, as has Bruce Shuttlewood of Originator. At Praxis Publishing, Clive Horwood and all of the Praxis staff have given their usual unending assistance and support. I want to thank them all for their assistance. Finally, I have to thank my wife, Sue, for her patient understanding and support while I was chained to the computer, writing.

John Catchpole

Conversion table

Multiply	*By*	*To obtain*
	Distance	
Inches	25.4	Millimetres
Inches	2.54	Centimetres
Feet	304.8	Millimetres
Feet	0.3048	Metres
Yards	0.9144	Metres
Metres	3.281	Feet
Kilometres	3,281	Feet
Kilometres	0.6214	Statute miles
Statute miles	1.6093	Kilometres
Nautical miles	1.852	Kilometres
Nautical miles	1.1508	Statute miles
Statute miles	0.8689	Nautical miles
Statute miles	1,760	Yards
	Liquid	
Gallons	3.785	Litres
Litres	0.2642	Gallons
	Weight	
Ounces	28.35	Grams
Pounds	0.4536	Kilograms
Kilograms	2.205	Pounds
Metric tonne	1,000	Kilograms
Short ton	907.7	Metric tonne

(*continued*)

Multiply	*By*	*To obtain*
	Pressure	
Pounds/square inch	6.895	Kilonewtons/square metre
Pounds/square inch	6.895	Kilopascals
Pounds/square inch	70.31	Grams/square centimetre
Pounds/square inch	51.75	Millimetres of mercury
Millimetres of mercury	133.32	Newtons/square metre (pascals)
	Thrust/Force	
Ounces	0.278	Newtons
Pounds	4.448	Newtons
Newtons	0.225	Pounds
Kilograms	9.807	Newtons
	Cubic measure	
Cubic inches	16.387	Cubic centimetres
Cubic feet	28,317.0	Cubic centimetres
Cubic feet	0.02832	Cubic metres
Cubic yards	0.7646	Cubic metres
Cubic centimetres	0.06102	Cubic inches
Cubic metres	610,233,770.0	Cubic inches
Cubic metres	35.314	Cubic feet
	Liquids	
Fluid ounces	29.57	Millilitres
Fluid ounces	0.0296	Litres
Fluid quarts	0.9464	Litres
Gallons	3.7854	Litres
	Flow rate	
Cubic feet/minute	0.283	Cubic metres/second
Gallons/minute	3.7854	Litres/minute
Pounds mass/hour	0.4536	Kilograms/hour
Pounds mass/minute	0.4536	Kilograms/minute
Pounds mass/second	0.4536	Kilograms/second
Pounds/cubic foot	16.02	Kilograms/cubic meter
	Square measure	
Square inches	6.452	Square centimetres
Square feet	929.03	Square centimetres
Square feet	0.0929	Square metres
Square yards	0.8361	Square metres
Square miles	2.59	Square kilometres
Square centimetres	0.15499	Square inches
Square metres	1,549.9	Square inches
Square metres	10.763	Square feet
Square kilometres	0.386	Square miles

Multiply	*By*	*To obtain*
	Velocity	
Feet/second	0.3048	Metres/second
Metres/second	3.281	Feet/second
Metres/second	2.237	Statute m.p.h.
Feet/second	0.6818	Statute m.p.h.
Feet/second	0.5925	Nautical m.p.h.
Statute miles/hour	0.447	Metres/second
Statute miles/hour	1.609	Kilometres/hour
Nautical miles/hour (knots)	1.852	Kilometres/hour
Kilometres/hour	0.6124	Statute m.p.h.
	Temperature	
Degrees Fahrenheit	Minus 32/Divide by 1.8	Degrees Celsius/Centigrade
Degrees Celsius/Centigrade	+273	Degrees Kelvin

1

Early construction

WHAT IS THE INTERNATIONAL SPACE STATION?

Today, the National Aeronautics and Space Administration (NASA) is proud to boast that the American-led International Space Station (ISS) is the result of years of co-operation between 16 countries. It was not always so, the American-led effort to build a space station with its political allies in Europe and Japan began as an attempt to construct a station that would be better than the Soviet-launched Mir space station. President Reagan called that effort Space Station Freedom, claiming that political freedom was something that America, Europe, and Japan shared while the Soviet Union suffered under a Communist dictatorship. Space Station Freedom represented the dying throw of America's effort in the Cold War-driven Space Race that had begun with Sputnik and ended with Mir. Whilst America had stunned the world by landing six Apollo flights on the Moon, that effort was not sustained. The Space Shuttle was an iconic flying machine, but until Space Station Freedom was authorised it had attempted to be all things to all people and its design had suffered as a result.

The Space Station Freedom effort was driven by the desire to prove American technological superiority over the Soviet Union. With the long-duration Mir (a modular space station that was built up gradually with extra modules being added to it in orbit and permanent occupation being supported by a regular supply of fresh Soyuz spacecraft and Progress cargo vehicles) as the benchmark, America strove to design a station that would hold a six-person crew (twice as many as Mir's three cosmonauts) and produce top-quality science. To that end, America, Europe, and Japan would each construct a scientific laboratory to be added to the station. Over the years the Space Station Freedom design grew beyond what was realistically sustainable, with pressurised modules docked to a huge girder-like structure and even a large repair shop where malfunctioning satellites would be recovered and repaired on-orbit, before being released to continue their missions. Individual elements of the station would be delivered by the Shuttle. Science would be automated when no Shuttle was present, but tended experiments would be performed by visiting Shuttle crews. The station would not be permanently occupied until the six-person Habitation Module, the last element, was installed and a six-person American Crew Return Vehicle had been developed to allow the crew to evacuate the station in the event of an emergency when no Shuttle was present.

Billions of dollars were spent on the design process before any metal was cut to begin construction of the station's modules. Budget restraints finally bit hard, and Space Station Freedom was continually redesigned in an attempt to bring spending under control. One such redesign saw the size of the pressurised modules halved in length to save money. In another, the satellite repair shop was discarded. The huge, square truss was replaced by a much smaller version. European plans for an Ariane-V launched free-flying laboratory, called Columbus, that could be docked to Space Station Freedom when required, and serviced by a European mini-Shuttle called Hermes were shelved due to budget restrictions. A smaller Columbus would be launched on the American Shuttle in return for services rendered, and would remain permanently attached to the station. Hermes would not be built. Likewise, the Japanese plans to develop the H-II Orbital Plane Experimental (HOPE) mini-Shuttle came to nothing.

Space Station Freedom was redesigned almost yearly in an attempt to bring the spending under control, while all the time eating further into its initially allocated budget. On one occasion, a Congressional vote to continue building the space station, rather than scrap it completely, was passed by a single vote. Meanwhile, Mir taunted the Americans from on high, as the station that was designed to operate for just 5 years was permanently occupied for the majority of its 15-year operational life.

When Soviet Communism collapsed in Eastern Europe it was America's President Clinton who insisted that the Russians be invited to join the space station effort, in order to prevent their missile experts selling their knowledge to states that were unfriendly towards America. With the Russians now included in the team, the station's name changed to the International Space Station, a name that had been used in some quarters from the time when the Europeans and Japanese had first joined the effort.

In the meantime, American astronauts began training in Russia before performing long-duration flights on Mir with Russian cosmonauts. The first NASA astronaut was delivered to Mir on a Soyuz and recovered by the Shuttle. The remaining Americans were delivered to Mir and returned to Earth on the Shuttle. The experience of the American astronauts on Mir was not a happy one, with several suffering symptoms of isolation and depression. Many of their post-flight writings show a typical Cold War American attitude towards their old enemies. Nothing the Russians possessed or did was good enough for certain members of the American Shuttle–Mir team! Two outstanding exceptions to this were British-born Michael Foale and female astronaut Shannon Lucid.

A new design was established for ISS, with a number of Russian modules in the critical line. A TKS-style module would provide attitude control and electrical power during the earliest phase of the construction. A second module, similar to the Mir base block, would provide early sleeping quarters and an operations centre. Employing the Soyuz/Progress routine, proven on the late Salyut stations and Mir, would allow permanent occupation and science to be undertaken from the earliest stages of construction, rather than only towards the end, as envisaged for Freedom. A Docking Module/airlock would provide support for the Russian Orlan-M EVA suit.

Alongside the Russian modules the Americans would install Node-1, a berthing station for later modules, and an electrical attitude control system, consisting of four gyrodynes in the Z-1 truss, mounted on the zenith of Node-1. The Port-6 (P-6) truss element, a single piece of the Integrated Truss Structure (ITS), would be temporarily mounted on the Z-1 truss and its Solar Array Wings (SAWs) deployed to provide additional electrical power while the American sector of the station was built up. A temporary ammonia cooling system was also established using the radiator on the P-6 ITS. The American science laboratory would become the centre of the American sector of the station, taking over many of the control operations previously performed by the Russian modules, and the American Joint Airlock would provide support for both the Russian Orlan-M EVA suits and the American Extravehicular Mobility Unit (EMU). The massive pieces of the ITS would then be delivered and constructed in orbit, with their SAWs and cooling radiators being deployed as the individual ITS elements were added to the station. As the ITS progressed, the station's electrical and cooling system would be reconfigured into its permanent configuration. The P-6 ITS would be relocated to its final position, as the final element on the port side of the ITS. Then Node-2 would be delivered to serve as the mounting for the European and Japanese science laboratories, which have still to be delivered to the station as this manuscript ends. Finally, Node-3 would bring additional sleeping quarters, allowing the station's crew to finally be raised to six people.

October 4, 2007 was the 50th anniversary of the launch of Sputnik, the first artificial satellite. It has required every second of those 50 years of experience to allow the development, construction, and occupation of ISS to take place. In those 50 years both America and Russia have developed their own engineering methods and infrastructure that has made their spaceflight programmes possible on the grand scale that they have followed. As a result, two extremely proud cultures clashed when America invited Russia to join the ISS programme. The Russians felt that they had begun the Space Age and had achieved many of the major "firsts" in the early years. Having lost the race to the Moon they had concentrated on developing space stations and the infrastructure required to support them. They initially feared that the Americans were intending to strip them of their knowledge relating to long-duration spaceflight and then discard them. All training of American personnel in Russia took place in the Russian language, requiring the Americans to learn that language before they could advance to detailed spaceflight training. All training manuals were produced in the Russian language, with no English translations. Despite repeated requests from their new colleagues, the Russians refused to produce English translations of vital documents unless the Americans paid heavily inflated prices for them, in US dollars. Trust between the old enemies came very slowly, usually only after personal friendships had been developed between individuals performing the same role on both teams. Facilities in Russia were often decrepit when compared with similar facilities in America. In many cases, Russian training facilities were of a lower fidelity than their American equivalents. Even the living quarters were of such poor quality, when compared with homes in America, that NASA shipped in a number of houses in kit form for their personnel to live in. The houses were assembled by a Russian workforce. In the basement of his new home, future Expedition-1 Commander William Shepherd established "Shep's Bar" and gymnasium, which became a focal point where American astronauts and Russian cosmonauts could relax in each other's company after the working day finished. Shep's Bar and gymnasium remained in place and continued to function many years after Shepherd had completed his occupation of ISS. Several American astronauts have mentioned Shep's Bar in their memoirs and have credited it with being one of the most important elements responsible for breaking down the barriers that originally stood between the Americans and Russians.

When the first female American astronauts began training in Russia they were disrespected by their hosts, many Russians joked that Shannon Lucid was flying to Mir, because it was dirty and required cleaning. The Russians had even stated publicly that when Russian women return to space it will be "in the strong arms of their male colleagues." It was an attitude that did not go down well with the American women. Over the years the Russian attitude has changed, as the American women, many of them military officers or doctors of various scientific disciplines, proved beyond all doubt that they had the educational standards, technical know-how, and physical capabilities required to meet all of the demands of the Shuttle–Mir and ISS programmes. Today a professional respect exists to the point that, while one female Russian cosmonaut has been removed from two Soyuz crews in order to fly commercial spaceflight participants (SFPs), male Russian cosmonauts have served under American female Shuttle commanders and the first female ISS commander.

Regretfully, given the dire state of the budget for the Russian space programme, this proud organisation now finds itself reduced to little more than the role of delivery driver, with its Soyuz spacecraft delivering crews to ISS and returning them to Earth, and their Progress vehicles regularly re-supplying consumables on the station. Even then, the transfer to a capitalist economy in Russia and the reluctance of the Russian government to fund the space programme has led to regular disruptions in the large Space Station module, Soyuz, and Progress production lines as contractors withhold vital spacecraft equipment until they are paid for in full. Time and again, NASA has been required to provide additional money to Moscow in order to help the Russians meet their commitment to ISS. This has been necessary primarily because all of the Russian modules were in the critical line of ISS development and occupation. Without the first Russian module there was no early electrical power, no attitude control, and nothing to add the American modules to. Without the second large module there was no onboard oxygen production system, crew living quarters, or orbital re-boost capability. Without Soyuz, the crew had no emergency escape system and therefore could not permanently occupy the station. Progress was vital in re-supplying the propellant, oxygen, and water supplies in the Russian modules as well as providing fresh food, dry goods, and vital personal items and letters from the crews' families. In short, without the Russians there was no ISS in its present form.

America trailed behind the Russians in the early years of the space programme, only overtaking early Soviet achievements in late 1965. Project Apollo landed 12 men on the Moon and then led to the Skylab prototype space station. Skylab was a mixed blessing. It gave America its first experience of long-duration spaceflight, and the endless onboard maintenance required to occupy a space station. It also led to the first strike by astronauts in space, when the Skylab-4 crew rebelled against the workload placed upon them in the early part of their flight.

The Shuttle was designed to be America's only launch vehicle, planned to replace all one-shot launch vehicles, as demanded by President Richard Nixon. Its final shape and size were dictated by the large reconnaissance satellites that it had to be capable of launching. As a result, the Russian Salyut stations and Mir's individual modules would fit in the Shuttle's payload bay with room to spare. From the beginning, NASA wanted the Shuttle to be used to launch the elements of an American space station, but for many years no funding was forthcoming for that station.

The overall success of America's space programme meant that NASA's name has come to represent excellence in western culture. This attitude only began to fade as several Shuttle-launched satellites failed to achieve the correct orbit and the Hubble Space Telescope was launched with ineffective optics. A number of rescue and repair flights went partway to recover NASA's reputation and the loss of several probes to Mars owing to efforts to fly "cheap" missions has been balanced by the more recent success of several robotic rovers used to explore the Martian surface. As a result, it was a confident, if not over-confident, organisation that clashed with Russian pride when budget restrictions dictated that the two superpowers pool their spaceflight experience.

The clash of cultures was perhaps felt most by the American astronauts, many of whom, on being removed from their families and relocated to Russia felt isolated. American astronauts are pilots, engineers, and scientists. They are encouraged to get involved in the development of the spacecraft they will fly and the experiments they will perform. Their input is often responsible for the redesign of equipment or procedures. In training they work closely with NASA's Flight Operations Directorate, which writes their Flight Plan, runs their training regime, and operates Mission Control. In the event of a malfunction or other problem in flight, the astronauts' input is as important as that of the engineers on the ground. In Russia they found that the cosmonauts had no input into the design of their vehicles, flying whatever hardware was delivered to the launch pad by the contractor. In Russia, mission control dictates what happens on a spaceflight, to the point that Americans working in Russia during the Shuttle–Mir programme found that Russian cosmonauts were not offended by a poster that showed mission control operating the Mir space station, which hung from a set of puppet's strings. The Americans, astronauts and controllers alike, found that the poster offended their professional respect for each other. While American astronauts received a standard Civil Service salary, based on their military rank, or civilian Civil Service grade, irrespective of whether or not they were in flight, Russian cosmonauts received a basic wage, much less than the Americans, and a basic sum when they were launched into space. They were then able to earn bonus payments if they completed individual in-flight goals, such as EVAs, or a manual docking of the Soyuz spacecraft in the event of the automated system failing. It was also obvious to the Americans that, if anything went wrong on a Russian spaceflight, the blame was usually laid at the feet of the cosmonauts, rather than in mission control, or at the door of the contractor who built the item concerned.

Another major difference that was obvious to the American astronauts was Russia's reliance on automated systems in their spacecraft. With the exception of the Soyuz spacecraft, all Russian ISS modules were launched into orbit and thereafter used their own guidance and attitude control to complete their automated flight regime. Even the primary systems on the Soyuz were automated, with the manual override intended for use only if the automation failed. In America, the astronaut was part of the control loop in the first several generations of spacecraft. All major in-flight manoeuvres, including rendezvous and docking, were performed manually. The Apollo lunar landings were performed with a pilot at the controls, as were all Shuttle landings, although the latter make maximum use of standard airport automated landing systems. No American spacecraft after Mercury would have been capable of performing its mission without the onboard computer(s), but even then computer software was more often than not changed and even updated in flight by the human crew. The Russian preference for automation was equalled on ISS by America's preference for robotics and human input. Where the Russians would have docked the major ISS elements together using automated systems, the Americans preferred to launch inert ISS modular elements on the Shuttle and then use the Canadian built RMS, operated from within the Shuttle's aft flight deck, or the SSRMS, operated from inside the American Laboratory Module, and human astronauts operating outside the station on EVAs to attach the new elements to the station.

While the individuals in Russia and America's space programmes may have learnt how to work together, pooling their individual talents and national expertise, there will always be political manoeuvring behind the scenes, and politicians in both countries, eager to drive their own agendas, who will blame the other country for all of the ISS programme's ills. No doubt the accusations will continue in both directions as long as Russia and America continue to co-operate on space exploration.

Creating the International Space Station, the first book in this series, described the convoluted history of ISS, along with the Shuttle–Mir programme (ISS Phase I), and the on-orbit construction of ISS Phase II and ISS Phase III in detail and in the correct chronological order. The account that follows is a very brief description of the flights covered in that volume. While these descriptions are in the correct chronological order, they contain very little detail. Even so, the account should remind the reader how the early ISS modules were delivered into orbit and joined together.

The first Russian ISS module, Zarya, was launched into orbit by a Proton launch vehicle, on November 20, 1998. It was followed by an American Shuttle, STS-88, launched in December 1998. STS-88 carried "Unity", the first of three nodes, with a Pressurised Mating Adapter (PMA) on either end. The Shuttle's RMS was used to manoeuvre Unity from its mounting in the payload bay and dock it (via PMA-2) to the Shuttle's own docking system. The RMS was then used to take hold of Zarya and dock it, via PMA-1, to the free end of Unity. Following several days of operations inside the station, Unity was undocked from the Shuttle and the Zarya–Unity combination commenced solo flight. Future Shuttles would dock to PMA-2 on Unity's exposed end.

With the next Russian module, Zvezda, delayed by lack of funding from the Russian government, STS-96 in May–June 1999 and STS-101 in May 2000, both with international crews, docked to the Station carrying pressurised SpaceHab modules full of logistics. These modules remained in the Shuttle's payload bay and were unloaded through the docking system, with a pressurised tunnel giving access to the SpaceHab. Meanwhile, NASA provided additional funding to the Russians, to ensure the completion and launch of Zvezda. In return, the Russians agreed that much of the time that had been allocated for their cosmonauts to perform Russian experiments on ISS would now be spent with those Russian cosmonauts performing American experiments.

Two years later, in July 2000, Zvezda was carried into orbit by a Proton launch vehicle, and completed an automated rendezvous using its KURS guidance system, then docked with Zarya at the opposite end to Unity. This module formed the centre of the Russian segment of ISS. It contained flight control systems, manoeuvring thrusters, life support systems, experiments, and exercise machinery. Zvezda also contained three Soyuz and Progress spacecraft docking drogues, at the wake, nadir, and zenith. Zvezda's crew quarters, including individual sleeping quarters for two people, allowed ISS to be permanently occupied, in the same manner as the Soviet/Russian Mir space station that had preceded it.

Once occupied, ISS could be re-supplied by Progress cargo vehicles, and the first, Progress M1-3, was launched in August 2000 and docked to Zvezda's wake. Individual Progress vehicles were launched by a Soyuz launch vehicle and employed their KURS guidance system to complete an automated rendezvous and docking with Zvezda. In the event that the KURS failed, the crew on ISS could use the TORU controls mounted inside Zvezda to manually guide the Progress to docking. Once docked, internal hatches were opened to allow the crew to unload the dry cargo delivered by the Progress. Propellants, water, and later oxygen were held in tanks in the unpressurised section of the Progress spacecraft and were pumped to the relevant tanks in Zvezda and Zarya. Zvezda's wake and nadir docking systems contained plumbing to accept Progress-delivered liquids and oxygen. Following unloading of the dry cargo, the pressurised section of the Progress was loaded with rubbish and items that were no longer required on the station. With the hatches closed once more, Progress was undocked and commanded to enter Earth's atmosphere, where it was destroyed by frictional heating.

In September 2000, STS-106 carried a further SpaceHab double module full of logistics to the station. This was followed by STS-92, in October 2000, which delivered the Zenith-1 (Z-1) Truss. This housed the station's primary attitude control system, consisting of four 98 kg Control Moment Gyroscopes (CMGs), steel wheels that spun at up to 6,600 revolutions per minute. The combination of their large mass and high rotation rate allowed them to store angular momentum. When the CMG was commanded to "speed up or slow down" the resulting force caused ISS to rotate.

The rate of spin was a measure of the momentum that they held; speeding up or slowing down would cause the station to rotate on that axis in response. By the application of multiple CMGs in this manner ISS could either be maintained in orientation, or manoeuvred to a new orientation. When any planned movement of the station's attitude required momentum beyond that available in the CMGs, rocket motors on Zvezda, or on the Shuttle or Progress cargo vehicles, could be used to provide that momentum. The Z-1 Truss also supported a number of communications antennae for NASA's Tracking and Data Relay Satellites (TDRSs), and served as a temporary mount for PMA-2 during the installation of the next pressurised module and later for storing PMA-3. Progress M1-3 was undocked in October and de-orbited in November 2000.

Expedition crews of three professional cosmonauts/astronauts could now be delivered by Soyuz spacecraft, and the first arrived on Soyuz TM-31 in November 2000. Command of alternate Expedition crews was rotated between American astronauts and Russian cosmonauts. When the Expedition crew Commander was an American astronaut the Russians insisted that their cosmonaut serve as Soyuz Commander when the Soyuz spacecraft was in solo flight, which was reasonable. If the Expedition crew was commanded by a Russian cosmonaut then this person served in both roles. Whenever there was an Expedition crew onboard ISS, there was always a Soyuz spacecraft docked to the station, ready to serve as a Crew Return Vehicle (CRV), a lifeboat in an emergency.

Progress M1-4 was launched in November 2000, and unloaded following docking to Zvezda's nadir. It was subsequently undocked and re-docked in the same location during December 2000. During the original docking a coating of ice on the spacecraft's camera had led to a failed automated approach. The undocking and re-docking was to prove that the problem had been resolved. Progress M1-4 was finally undocked and de-orbited in February 2001.

Figure 1. America's Space Shuttle was the main cargo haulage vehicle for the major elements in the American sector of the International Space Station. It was launched vertically from Launch Complex 39, Kennedy Space Centre, Florida.

Figure 2. After re-entry through Earth's atmosphere the Shuttle orbiter approaches the landing strip along a trajectory that is far steeper and faster that a commercial airliner.

The next Shuttle launch was STS-97, at the end of November 2000, which docked to Unity's ram and used its RMS to install the Port-6 (P-6) ITS on the Z-1 Truss. From there, its two Solar Array Wings (SAWs) would provide sufficient electricity to support the initial enlargement of the American sector of ISS. Installing the P-6 and deploying its SAWs required three EVAs by the Shuttle's crew.

STS-98, launched in February 2001, docked to Unity's nadir. The Shuttle's RMS was used to remove PMA-2 from Unity's ram and mount it on the Z-1 Truss. The RMS was then used to lift the Destiny American laboratory module out of the Shuttle's payload bay and mount it on Unity's ram. PMA-2 was subsequently retrieved and mounted on Destiny's ram for use in future Shuttle dockings. Destiny would become the new heart of the station, taking over many of Zvezda's control operations as well as serving as the principal American scientific laboratory on ISS.

Having docked their spacecraft to Zvezda's wake, the Expedition-1 crew had to move it to Zvezda's nadir in February 2001. This cleared Zvezda's wake for use by Progress M-44, which flew in February 2001. Having delivered its cargo, it carried away rubbish for disposal in April 2001.

While early Shuttle flights had carried their cargos in SpaceHab pressurised modules, which remained in the Shuttle's payload bay, only cargo that could pass through the narrow circular docking system could be transferred. Later Shuttle logistics flights carried the Italian-built Multi-Purpose Logistics Module (MPLM). This was lifted from the payload bay using the Shuttle's RMS and docked to Unity's nadir. Thereafter, large square internal hatches were opened to allow internal access for unloading of bulky items (in particular, racks of equipment to outfit Destiny). Once empty, items to be returned to Earth were loaded into the MPLM, which was then sealed, undocked, and returned to the Shuttle's payload bay for return to Earth.

The first MPLM, Leonardo, flew on STS-102 in March 2001. That flight also carried the Expedition-2 crew up to ISS and recovered the Expedition-1 crew at the end of their occupation.

The second MPLM, Raffaello, flew on STS-100 in April 2001. That flight also delivered the SSRMS, which was initially mounted on the exterior of Destiny, and was capable of moving end over end to manoeuvre around the exterior of the American sector of the station. It would ultimately be mounted on the Mobile Base System, on the ram face of the ITS, and would be vital to the construction of the outer segments of the ITS itself. The SSRMS was operated by crew members from a work station inside Destiny.

Expedition crew occupations were planned to last from 4 months to 6 months, which was within the operational limit of a single Soyuz spacecraft. Despite this fact, the Russians preferred to replace the Soyuz several months before its 6-month operational life expired. This resulted in Mir-style, 10-day duration, Soyuz "taxi" flights, which delivered a new Soyuz to ISS. These taxi flights were usually flown by two Russian cosmonauts and a paying passenger, either a professional European Space Agency astronaut, or a commercial Space Flight Participant (more commonly referred to as a "space tourist"). The taxi crew returned to Earth in the original Soyuz, leaving the fresh vehicle for the Expedition crew. Soyuz TM-31 was replaced by Soyuz TM-32 in April 2001. This flight was also the centre of much controversy, when the Russians announced that they had contracted to launch the first commercial passenger, an American multi-millionaire, to ISS without clearing their plans with NASA beforehand. Progress M1-6 docked to Zvezda's wake in May 2001.

STS-104 delivered the Quest Airlock in July 2001. After docking the Shuttle to Destiny's ram, Quest was lifted out of the payload bay and docked to Unity's starboard side. Two EVAs from the Shuttle's airlock and one from Quest itself were required to support the installation of four gas tanks on the exterior of the new airlock. Quest would support EVA by crew members wearing either the American EMU or Russian Orlan-M pressure suits. The following month, August 2001, STS-105 docked to Destiny's ram and installed MPLM Leonardo on Unity's nadir for unloading, and re-filling before it was returned to the Shuttle's payload bay. STS-105 also delivered the Expedition-3 crew and recovered the Expedition-2 crew. Progress M1-6 was undocked from Zvezda in August, and was replaced at Zvezda's wake by Progress M-45 the following day.

Soviet budget restrictions had long ago caused the cancellation of all remaining Russian ISS modules except one, the Pirs Docking Module-1. This was launched by a Soyuz launch vehicle in September 2001, and carried out an automated rendezvous and docking using a Progress tug, which was jettisoned following docking and burned up in Earth's atmosphere. Pirs contained a docking drogue for Soyuz spacecraft at its nadir, but it was also an airlock supporting EVA by crew members wearing Russian Orlan-M pressure suits. Two Russian Strela cranes (mechanically driven telescopic booms, extremely low-technology versions of the American RMS and SSRMS) were installed by EVA on the exterior of Pirs, to provide efficient access to the exterior of the Russian sector of ISS.

Progress M-45 was undocked from Zvezda's wake in November 2001.

Figure 3. A Soyuz-FG launch vehicle lifts off from Baikonur Cosmodrome in Kazakhstan.

Figure 4. A Soyuz spacecraft approaches the International Space Station.

Soyuz TM-33 was a taxi flight to change out the CRV attached to ISS. Launched in April 2001, it docked to Zarya's nadir. The crew returned to Earth in Soyuz TM-32 a week later. Occasionally, the Expedition crew's Soyuz CRV would have to be moved from one docking point to another, thus making way for a Progress or a replacement Soyuz at the original docking point. In order to do this, the Expedition Crew had to prepare ISS for automated, uncrewed operations before sealing themselves in the Soyuz and undocking. After manoeuvring and docking to the new location, the crew would then return to the station and re-activate everything. This long and complicated procedure, which could take up to a week, was used in case it were to prove impossible to re-dock the Soyuz spacecraft, in which case the crew would use it to return to Earth, leaving ISS unoccupied. The Expedition-3 crew followed this procedure before they undocked Soyuz TM-32 from Zarya's nadir and moved it to Pirs' nadir in October 2001.

Progress M1-7 flew in November 2001. Although it was able to soft-dock to Zvezda's wake, the hard-docking was prevented by a rubber seal, debris left behind by Progress M-45 as it left. The Expedition-3 crew made an impromptu EVA to remove the seal, and the Progress was able to hard-dock in December.

STS-108 brought the Expedition-3 occupation to an end in December 2001, when it delivered the Expedition-4 crew to ISS. The MPLM Raffaello was docked to Unity's nadir for unloading while the two crews carried out their week-long hand-over.

Representatives from several other nations involved in ISS have already flown to the station in orbit and new conflicts have arisen as individual cultures are integrated into the programme. In the coming years both the European and Japanese science laboratories and their unmanned cargo delivery vehicles will be integrated into the programme. They will bring with them their own astronauts, control centres, and management teams, all of which will add to the international melting pot that is the ISS programme. In 2007, China, only the third nation to develop an independent human spaceflight programme, also voiced an interest in becoming involved in ISS. On top of everything, Russia continued to offer commercial flights to ISS to anyone, regardless of nationality, who can afford the $20 million that it costs to buy the third seat on a Soyuz taxi flight. No doubt all of these nations will bring with them a new set of cultural clashes and a new set of lessons to be learnt. They will also bring with them newfound professional respect, both given and received, and newfound personal friendships.

In short, ISS is the most complex spacecraft ever built to support human beings in space. However, just as much as the incredible machinery, it is the clash of national cultures here on Earth, and the requirement for individuals to learn to work together and trust each other implicitly that is at the heart of the ISS programme. In the words of American astronaut Kenneth Bowersox,

> "The best thing is we've got Americans and Russians and international crews, they're not just working together they're actually getting along, making friends, building relationships. That's the visible part. The part that's much more important and not visible is the relationships being built on the ground. I tell people that 95 percent of what's important about Space Station happens on the ground: when American engineers and Russian engineers get together; when a Canadian meets a Russian and they talk about what life is like in their countries; when we send somebody from Houston over to Japan, and he talks to somebody at dinner. The relationships that we're building are building a stronger world, and that's just as important as building our Space Station."

PARTNERS

The American National Aeronautics and Space Administration (NASA) is fond of stating in its media releases that there are 16 partner countries participating in the ISS programme. While technically true, this is misleading. America and Russia are the principal partners in the ISS programme and both of them brought several decades of experience and achievement in human and robotic spaceflight to the programme. Japan, Canada, and Brazil also bring national programme experience to ISS. The remaining 11 nations are all involved in the ISS programme by way of their membership of the European Space Agency. Even then, five ESA member states chose not to participate in the ISS programme. Brazil became involved in ISS by agreeing to construct an external experiment rack, in return for a flight by a Brazilian astronaut to the station.

America

NASA was established in October 1958 and has maintained an Astronaut Office since 1959. In the intervening five decades NASA has established a technical superiority in human spaceflight that is second to none. The one-man Mercury spacecraft allowed the Americans to learn how to reach and survive in space. Gemini, with two astronauts, gave experience in orbital rendezvous, docking two vehicles together, and EVA. All of that experience made it possible for the three-man Apollo spacecraft to support six two-man landings on the Moon, while the third astronaut remained in lunar orbit. Apollo hardware was then developed into the three-man Skylab prototype space station, where NASA learned many of the lessons that it is now applying to the International Space Station. Meanwhile, in 1969 NASA had decided to develop a partially re-usable spacecraft, the Space Shuttle, with a crew of up to seven people. The Shuttle programme was when NASA introduced the first female, and ethnic astronauts. They now select small groups of astronauts bi-yearly, as required.

From the beginning, NASA had hoped that the Shuttle would be used to build a large, permanent space station in Earth orbit. Although development of the Shuttle began in 1972, the Presidential call to develop a space station did not come until 1984. Today, America talks proudly of the international co-operation involved in the ISS programme. What NASA carefully forgets to mention is that the original programme, which led to the current ISS, was an American-led programme to build a space station with the principal goals of being bigger and better, and producing better science than the Soviet Mir space station. Europe, Japan, and Canada all agreed to participate in Space Station Freedom.

Despite lots of misleading promises of what might be achieved on Freedom, almost annual budget cuts meant that the bold American designs came to nothing and the whole concept was continually downsized. Two complete sets of Solar Array Wings (Port-2 and Starboard-2) were deleted completely from the design while the pressurised modules were halved in size and minimised in number. Like its predecessors, the new, much more conservative design would be constructed in a modular format, using the Space Shuttle to deliver individual elements to orbit. Freedom would be used to run experiments requiring a human presence when the Shuttle was present and automated experiments when it was not. The station would not be permanently occupied until the last element, the Habitation Module, was delivered and put in place. Thereafter, permanent crews of six astronauts would perform maintenance and experiments, being relieved regularly by Shuttles that would deliver new crew members and take away those that had completed their time on the station. A new American Crew Return Vehicle (CRV) would ensure their safety when the Shuttle was not present. The CRV was added to the specification after the loss of STS-51L in 1986, when it was realised that crews on Freedom would need an emergency escape system if the Shuttle fleet were to be grounded.

With the collapse of Communism in the Soviet Union, America's President Clinton insisted that the Russians be brought into the Space Station programme, in an attempt to prevent their missile engineers taking their experience to nations that were politically unfriendly towards America and her allies. This led to yet another new design, the one that is now being constructed in orbit. Since then, severe budget over-runs have led to the unilateral decision to ignore the Memorandum of Agreement, a legally binding contract that America signed with the other ISS partners, and cancel the American Habitation Module (reducing future Expedition crews from six to three people) and the American Crew Return Vehicle (causing the Americans to have to rely on the Russian Soyuz for this facility). Despite this decision America still insists that all of the other ISS partners stick to the letter of the Memorandum of Agreement, and provide everything that they promised they would.

Nevertheless, despite the cancellation of the hardware listed above, America has still provided most of the major elements of ISS:

- Zarya was built by the Russians under contract to Boeing, America's prime contractor for ISS. The Russian TKS-style module provided attitude control prior to the launch of Zvezda.
- Unity was the first of three Nodes (pressurised junctions, gateways) providing docking systems for and access hatches to the different modules of the station. Unity was docked to Zarya's ram and served as the gateway between Zarya and Destiny.
- Destiny was the American science laboratory at the heart of the American sector of ISS. It contained 24 standard equipment racks of which half hold experiments and the other half contain vital systems. Destiny also housed a single sleeping quarter. The Pressurised Mating Adapter (PMA) on Destiny's ram was the standard docking location for visiting Shuttles, until Harmony was added. Destiny was docked to Unity's ram.
- Z-1 Truss provided an early mounting point for the Port-6 Solar Array Wings, which provided electrical power to the American sector of ISS during its construction. The Z-1 Truss's principal role was to house the station's four Control Moment Gyroscopes, the electrically driven primary attitude control system.
- Quest Airlock provided extravehicular activity access to the exterior of ISS from the American sector. It was capable of supporting both the American Extravehicular Mobility Unit and the Russian Orlan-M pressure suits, but was used primarily with the American suit.
- Integrated Truss Structure (ITS) was the huge cross-beam, assembled from nine pieces (Port-3/4 and Starboard-3/4 were both launched as single pre-joined units) that held the eight solar array wings to provide electrical power and the radiators to provide cooling for the American sector of ISS. The central ITS element was mounted on Destiny's Zenith.
- Harmony was constructed for NASA by the European Space Agency (ESA), at the Europeans' expense, in return for a free launch of Columbus. It was a utilities centre to provide resources from the ITS and Destiny to Columbus and Kibo, the International Partner modules that were to be docked to it. Harmony was docked to Destiny's ram, with PMA-2 being placed on its own ram.
- Node-3 ( ) was constructed for NASA by ESA. It would serve as a replacement for the cancelled American Habitation Module, allowing the standard three-person Expedition crew to be raised to six people. It also contained

much of the Life Support system and environmental Control system that was originally intended for the cancelled American Habitation module. Node-3 will be docked to Unity's nadir.
- Cupola was a seven-window observation portal mounted on Node-3, where it will provide all round observation for the astronauts operating the Space Station Remote Manipulator System. It was built by ESA under a NASA contract.

Most of NASA's field centres were involved in the development of the various ISS modules and ITS elements, co-operating with Boeing, NASA's prime contractor. Five principal centres were involved in the on-orbit operations of ISS:

- NASA Headquarters (NASA HQ), Washington DC, was at the centre of the Administration, providing managerial oversight of the field centres.
- Lyndon B. Johnson Space Centre (JSC), Houston, Texas, was where the American Astronaut Office was based and where Shuttle and Expedition crew members were trained. It was also the principal centre for human spaceflight, crewed spacecraft design, development, and operation, and the location of the Shuttle flight control room.
- John F. Kennedy Space Centre (KSC), Florida, was the location where all American, Canadian, European, and Japanese ISS modules were prepared for flight. The Shuttle was launched from Launch Complex-39, using facilities converted from those used for the Apollo Moon landings. ISS modules and ITS elements were prepared for launch in the Space Station Processing Facility (SSPF).
- George C. Marshall Space Flight Centre (MSFC), Huntsville, Alabama, was NASA's principal propulsion and launch vehicle development centre. MSFC was responsible for overseeing the development of most of the American ISS modules. It was also the location of the Payload Operations Centre (POC), the control centre for American experiments and daily operations onboard ISS.
- Four NASA field centres, including JSC and MSFC, also house Tele-science Support Centres, capable of performing automated scientific experiments in the American sector of ISS. The other two centres were the Ames Research Centre (ARC), Moffett Field, California and the John H. Glenn Research Centre (GRC), Cleveland, Ohio.

Russia

Russia, or at least the Soviet Union, placed the first cosmonaut into space in April 1961, and has maintained a human spaceflight programme and cosmonaut group since that time. While NASA's space programme is civil in nature, with considerable assistance from the US military, Russia's space programme has been run by the military from the beginning.

Vostok was a one-man spacecraft, which introduced Russian cosmonauts to spaceflight. After six flights the propaganda requirements of Nikita Khrushchev dictated that Vostok was stripped out in order to allow it to carry three cosmonauts, and then to take two cosmonauts and an extendable airlock to allow for the first EVA before the Americans flew the first two-man crew and attempted an EVA in Project Gemini. The refitted Vostok spacecraft was called Voskhod to give the impression that it was an entirely new design. Vostok/Voskhod was replaced by Soyuz, a spacecraft designed to support an Earth orbital programme, or a human lunar landing programme, launched by the N-1. All of these programmes were developed by OBK-1, Sergei Pavlovich Korolev's design bureau.

The earliest years of the Soviet human space programme were highlighted by a series of politically driven propaganda events, the first man in space, the first woman, the first three-man crew, the first EVA, and then it all went wrong. When the Soviet Premier changed in October 1964 and Korolev, the man in charge of the Soviet space programme, died on the operating table in January 1966, everything changed. The new premier, Leonid Brezhnev, had less interest in the space programme and Korolev's replacement Vasili Mishin was not up to the job, faced as he was with racing America to the Moon, having started several years behind the Americans.

Meanwhile, a second design bureau, led by Valentin Chelomei, had designed the Proton launch vehicle and OKB-1 had designed the Soyuz-derived Zond to be launched on a Proton and carry a single cosmonaut on a high orbit that would pass around the far side of the Moon and fall straight back to Earth. When Apollo won the Moon race Russia cancelled both of its human circumlunar and lunar landing programmes.

Chelomei had begun a space station project in the mid-1960s, consisting of a crew/cargo ferry designated TKS and a station element designated OPS Almaz (despite this distinction, large Russian space station size modules based on the OPS Almaz element of this vehicle have generally come to be identified as "TKS modules" and this is the way in which the designation is used in this book). In December 1969, a decision had been made to have OKB-1 modify the OPS Almaz to operate with the Soyuz, and to serve as a scientific station rather than as a military reconnaissance platform. The scientific version of the Almaz station was originally called Zarya. Ultimately, both Almaz (military) and Zarya (scientific) versions of the station later flew under the cover name Salyut.

The first seven Salyut stations (two malfunctioned before they were occupied) supported a single Soyuz spacecraft, with their crews performing a series of record-breaking long-duration flights, but the stations were left unoccupied between crew visits, just as the American Skylab prototype space station would be. Salyut-6 introduced two docking ports, one at each end of the station's long axis. This allowed two Soyuz spacecraft to dock at the same time, leading to permanent occupancy and crew relief on-orbit. When crew occupations surpassed the 6-month guaranteed life of a Soyuz spacecraft the Soviets introduced 10-day Soyuz taxi flights, where two-man crews delivered a new spacecraft to the station and returned to Earth in the old one. Salyut-6 also saw the introduction of the Progress delivery vehicle carrying dry cargo in a pressurised compartment as well as liquid water and rocket propellant. Once the new cargo had been transferred to the station the pressurised compartment in the Progress was filled with rubbish and the spacecraft was undocked and commanded to re-enter the Earth's atmosphere, where it was heated to destruction. Salyut-7 also received several TKS modules, each of which completed an automated rendezvous and docking, a precursor to the construction methods used to build the next generation of Soviet space stations. Indeed, in Salyut-6 and Salyut-7 the Soviets had rehearsed everything required for their third generation of space stations.

The Mir base block, launched in 1986, was the beginning of a new space station. The docking system at the module's wake received Soyuz spacecraft, but also included the plumbing to support the liquid cargo deliveries from Progress spacecraft. At the module's ram, the spherical node contained five docking systems. The one at the module's ram was used for Soyuz spacecraft. When there were no Soyuz spacecraft docked, the system was also used to dock automated TKS-style modules, which were then moved to the radial ports and were accessed by the crew internally, from the node. When Communism collapsed in the Soviet Union, only two of the four Mir science modules had been launched. The remaining modules were only launched after they were fitted out using American money.

After the Russians had signed up to ISS, Mir became the location for a co-operative programme with the Americans, allowing their astronauts to gain long-duration flight experience. Prior to this, Mir had been hosting European astronauts, ESA having grown frustrated by the delays in creating an American-led station. This Shuttle–Mir programme was designated Phase 1 of the ISS programme. By the time the first ISS module was launched the Russians already had nearly 30 years of space station, long-duration flight experience.

The Russians have provided three major ISS modules:

- Zarya was built by the Russians under contract to Boeing, NASA's primary ISS contractor. The module provided attitude control until later American modules were docked to it, after which it became a storage area. Zarya was designated as an American module, although it is now seen as part of the Russian sector of ISS.
- Zvezda was originally an all-Russian module, but lack of funding from the Russian government meant that it was only completed, two years late, after an injection of NASA's cash. Basically similar to the Mir base block, and the Salyut stations that had preceded it, Zvezda was the control centre of the Russian sector of ISS, and the social centre of the station, as it contains the food preparation area and a galley table, as well as a waste management facility (toilet). Zvezda allowed the permanent occupation of ISS from the earliest days of its activation. Under the original plans for Space Station Freedom the station would not have been permanently occupied until the very last module was in place. Zvezda re-wrote the flight plan, but only with help from the other two vital elements in the Salyut/Mir programme!
- Soyuz was Russia's human-carrying spacecraft. As such, it could deliver crews to ISS, docking to the Russian modules. A Soyuz spacecraft always remained docked to ISS, serving as a Crew Return Vehicle (CRV), in case of an emergency evacuation. Ten-day taxi flights, often with commercial occupants in the third couch, replaced the Soyuz attached to the station approximately every six months. The ISS was originally serviced by the Soyuz TM spacecraft, but this was replaced by the upgraded Soyuz TMA.

Figure 5. A Soyuz-U launch vehicle and Progress cargo spacecraft are prepared for launch at Baikonur Cosmodrome in Kazakhstan. The snow on the ground and overcast sky highlight the winter conditions at the Kazakhstan launch site.

Figure 6. A Progress cargo spacecraft approaches the International Space Station. The similarities with the Soyuz spacecraft are obvious, but are in fact only superficial.

- Progress carried dry cargo, propellants, water, and, in the Progress M, air (oxygen and nitrogen) to ISS. It brought food, spare and replacement parts, and personal effects to the Expedition crews on the station. The propellants Progress carried allowed the thrusters on Zvezda to be refuelled, and thus remain operational, maintaining the station in the correct attitude, when the CMGs in the American Z-1 Truss became momentum-saturated.

  Zvezda, Soyuz, and Progress, are the three Russian elements that made it possible to permanently occupy ISS at the earliest opportunity, but without the American Shuttle there would be very little ISS to occupy.
- Russian Docking Module-1, Pirs, was the final Russian module to receive funding from the Russian government, the planned science modules and power module were never built. Pirs docked automatically to Zvezda's nadir to provide an airlock supporting EVA by crew members wearing Russian Orlan-M pressure suits while retaining a docking port for Soyuz and Progress spacecraft on the nadir. Two Strela cranes were later mounted on the exterior of Pirs.

The principal Russian Space Agency centres involved in the ISS programme were

- S. P. Korolev Rocket and Space Corporation (RSC) Energia, Korolev, Moscow, manages the Russian sector of the ISS programme and was responsible for the integration of Russia's space station modules, the Soyuz and Progress spacecraft, and their respective launch vehicles.
- Yuri Gagarin Cosmonaut Training Centre, Zvezdny Gorodock, was where Russian cosmonauts, their international partners, and commercial Space Flight Participants are trained.
- Khrunichev State Research and Production Space Centre, Khrunichev, Moscow, was responsible for developing and constructing the Zarya (under contract to the American Boeing Company) and Zvezda modules and the Proton launch vehicle.
- Korolev Mission Control Centre (TsUP), Korolev, Moscow was the main Russian control centre for ISS operations.
- Baikonur Cosmodrome, Kazakhstan, was Russia's launch facility. Its facilities oversaw the integration of all crewed and uncrewed spacecraft and their launch vehicles, before transporting them by rail to the launch pad, where they were erected and launched. (Baikonur Cosmodrome is named after the Baikonur region in which it lies, and not the town of Baikonur which is several hundred kilometres away. In the 1970s one NASA engineer explained facetiously that this "... is like calling Kennedy Space Centre Tampa Spaceport.")

Canada

Canada became only the third nation to launch a satellite, in 1962. Despite this, Canada maintains no launch facilities of its own and uses NASA facilities within America's national borders. The civilian Canadian Space Agency (CSA) (in French *L'Agence Spatiale Canadienne* or ASC) was not established until 1989. Canada co-operates regularly with NASA in America and is a "Co-operating State" with the

European Space Agency. By paying into the ESA budget, Canada is given a position on the main committees of the ESA. Canada also provides instruments to fly on European satellites and deep-space probes.

Canada has provided two elements that are vital to the construction of ISS, plus a third which is vital to Space Shuttle survivability in the final years of its service. All three were developed and built under NASA contracts:

- Remote Manipulator System (RMS): Development of the Shuttle's RMS, popularly referred to as the "Canadarm", began in 1974, when Canada agreed to develop and build a single RMS for the Shuttle orbiter Columbia. NASA subsequently ordered four more, for the other orbiters. The RMS *was* 15 m long and had two rotating joints (pitch and yaw) at the shoulder, one joint (pitch) at the elbow, and three joints (pitch, yaw, and roll) at the wrist. The two booms were made of graphite epoxy. The upper boom was 5 m long, and the lower boom was 5.8 m long, both were 33 cm in diameter. A single end effector, on the free end, housed three wires that were used in conjunction to grasp special grapple fixtures on the items to be lifted. These wires pulled the payload snug against the end effector and allowed it to be moved around. Shuttle Mission Specialists operated the RMS from the orbiter's aft flight deck, using either the Shuttle's own computers to translate hand controller commands into smooth RMS movements, or manually, by commanding each rotation joint individually. The RMS was permanently fixed to the Shuttle's payload bay door hinge-line and is returned to Earth with the orbiter at the end of each flight.
- Space Station Remote Manipulator System (SSRMS), affectionately called "Canadarm-2", was a more advanced tool than the Shuttle RMS. It was originally designed to grasp the Shuttle and pull it in to dock with Space Station Freedom. The unique feature of the SSRMS was the Latching End Effector (LEE) at each end, which allowed either end to mate to a Power-Data Grapple Fixture (PDGF) on the exterior of ISS, while the free end performed the lifting tasks. This feature also allowed the SSRMS to be "walked" end over end across the exterior of the American sector of ISS, from one PDGF to another. The RMS on a docked Shuttle and the SSRMS were capable of working together, either lifting items out of the Shuttle payload bay, or handing items from one to the other.

  The SSRMS was also designed to be mounted on the Mobile Base System, a small cart that could translate along rails mounted on the ram face of the ITS. This additional mobility allowed the SSRMS to support the construction of the ends of the ITS, while being mounted on and travelling along the face of those ITS elements already in place. The reach of the SSRMS was dictated by the necessity to move the Port-6 ITS from its temporary position on the Z-1 Truss and install it on the exposed end of the Port-5 ITS. The SRMS was designed for on-orbit replacement of its major parts and was not expected to return to Earth.

  An extension tool, to be used with the SSRMS was the "Dextre" manipulator system. This consisted of two smaller Remote Manipulator Systems that could be used to complete more delicate work on the exterior of ISS without the requirement for crew members to perform EVAs.

- Orbiter Boom Sensor System (OBSS): As a result of the STS-107 tragedy in February 2003, the Columbia Accident Investigation Board recommended that NASA develop a method of inspecting areas of the orbiter that had previously been inaccessible to the crew in flight. The OBSS provided an extension to the Shuttle's RMS and the cameras and laser sensors on the OBSS allowed the crew to inspect those areas of their spacecraft that were not readily visible from the flight deck windows, or with the cameras on the un-extended RMS. On early flights the OBSS was mounted on the opposite payload bay door hinge-line to the RMS, making it easy for the latter to pick up. It was then returned to that location and carried back to Earth at the end of each flight. In 2007, STS-118 astronauts installed a mounting to allow the OBSS to be stored on the exterior of ISS when the Shuttle is retired in 2010. This would mean that the OBSS was still available to examine the exterior of the ISS for meteorite strikes, or other damage, even when the Shuttle is no longer flying. The Shuttle RMS, SSRMS, and OBSS were all developed and built by Macdonald, Dettwiler & Associates, Limited, Brampton, Ontario.

The first Canadian astronaut flew on the Shuttle in 1984 and was followed by a further 7 Canadian nationals who have taken part in a total of 13 Shuttle flights through the end of 2007. Two further Canadian astronauts have retired without flying in space.

Canada maintains a number of space centres:

- John H. Chapman Space Centre, Saint-Hubert, Quebec was the CSA's Headquarters and oversees the management of the national space programme.
- David Florida Laboratory, Ottawa, Ontario, an engineering facility.
- Mobile Servicing System Operations Complex (MOC), Longueluil, Quebec, prepared the Canadian systems and provided astronaut training on Shuttle RMS and SSRMS systems. It also provided full support for all RMS and SSRMS engineering and operations.

Europe

Post-World War II European space efforts were originally split between two organisations. The European Launch Development Organisation (ELDO), to develop a space launch vehicle, and the European Space Research Organisation (ESRO), to develop and exploit payloads for those launchers. The European Space Agency (ESA) was established in 1974 by merging these two organisations. ESA's Headquarters building was in Paris, and oversaw the activities of 16 member states, 11 of which are participating in the ISS programme. Most member states also have national space agencies and some have thriving national space programmes. In the intervening years ESA has developed a reputation for reliable satellites and deep-space probes. The French/ESA Ariane launch vehicle programme has developed through 14 different configurations, of which the latest, the Ariane-V ES-ATV, is the only one involved in the ISS programme. Ariane has established a reliable series of launch vehicles, placing ESA at the forefront of the worldwide commercial launch vehicle market. In 2007, two out of every three commercial payloads then in Earth orbit had been launched by Ariane launch vehicles.

ESA originally co-operated widely with NASA, but information-sharing restrictions in America in the 1990s led to a change. In 2007, ESA considered Russia to be its principal partner in space exploration. For ISS Russia had an agreement with ESA whereby if the third couch on a Soyuz spacecraft launched to ISS was not taken by a commercial Space Flight Participant, it would be offered to ESA on a commercial basis, before Russia put one of their own cosmonauts in it. ESA has also co-operated with China.

Human spaceflight has never been a major part of ESA's mission. As a result, although ESA maintains an astronaut cadre, it has developed no crewed spacecraft of its own. A French national project, the Hermes spaceplane, was designed to service the original concept of a free-flying Columbus module, but was not developed. Instead, European astronauts are launched on American Shuttles and Russian spacecraft on a commercial basis. The first ESA astronaut flew on the Shuttle in 1983 and, as of 2007, ESA has between 10 to 15 astronauts in training, several of whom have flown solo Shuttle flights and both Shuttle and Soyuz flights to ISS. So far, only one ESA astronaut has served on an ISS Expedition crew.

ESA's main participation areas in the ISS programme were

- Multi-Purpose Logistic Modules Leonardo, Raffaello, and Donnatello were Europe's major input to ISS during the early years of its construction. The three pressurised cargo modules were carried into orbit in the Shuttle's payload bay. Following docking they were lifted out of the bay by the Shuttle's Remote Manipulator System and docked to Unity, from where they were accessed internally and unloaded. After re-filling with items to be returned to Earth they were undocked and returned to the Shuttle's payload bay, allowing them to be returned to Earth and re-used. This was the principal method of returning major items to Earth as all three alternative ISS cargo delivery systems were designed to be destroyed during re-entry into Earth's atmosphere.
- Columbus was the principal ESA ISS element. The pressurised laboratory module would be launched by STS-120 and docked to Harmony. The laboratory would house ten standard experiment racks housing the following experiments:
  1. Fluid Science Laboratory (FSL)
  2. European Physiology Modules (EPMs)
  3. Biolab
  4. European Drawer Rack (EDR)
  5. European Stowage Rack (ESR)

  The ten racks would be split 51%/49% between European and American experiments. The Columbus module would also have an external facility to allow for the fixture of experiments that required exposure to the space environment. The

Figure 7. A Shuttle orbiter approaches the International Space Station carrying a Multi-Purpose Logistics Module in its payload bay.

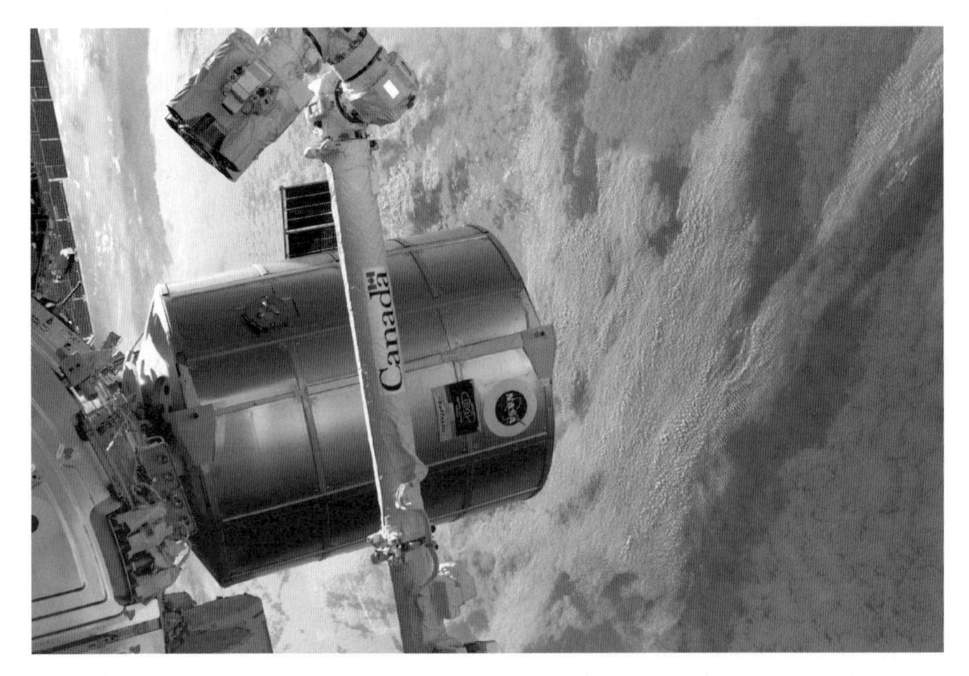

Figure 8. Multi-Purpose Logistics Modules were removed from the Shuttle's payload bay and docked to Unity before being unloaded via an internal hatch. The empty module was then filled with items no longer required on the station and returned to the Shuttle's payload bay for return to Earth.

first two external experiments were

a. European Technology Exposure Facility (EuTEF)

b. Solar Monitoring Observatory (SOLAR)

Two additional external experiments were also planned:

c. Atomic Clock Ensemble in Space (ACES)

d. MISSE-6 (NASA)

- Automated Transfer Vehicle (ATV) is a delivery vehicle for dry cargo, propellant, water, and air. It will be launched by Ariane-V ES-ATV and then perform an automated rendezvous and docking at the rear of Zvezda. The ATV's Kurs rendezvous equipment and Soyuz docking probe were supplied by RSC Energia, in return for the European Command and control computers in Zvezda, which were the same as those in the ATV. After unloading it would be filled with rubbish before being detached from the station, manoeuvred clear, and commanded to re-enter the atmosphere, where it would be heated to destruction.
- Ariane-V ES-ATV is a heavy-lift launch vehicle, is a mixture of the Ariane-V ECS first stage and the Ariane-V G second stage, and would be used to launch the ESA Automated Transfer Vehicle to ISS at 12-month to 15-month intervals. It is launched from the Guiana Space Centre, in South America.
- Node-2 (Harmony), Node-3 ( ), and the Cupola have all been described previously. They were constructed by ESA under a NASA contract that provided for the launch of Columbus on an American Shuttle in return for their construction.

ESA maintains six major centres related to their space programmes. In addition, member states maintain national centres to support their own activities related to ESA programmes:

- European Space Agency Headquarters, Paris, France, oversees and manages all ESA programmes.
- European Space Research and Technology Centre (ESTEC), Noordwijk, Holland, is ESA Headquarters and is where most ESA programmes are developed and managed.
- European Astronaut Centre, Cologne, Germany, is where the European astronaut group is trained.
- Columbus Control Centre (COL-CC), Oberpfaffenhofen, Germany, is the control centre for the Columbus Laboratory Module and the operation centre for European experiments on ISS.
- Guiana Space Centre (GSC), was originally established by the French Space Agency as the launch site for their Ariane series of launch vehicles. Today it is operated jointly by France and ESA. For the ISS programme the GSC will be used to launch the Ariane-V, carrying ATVs to the station.
- Automated Transfer Vehicle Control Centre (ATV-CC), Toulouse, France, would operate the ATV in flight.

Japan

Japan's National Aeronautical Laboratory was established in 1955, gaining a new Aerospace Division and changing its name to the National Aerospace Laboratory (NAL) in 1963. The Institute of Space and Aeronautical Science (ISAS) was also established in 1955, within the University of Tokyo. This led to co-operation between a number of Japanese universities on aerospace projects, including Japan's infant space programme. In 1984 the ISAS name was changed to the Institute of Space Aeronautical Science. Also, the National Space Development Agency of Japan (NASDA) was formed in 1969 and assumed overall responsibility for the national space programme, including development of launch vehicles, payloads, and a launch, tracking, and support infrastructure. Finally, on October 1, 2003, these three organisations were merged to form the Japan Aerospace and Exploration Agency (JAXA). Japan has maintained an autonomous space programme, developing a series of sounding rockets and space launch vehicles since the mid-1960s. In the intervening decades it has developed a series of launch vehicles, initially of American designs under licence, but then of its own design, the latest of which is the general-purpose H-IIB, and it has also built a wide variety of satellites and deep-space probes. Despite plans for development of the HOPE mini-Shuttle, which was the victim of budgetary constraints, the Japanese have never developed their own crewed spacecraft, preferring to co-operate with the Americans from the earliest years of Space Station Freedom. From the outset Japan announced that it would develop a laboratory module for ISS, which would be launched and installed on the station by American Shuttle crews. In 2007, Japan maintains an astronaut group of six men and two women. The first Japanese astronaut flew on the Shuttle in September 1992. Since then four other Japanese astronauts have made Shuttle flights. Two more have been assigned to Shuttle flights relating to the launch and installation of the various sections of the Kibo laboratory module to ISS, and the first Japanese Expedition crew member has been named. Koichi Wakata would fly as part of the Expedition-18 crew.

Kibo would require three separate Shuttle launches before all of its separate parts were installed in ISS. STS-123 would launch the Experiment Module and place it in a temporary location on the exterior of ISS; STS-124 would deliver the Kibo laboratory module and install it on Harmony. After the Shuttle's departure the experiment module would be positioned on Kibo's zenith. Finally, STS-127 would deliver the Exposed Facility. All of the Kibo elements were at Cape Canaveral, Florida, awaiting launch when this was written in mid-2007. Kibo arrived at the Space Station Processing Facility in May 2003 and was joined by the Exposed Facility in March 2007. Despite being the smallest module in the Original Space Station Freedom design, the dimensions of Kibo have not changed, while the size of the American and European modules has been decreased. Kibo is now the largest of the three laboratory modules in the American sector of ISS.

In the first instance Kibo would house three major experiments:

- Monitor of All-sky X-ray Image (MAXI)
- Superconducting Sub-Millimeter-wave-limb Emission Sounder (SMILES)

Figure 9. Russian stage extravehicular activities were completed from the Pirs Docking Module by astronauts wearing Russian Orlan-M pressure suits.

- Calorimetric Electron Telescope (CALET)

Japan's second major input into the ISS programme would be the H-II Transfer Vehicle (HTV). The automated logistics carried would be launched on a Japanese H-IIB launch vehicle and would carry logistics to ISS. Following launch, the HTV would carry out an automated rendezvous and station-keeping with ISS, before being grasped by the Space Station Remote Manipulator system and being docked to Unity for unloading. Following unloading, the HTV would be filled with rubbish and unwanted materials before being undocked from Unity and performing a separation manoeuvre. Once clear of ISS the HTV would perform a retrofire manoeuvre and re-enter Earth's atmosphere, where it would be heated to destruction.

JAXA maintains three principal centres for their operations relating to ISS.

- JAXA Headquarters is in Tokyo and oversees the management of the Japanese space programme.
- Tsukuba Space Centre (TKSC), where Kibo was developed, constructed, and tested. It is also the location of the Kibo Control Centre.
- Tanegashima Space Centre contains the Osaki Range where the H-II launch vehicles for the HTV will be launched from.

Figure 10. American stage extravehicular activities were made from the Quest Joint airlock by astronauts wearing American Extravehicular Mobility Units.

Brazil

The Brazilian space programme began in 1961, under the command of the national military authorities. It has developed a national launch vehicle programme, beginning in the 1970s. In 1994, the programme was civilianised, under the charge of the Ministry of Technology and Science and the Brazilian Space Agency (in Portuguese the *Agência Espacial Brasiliera* or AEB) was formed, with its launch site in Alcântara. Early co-operation with the Americans was hindered by American laws relating to technology transfer, and Brazil now co-operates with a number of other nations.

AEB originally signed a contract to provide an experiment rack, to be launched on the Shuttle and mounted on the exterior of ISS. In return a Brazilian astronaut would fly to ISS on a Shuttle mission, as part of NASA's allocation of ISS utilisation time. Budgetary constraints prevented the Brazilians manufacturing the experiment rack, but the flight of their astronaut has taken place.

Brazil operates several AEB centres:

- National Institute for Space Research is responsible for the oversight of all of Brazil's space programmes and the development of relevant hardware.
- Institute of Aeronautics and Space (IAE) oversees the development of Brazil's launch vehicles.
- Alcântara Launch Centre.

2

The ISS Management and Cost Evaluation Task Force

On October 16, 2001, a tearful Dan Goldin announced his resignation as NASA's longest serving Administrator. His term of office had begun in March 1992 and lasted 8.5 years. Goldin had overseen the redesign of ISS to meet President Clinton's demands. On his President's instructions, it had been Goldin who invited the Russians to join the ISS programme. He had also overseen the initial privatisation of the Shuttle programme and the beginning of the ISS flight programme with its associated delays and massive budget overruns. In the end, it would be the budget overruns and the hardware cancellations that brought Goldin's tenure to a close. In his retirement speech he told the audience proudly, "I had a lease on this programme. I am handing back that lease."

In July 2001, even as he prepared for his retirement, Goldin had established the International Space Station Management and Cost Evaluation Task Force (IMCE) under the leadership of Thomas Young. Predictably, it became known as the Young Committee. The Task Force was established to answer the criticism surrounding NASA's management and funding of the ISS programme, with its huge overspend. The programme had overspent by $13 billion in just four years.

In setting up the Committee, Goldin told the media:

> "In the last year we have successfully carried out all of the scheduled assembly missions to the International Space Station. We did so with unbelievable precision and execution, completing the second phase of Space Station construction. It's an incredible management and engineering achievement, but we must ensure it is carried out in a more efficient and effective manner.
>
> "Since April, we've been working to select a team of outstanding innovators in the fields of science, engineering, finance and business to advise NASA and the Administration how to maximise the scientific returns on the Station while living within the guidelines of the President's budget ... The financial management of the International Space Station needs an overhaul, but we're going to do it in a

way that doesn't sacrifice safety ... This panel has been empowered to leave no stone unturned. We have experts in all fields that have the capacity to dig deep to help us restructure the business and financial approach to the programme."

Goldin's resignation was effective from November 15, 2001 and the Final Report of the Young Committee was dated six days later. It read, in part:

ISS Management and Cost Evaluation Task Force

Terms of Reference

These Terms of Reference establish the International Space Station (ISS) Management and Cost Evaluation (IMCE) Task Force of the NASA Advisory Council (NAC). The IMCE Task Force is chartered to perform an independent external review and assessment of cost and budget and provide recommendations on how to ensure that ISS can provide maximum benefit to the U.S. taxpayers and the International Partners within the Administration's budget request.

In addition there are reviews of the ISS financial management tools being conducted by the IMCE Financial Management Team (FMT) to identify and recommend Agency-wide improvements in these tools. The report of the FMT will be integrated into the report of the IMCE Task Force ...

The integrated final report is to focus specifically on the following items

- Assess the quality of the ISS cost estimates approved for the ISS Program, including identification of high-risk budget areas and potential risk mitigation strategies.
- Ensure that the program can remain within its available budget, assess program assumptions and requirements—specifically those that led to significant cost growth relative to FY 2001 budget estimates—and identify options for smaller growth and/or budget savings and efficiencies that offset any additional spending recommended by the Task Force and approximately $500 million in unfunded cost growth.
- Review the management reforms in the ISS Program Management Action Plan—particularly cost estimating and reporting issues, early warning of potential growth, and managing program reserves—and make recommendations for additional and/or refined management reforms. Integrate results from the FMT.
- Identify opportunities for maximizing capability to meet priority research program needs within the planned ISS budget and International Partner contributions.
- In addition, assess cost estimates for potential U.S.-funded enhancements to the core station (e.g. providing additional crew time for enhanced research) and recommend refinements as necessary to achieve high-confidence estimates.

SPECIAL FINDINGS

1. The ISS Program, while taking a conservative approach and making safety paramount, has achieved excellent progress in integration of diverse international technologies.
2. NASA has not accomplished a rigorous ISS cost estimate. The program lacks the necessary skills and tools to perform the level of financial management needed for successful completion within budget.
3. The cost to achieve comparable expectations at assembly complete [original design complete with American Habitation Module and American CRV] has grown from an estimate of $17.4B to over $30B. Much of this cost growth is a consequence of underestimating cost and a schedule erosion of 4+ years.
4. A cost of $8.3B (FY02–06) is not credible for the core complete baseline [No American Habitation Module, no Node-3, and no American CRV] without radical reform.
5. The management focus is on technical excellence and crew safety with emphasis on near-term schedules, rather than total program costs.
6. The Program is being managed as an "institution" rather than as a program with specific purpose, focused goals and objectives, and defined milestones.
7. The financial focus is on fiscal year budget management rather than on total Program cost management.
8. Lack of a defined program baseline has created confusion and inefficiencies.
9. Current research support funding represents a 40-percent reduction in buying power from that originally planned.
10. The Office of Biological and Physical Research (OBPR) is not well coordinated with the Office of Space Flight (OSF) or the program office for policy and strategic planning. The scientific community representation is not at an effective level in the program office structure.
11. A centrifuge is mandatory to accomplish meaningful biological research. Availability as late as FY08 is unacceptable.
12. There are opportunities to maximize scientific research on the core station with modest cost impact.
13. Cost estimates for the U.S. funded enhancement options need further development to assess credibility.

SPECIAL RECOMMENDATIONS

A. Establish the ISS Program Office separate from, but residing at JSC, reporting to a new Associate Administrator (AA) for ISS.
B. Consolidate prime and non-prime contracts into a minimum number of resulting contracts all reporting to the program office.
C. Develop a life cycle technical baseline and manage the ISS Program to total cost and schedule as well as fiscal year budgets.
D. Consider revising the ISS crew rotation period to 6 months and reducing the

Space Shuttle flight rate accordingly. The result would be a delay in U.S. core complete assembly sequence by up to 2 months. Target cost savings: $668m, and continue to examine Strategic Resources Review (SRR) and institutional cost reductions. Target cost savings: $350M–$450M.

E. Develop a credible program road map starting with core complete and leading to an end state that achieves expanded research potential. Include gate decisions based on demonstrated ability to execute the program and identify funding to maintain critical activities for potential enhancement options.
F. Establish research priorities. The Task Force is unanimous in that the highest research priority should be solving problems associated with long-duration human space flight, including the engineering required for human support mechanisms, and provide the Centrifuge Accommodation Module (CAM) and centrifuge as mandatory to accomplish top priority biological research. Availability as late as FY08 is unacceptable, and establish a research plan consistent with the priorities, including a prudent level of reserves, and compliant with the approved budget.
G. Provide additional crew time for scientific research through the use of extended duration Shuttle and overlap of Soyuz missions.
H. Create a Deputy Program Manager for Science position in ISS Program Office. Assign a science community representative with dual responsibility to the Program and OBPR.
I. "The IMCE Report proposes a strategy to restore confidence in the ISS Programme"
J. "The goals of the US International Space Station Programme are not well-defined"
K. "The IMCE strategy raises serious issues for the ISS International Partners"
L. "NASA cannot afford to delay"
M. Manage strategically
N. Provide aerospace products and capabilities
O. Generate knowledge
P. Communicate knowledge.

The IMCE addressed the two major concerns over the ISS programme. The problem of severe cost overruns was responsible for the introduction in the report of the new concept of bringing ISS to "American Core Complete", rather than the original, legally agreed, "Station Complete". Core Complete was a unilateral American decision to save money by reducing American involvement in ISS while retaining America's role as the controlling partner in the alliance. Core Complete deleted the American Habitation Module, the American CRV, and Node-3 from the ISS design, without any negotiation with Russia and the International Partners. The decision to eliminate the Habitation Module effectively limited future Expedition crews to just three people for the foreseeable future, which would severely constrain the European and Japanese agencies' access to their own laboratories.

The cancellation of the X-38 CRV development programme meant that the Space Shuttle would remain NASA's only access to ISS for the foreseeable future, given that American access to the Russian Soyuz spacecraft was limited by the Iran Non-proliferation Act at that time. Even without that legislation, the idea of NASA paying the Russians to carry American astronauts to and from space had never been very palatable to the Americans, even though it would be cheaper than continuing to fly the Shuttle. Attempts to replace the Shuttle were plagued by lack of long-range goals and a too narrow focus on ISS. Problems developing the X-33 Flight Test Article proved that Lockheed-Martin's bold talk of the VentureStar vehicle were nothing but hyperbole. The Orbital Space Plane (OSP) would be criticised as being too poorly defined and too narrowly focused on the ISS CRV role. Finally, in the wake of disaster (STS-107, with seven people onboard would be lost in February 2003), NASA would be set a new goal and the definition of a new spacecraft would become easier to complete.

The second area that the IMCE addressed was the lack of science that would be able to be completed on a station that was restricted to the new Core Complete configuration. One recommendation that the committee made was one-month-long Soyuz taxi flights. In this scenario one Expedition crew would already be in space. Their relief would be launched one month before the original crew's occupation came to an end. The six astronauts would then work together for one month with two Soyuz CRVs docked to the station. At the end of that month the original crew would return to Earth, leaving the new crew in orbit. In turn their relief crew would be launched one month before the end of their occupation and the two crews would work together for one month before the second crew returned to Earth, leaving the third crew alone in orbit. Due to restrictions on consumables, Shuttle flights would only visit the station during the periods when a single Expedition crew was in occupation. This recommendation would not be acted on after Node-3 was re-instated with living quarters for a further three occupants.

"Pioneer the future"

Goldin's replacement as NASA Administrator was the former Deputy Director of the White House Office of Management and Budget Sean O'Keefe. On taking up his new post on December 21, 2001, O'Keefe stated that he intended to adhere to the recommendations of the Young Committee on how to bring the ISS budget back under control. The report had suggested that NASA be placed on probation until ISS reached "Core Complete", with a three-person crew performing both maintenance and science. O'Keefe had said that, if NASA brought the ISS budget under control, then consideration might be given to going beyond "Core Complete", such as reinstating an American Habitation Module and the X-38 CRV. He stated that the question of crew size was vital to the programme. The new Administrator was blunt and warned, "If NASA fails to meet the standards, then an end-state beyond 'Core Complete' is not an option."

O'Keefe also stated that he would ensure that ISS did not dominate NASA's programmes "at the expense of everything else this organisation does." The new Administrator said that he believed the civil and military sectors should increase their co-operation on space programmes, especially on the development of any future Shuttle replacement.

In the weeks that followed, the management of ESA demanded a meeting with O'Keefe. The Europeans were unhappy at NASA's unilateral decision not to construct and launch the Habitation Module and the X-38 CRV, thereby restricting the ISS crew to three people and severely limiting the amount of scientific research that could be performed on the station. The Europeans considered that all of the ISS partners, including America, had signed legal documents that committed America to constructing and launching a Habitation Module and thereby supporting an Expedition crew of up to seven people. However, NASA and the American government now considered that their commitment ended when Node-2 was launched, thereby allowing the European and Japanese Science modules to be launched and docked to ISS. Ultimately, NASA negotiated with ESA to have Node-3, which was being constructed in Italy, reinstated and outfitted with additional sleeping quarters and life support equipment.

At 14:00, April 12, 2002, while STS-110 was docked to ISS, O'Keefe made a public address at the Maxwell School of Citizenship and Public Affairs, Syracuse University, during which he voiced his vision of the future of NASA under his leadership. In his speech he described NASA's mandate as:

> "... to pioneer the future, to push the envelope, to do what has never been done before."

He called it, "An amazing charter indeed," and continued, "Our greatest asset in fulfilling this demanding charter is the excellence of our people."

O'Keefe stated NASA's mandate under his charge in three simple terms.

- To improve life here.
- To extend life to there.
- To find life beyond.

He expanded each of these three goals, but his vision for NASA was summed up by the headings that he gave to those expansions.

- To understand and protect our home planet.
- To explore the Universe and search for life.
- To inspire the next generation of explorers ... as only NASA can.

The Young Report would become a major planning tool by which NASA's future involvement in the ISS programme would be ruled.

3

Commencing the Integrated Truss Structure

STS-108, and the beginning of the Expedition-4 occupation of ISS, was where *Creating the International Space Station* ended. This manuscript returns to STS-108 as it provides a natural starting point for the original flight accounts that occupy the remainder of this volume.

STS-108 BEGINS EXPEDITION-4

	STS-108
COMMANDER	Dominic Gorie
PILOT	Mark Kelly
MISSION SPECIALISTS	Linda Godwin, Daniel Tani
EXPEDITION-4 (up)	Yuri Onufrienko (Russia), Daniel Bursch, Carl Walz
EXPEDITION-3 (down)	Frank Culbertson, Vladimir Dezhurov (Russia), Mikhail Tyurin (Russia)

In Florida preparations progressed, aiming to launch STS-108, Endeavour, on November 29, 2001. Following the Russians' difficulty confirming Progress M1-7's hard-docking to ISS the launch was delayed for 24 hours. On November 30, the launch was held at $T-11$ hours, ultimately for 96 hours. The Rotating Service Structure was put back around the vehicle to protect it as it remained on the pad in a powered-up, flight-ready condition. The countdown would be resumed at $T-11$ hours on December 4.

On that date, the countdown for STS-108 progressed until $T-5$ hours when it was stopped and the launch cancelled due to bad weather. Because the fuelling of the ET had not started, it was possible to initiate a 24-hour recycle.

Figure 11. STS-108 crew (L to R): Mark Kelly, Linda Goodwin, Daniel Tani, Dominic Gorie. These four astronauts were joined by the Expedition-4 crew during launch and the Expedition-3 crew during landing.

On December 5, the countdown was stopped due to software difficulties in the orbiter. New software was loaded and the countdown was re-cycled to $T-6$ hours and restarted. Endeavour lifted off at 17:19 (all times US Eastern), just 11 seconds before the launch window closed, and climbed into orbit without event. Once in space, the payload bay doors were opened. Gorie and Kelly entered the on-orbit software and set about the first manoeuvres in the rendezvous with ISS.

On the NASA Human Spaceflight website the Shuttle's rendezvous with ISS was described in the following terms by Nicholas O'Dosey, a NASA Shuttle Rendezvous Guidance Procedure Officer:

> "The operations that the Shuttle does from 'go for orbiter ops' to docking are as follows. The Shuttle does a series of burns to catch up to the station. These burns NC-1, NC-2, NP-C and NC-3, all use ground tracking to target a point 40 miles [74 km] behind ISS. Ground tracking is the process of using ground-based radars, like C-Bands or S-Bands (big radar dishes), and TDRS satellites to tell the position and velocity of the Shuttle or station at a certain time to the Earth.

Ground tracking is accurate to within a couple of hundred feet of the position. The final ground-targeted burn is done at NC-4, which is nominally 40 nautical miles [74 km] behind the station, sending the Shuttle to an 8-nautical-mile [15 km] point were the TI burn is done. The TI burn sends us on a course to roughly 2,000 feet [610 m] away from the station. Now, the TI burn and the midcourse correction burns are done using Onboard Navigation. Onboard Navigation uses either a Star Tracker or the Ku-Band radar to track the station from the Shuttle. Now, Onboard Navigation gives the relative position and velocity between the Shuttle and ISS. Onboard Navigation can tell the distance between the Shuttle and ISS to a couple of feet [0.6 m]. The last midcourse burn aims the Shuttle to arrive below the station. From there, the crew does a quarter-circle around the ISS to dock. Inside 2,000 feet, other sensors are used to measure distance and speed because ISS is too big for the Star Tracker and Ku-Band radar. The crew uses some laser sighting sensors, cameras with special overlays and the windows to pilot the Shuttle from 2,000 feet into docking with the ISS. The Commander has a tough job meeting the contact condition of bull's-eying a circle to within a couple of feet, but by using the overhead window and a special camera overlay to measure distance inside 15 feet [4.6 m], they do a great job. In general, these are the burns done for ISS rendezvous flights."

STS-108 would deliver the MPLM Raffaello to ISS, with 3 tonnes of additional equipment for the Expedition-4 mission. Raffaello would be temporarily docked to Unity while it was unloaded. Goodwin and Tani would perform a single EVA to install thermal blankets around the motors that drive the P-6 ITS photovoltaic arrays.

In Endeavour's payload bay two experiments were housed in a Multiple Application Customised Hitchhiker-1 (MACH-1) facility. The Capillary Pumped Loop Experiment (CAPL) was a multiple evaporator capillary-pumped loop system. The Prototype Synchrotron Radiation Detector (PSRD) measured cosmic ray data. Both experiments were active throughout the early days of the flight.

Following the crew's first sleep period, Day 2 was spent preparing for the rendezvous. Gorie and Kelly continued to oversee the rendezvous manoeuvres from the flight deck. The crew also tested the RMS and used its cameras to record a video inspection of the payload bay, including the exterior of Raffaello. Later, Godwin and Tani prepared their EMUs and the tools for their EVA. Godwin powered up the Shuttle's docking system and extended the docking ring. The Expedition-4 crew participated in the day's activities as well as overseeing and performing a number of experiments onboard Endeavour. The workday ended with a few hours of scheduled relaxation prior to the busy week ahead. On ISS, the Expedition-3 crew had spent the day unloading Progress M1-7, which had been launched on November 26 and had soft-docked to ISS two days later. Rubber debris on the ISS docking system had prevented hard-docking at that time, causing Dezhurov and Tyurin to make an unscheduled EVA to clear the debris on December 3, allowing the craft to hard-dock.

On December 7, Day 3, Gorie took manual control of Endeavour as his spacecraft followed the standard rendezvous, approaching ISS from below, before manoeuvring through 90° out in front of the station to place the orbiter with its docking system facing PMA-2 on Destiny's ram. Viewing the approaching Shuttle through the station's windows, Expedition-3 commander Frank Culbertson told Endeavour's crew, "It will be real nice to see you guys." Gorie replied, "It's great to hear your voice."

Docking occurred at 15:03. During the initial attempt the Shuttle's docking ring was not correctly aligned with PMA-2. Remaining in the soft-docked position, both vehicles were left to allow their vibrations to damp out before a second, successful hard-docking attempt was completed. Following docking, Godwin and Terry completed the necessary pressure checks before the hatches between the two vehicles were opened at 17:22, and the Expedition-3 crew welcomed their replacements to ISS, along with the crew that would take them home. The newcomers received the now standard safety briefing given to all visitors to the station.

The following day Culbertson, Dezhurov, and Tyurin removed their couch liners and Sokol re-entry suits from Soyuz TM-33 and transferred them to Endeavour. Culbertson's crew would now return to America on the Shuttle. The Expedition-4 crew, Onufrienko, Walz, and Bursch, moved their couch liners and pressure suits into the Soyuz. From that point onward Onufrienko and his crew were the new residents of ISS and would return to Earth in the Soyuz, landing in Kazakhstan. On NASA's website Walz explained that American astronauts

> "... have extensive theoretical and practical training in the Soyuz capsule. In the case of some dire emergency we could fly the Soyuz to a safe landing in accordance with flight procedures and with the help of the ground. Hopefully, that would never happen. Our Commander is a tremendous Soyuz pilot ... but we do have the training."

The two Expedition crews then began the hand-over briefings that would continue throughout the week of docked operations. Before launch, Bursch had described the hand-over procedures in the following manner:

> "I recently talked to Jim Voss ... he was on Expedition 2, and he put it a way that I thought was pretty interesting. He says that in hand-over, you have three sets of notes or questions that you need to hand over between the crews. One set is an ongoing set of items that the ground keeps track of, maybe a system we're operating in a different mode than we've been trained for, or ... it's had a failure so this is how we're operating it ... So that is one set of notes or questions. Another set might be just personal notes, from Expedition 3 that Frank and Mikhail or Vladimir will have of notes that they have seen that maybe surprised them when they got on station, that they didn't realize that were different, or different from what they trained, or maybe tips that they can give us ... So that's the second thing ... and then after we can see those first two notes, we'll probably come up with our own questions, or maybe even after reading the first list we'll come up with some questions ... so it's a list of our own personal questions. So, through those three lists I think that's how we conduct the hand-over."

For Culbertson the crew change meant that he would have to come to terms with returning to an America still caught up in the aftermath of the September 11, 2001 terrorist attacks. The only American not on Earth when the attacks happened reflected that, "The most common thing I hear is that the United States is different now, that I will be surprised."

During the day, Kelly operated the RMS, with Godwin's assistance, to lift Raffaello out of Endeavour's payload bay at 12:01 and dock it to Unity's nadir. The transfer was completed at 12:55, December 8. Following pressure checks the hatches giving access to Raffaello were opened at 19:30 and the crews began unloading the MPLM. Expedition-4 Flight Engineer Carl Walz had described the transfer of equipment between the two spacecraft:

> "I'm one of the loadmasters. So Linda Godwin and I will be working together to make sure that all the cargo comes off of Raffaello and then all the cargo that's going back down gets onto Raffaello. And so it's like moving into a house where the old occupant is moving out, you're moving in, but you're using one truck ... you have to make sure that you don't get your boxes mixed up. [W]e'll be working very hard to make sure that we don't do that. And then Frank Culbertson and his crew are going to have to help us out by getting their pre-packed items ready to go, so as the upcoming boxes come off, Frank's boxes will come in, and so it'll be kind of a constant logistical activity as we make sure that everything coming up comes up, everything going down gets stowed."

December 9 began with music from the Fire Department of New York Emerald Society Pipes and Drums. The music was given to Kelly when he visited the wreckage of the World Trade Centre with NASA Administrator Dan Goldin following the terrorist attack on September 11. The three crews took time out at 17:24 that afternoon to remember the fallen, their families, and the rescue workers of the events of that day. Under Dan Goldin's "Flags for Heroes and Families" initiative several US flags were flown on ISS and Endeavour during this flight. These included one flag that was recovered from "Ground Zero", the site of the two World Trade Centre towers, in New York, a Marine Corps flag that had been recovered from the Pentagon in Washington DC, and a US flag from the state of Pennsylvania, where the fourth hijacked aircraft had crashed. Endeavour also carried a New York Fire Department flag, 23 replica New York Police Department shields, and 91 New York Police Department patches for distribution following the Shuttle's landing. Six thousand small American flags stored in Endeavour would be distributed to the families of victims of the September 11 atrocities.

Endeavour's crew spent the day unloading Raffaello while the two Expedition crews continued their hand-over briefings. During the day, Gorie and Kelly oversaw the first series of orbital boost manoeuvres using Endeavour's thrusters. Godwin and Tani checked their EMUs and the tools that they would use during their EVA, which was planned for December 10. In the evening, the STS-108 crew and the Expedition-3 crew made their way back to the Shuttle. The hatches between Endeavour and Destiny were closed at 19:43, isolating the Expedition-4 crew on ISS for the first time. Endeavour's internal pressure was then lowered in preparation for the EVA. Following their evening meal the two crews went to bed.

The highlight of December 10 was Godwin and Tani's EVA. Godwin left Endeavour's airlock at 12:2. Tani followed, telling his colleagues inside the Shuttle, "I'm going for a walk. I'll be back in a couple of hours." Taking in the view of Earth he remarked, "Wow, look at that view." As Endeavour passed over Houston, Tani observed, "I see it, downtown Houston, Intercontinental Airport. I see Ellington Field and Clearlake. Beautiful."

Following the installation of the P-6 ITS on the Z-1 Truss, in November 2001, engineers in Houston had noticed that the P-6 Beta Gimbal Assembly (BGA) motors were drawing more electrical power than expected. Analysis of the problem had led to the conclusion that the additional power was being drawn because the motors were expanding each time the SAW passed from Earth's shadow into sunlight and contracting again each time it passed from sunlight, back into Earth's shadow. Each expansion caused the BGAs to bunch, creating additional friction, which required additional electrical power to overcome. Godwin and Tani would install thermal blankets in an attempt to equalise the temperature acting on the motors. Sally Davis, Lead Flight Director in Houston, described the thermal blanket installation in the following terms:

> "It's a matter of treating our hardware as carefully as we can. If a motor starts to stall and you don't start treating it a little more carefully, it could eventually stall in the wrong position ... The solar array would be in a position where we cannot generate electricity."

The two astronauts were lifted half-way up the P-6 ITS by the Shuttle's RMS, which was operated by Kelly. From there, they made their way to the top of the 30 m tall tower, where they placed the thermal blankets around the two drum-shaped BGAs that drove the 40 m long SAWs. The blankets were intended to stabilise the heat reaching the two electrically driven motors. On their way back down the P-6 ITS they stopped at a storage bin and removed a cover that had been placed in the bin after it had been removed from an antenna on an earlier EVA. The cover would be returned to Earth. They also performed a number of "get ahead" tasks, including positioning two switches on the exterior of the station, where they would be installed during an EVA on STS-110, planned for spring 2002. The two astronauts also recovered a number of tools, which they took inside ISS for use on the STS-110 EVA. The STS-108 EVA ended at 15:04, after 4 hours 12 minutes. Inside ISS, Onufrienko, Bursch, and Walz had spent the day unloading Raffaello.

December 11 was the 3-month anniversary of the atrocities in New York and was commemorated by President Bush's "Anthems of Remembrance" initiative. In the mission control centres at Houston and Korolev, as well as on Endeavour and ISS, the American and Russian national anthems were played in remembrance of those who had died in the terrorist attacks. The American anthem was commenced at 08:46, the time when the first aircraft impacted the World Trade Centre. During the day the American and Russian members of the various crews were given oppor-

Figure 12. Expedition-4: Daniel Bursch floats inside the Multi-Purpose Logistics Module docked to Unity's starboard side.

tunities to express their personal feelings to their respective control centres regarding the events of September 11. In Houston, caps honouring the New York Police Department, Fire Department of New York, New York Port Authority, and New York Office of Emergency Management were displayed in the control centre. Wayne Hale, Flight Director in Houston, told the astronauts,

> "More than 3,000 people perished this day three months ago, including more than 200 citizens from countries that are family members of the International Space Station programme—Canada, Italy, France, Germany, Japan and Russia."

During the morning Endeavour's crew were told that their flight had been extended by one day. The extra time would be used to assist in routine maintenance on ISS. STS-108 would now land on December 17. At 15:48 Culbertson made a ceremonial hand-over of command to Onufrienko. The remainder of the day was spent with all three crews participating in the unloading of Raffaello. Items already unloaded and moved to ISS included food, clothing, medical supplies, EVA equipment, experiments, and other crew provisions for use by the Expedition-4 crew.

December 12 was another day of cargo transfers. With most of the cargo already moved out of Raffaello, the crew began packing the near-empty module with items to be returned to Earth. Endeavour's crew also spent the day assisting the Expedition-4 crew by replacing some components on the ISS treadmill. The work went well and the task was completed several hours ahead of schedule. Hand-over briefings for the Expedition-4 crew also continued throughout the day. Gorie and Kelly completed the third series of orbital re-boost manoeuvres using Endeavour's thrusters. During the crews' sleep period, one of Endeavour's three Inertial Measurement Units (IMUs) suffered a transient problem. The unit was one of three, of which two were powered on and one powered off whilst Endeavour was docked to the station. IMU-1 and IMU-2 were on-line when the fault occurred, with IMU-3 off-line. IMU-2 was taken off-line and IMU-3 was placed on-line to replace it. Only two IMUs were required to fly the Shuttle, the third was carried as a redundancy measure. Although IMU-2 continued to perform correctly, controllers considered it to be "failed". Endeavour would return to Earth on IMU-1 and IMU-3.

The following day the STS-108 crew packed the last of 2 tonnes of cargo, including laundry, packaging foam, and equipment no longer required on the station inside Raffaello, while the Expedition crews continued their hand-over briefings. Culbertson noted, "Basically, everything's over there that should go today." He joked that the Expedition-4 crew would be glad "to get rid of those three extra bodies (the Expedition-3 crew)." At 16:09, the three crews gathered together to watch Culbertson officially hand over command of ISS to Onufrienko. Houston reminded the Expedition-3 crew, "In just a few days you'll be back on Earth, feeling the warm Sun on your faces."

Following the final loading of Raffaello, the hatches between the MPLM and Unity were closed at 11:00, December 14. After pressure checks Kelly and Godwin used the RMS to undock Raffaello, shortly after 14:20, and successfully placed it back in Endeavour's payload bay at 17:44. The Expedition crews continued with their final briefings and Dezhurov worked with Onufrienko to replace a faulty compressor in an air-conditioning unit inside Zvezda. During the day America's Space Command informed NASA that a Russian rocket upper stage from the 1970s would pass close to ISS and the decision was made during the night to use Endeavour's thrusters to raise the station's orbit before the Shuttle undocked the following day.

December 15 began at 05:17. The STS-108 and Expedition-3 crews said their farewells and returned to Endeavour. Culbertson presented the Expedition-4 crew with a small Christmas tree, candy canes, and individual presents consisting of small silver pins. He then saluted Onufrienko before shaking his hand and exchanging farewells. Culbertson, Dezhurov, and Tyurin had spent 125 days on ISS. By the time they landed on December 17, they would have been in space for 129 days including the days spent on the Soyuz spacecraft that carried them into space. The hatches between Endeavour and Destiny were closed and the usual pressure checks were completed. At 09:55 Endeavour's thrusters were pulse-fired over a 30-minute period to raise the station's orbit and move it clear of the spent 30-year-old Russian rocket stage.

Prior to undocking, Culbertson, now settled in the mid-deck of Endeavour, commented, "It's been a wonderful experience for all of us. The work continues, the research continues and will for many years to come."

Kelly undocked Endeavour at 12:28 and made a 90° fly-around of ISS before completing the separation burn. As Endeavour finally began to move away from the station Culbertson told Houston, "It feels wonderful to be heading home." He made it clear that he was looking forward to seeing his wife and five children, but also that he wanted a hot shower, and a bowl of ice cream covered in chocolate syrup.

Dezhurov was equally personal, saying, "After landing I want to meet with my family. Maybe after that, I will think of some food. I also want to go to the sauna and take a shower." Tyurin added his similar thoughts, "Mostly we are thinking about opportunities to see our families. But, also, I've said a big glass of cold beer would be fine."

Gorie's crew packed their gear in preparation for re-entry and then enjoyed a few hours free time before beginning their final night in space. On the station the Microgravity Acceleration Measurement System (MAMS) recorded the vibrations associated with the undocking. The Protein Crystal Growth-Single Thermal Enclosure System (PCG-STES) had continued to operate throughout the crew hand-over as had the PCG-STES Unit 10 experiment activated on December 6.

At 10:00, December 16, the last complete day in orbit, the STARSHINE satellite was launched from Endeavour's payload bay. The crew spent the day preparing Endeavour for re-entry. The final day began on Endeavour at 04:19, December 17. Endeavour's payload bay doors were closed at 09:10. The crew assumed their re-entry positions at 10:50 with Culbertson's crew strapping themselves into their reclining seats on the mid-deck. Retrofire occurred at 11:55 after which Gorie turned his spacecraft upright and nose forward for entry into the thick lower atmosphere. Endeavour landed at the Kennedy Space Centre at 12:55, after a flight lasting 7 days 21 hours 25 minutes. Following the routine making safe of the orbiter, Culbertson, Dezhurov, and Tyurin were removed from the mid-deck to begin their readjustment to life under 1*g*. They underwent bed rest and initial medical examinations for several days, before being transported to Houston for the official rehabilitation programme undertaken by all returning long-duration space station crews. The third occupation of ISS was at an end.

EXPEDITION-4

Prior to launch Bursch was asked how he viewed his flight with the Expedition-4 crew. He replied, in part,

> "I think early on, well, maybe after about a year of all the travelling and training, Carl and I were talking about it with Yuri and I think we all came to the realization that this mission is different. And Yuri has been very helpful and because of his experience already on Mir and the six months that he spent on Space Station, of letting us know that it is different ... the best analogy that I think I can come up with, it's kind of like the difference between a sprint and a long-distance race ... With a long-duration mission I think it's going to be more like a marathon, where we're going to have to pace ourselves ... we can't run at the

same pace that we're used to running on the Space Shuttle, and there'll have to be times where we'll have to help each other out just to say, hey, it's time for you to just take a break and sit by the window, look out, and take some time; have some time to yourself ...

However, there'll be times when a Shuttle comes to visit and we'll have six to eight very, very busy days of getting ready for them to come, while they're docked, helping with the EVAs, helping with installation of new equipment, and so, it'll vary from being very slow to very fast, almost a Shuttle-like pace, and then slowing down again. And what'll be different, also, is that the weekends, I haven't quite figured out how that's going to work; I know that on Saturday it will be a lighter duty, and Sunday in most respects we'll have much of the day off, but I can't imagine having the day off and not having anything to do ... Maybe ... I'll end up saying, OK, give me things that need to be done, because ... I've run out of personal time or, I don't want to read a book right now, I want to do something."

In the same interview Bursch explained how the Expedition-4 crew viewed their personal responsibilities,

"[J]ust as in Shuttle flights ... everybody has their own position and their own responsibilities ... so early on we ... divided up the responsibilities of the crew. Yuri, for a Commander it's different, in that he's responsible for everything ... he's tended to specialize more on the Russian segment and some of the Russian systems, especially because he had the experience on Mir and a lot of the systems ... are similar to Mir. Carl, with his experience on EVAs ... became the ... natural lead for any type of EVA activities that we would do, especially on the U.S. segment or using the U.S. spacesuit, or the EMU. Yuri's kind of, is more of an expert on the Russian Orlan, the Russian spacesuit ... I took on working with the Space Station Remote Manipulator System ... And that has been my prime responsibility and my prime system. There's also systems on the station that we divide up ... keep in mind, with only a crew of three, everybody has to know a little bit about everything ... however each one of us has tried to specialize in a certain part of the system to take care of that ... So we kind of had a big list of all the tasks and ... of all the equipment, and all the things that we need to do on space station, and we just basically divided them up."

On the subject of the crew's daily routine Bursch added,

"It'll basically be, wake up, get ready to go have some breakfast, read the morning mail, and have some time together, to talk about the upcoming day, what do we have planned for the day. So regular things like that, whether it's personal hygiene or reading or getting up-to-date is going to be somewhat normal. And then there'll be some weeks that I think the pace will definitely change . . .the pace will probably be toughest ... when the Shuttle is docked, and we're doing operations together

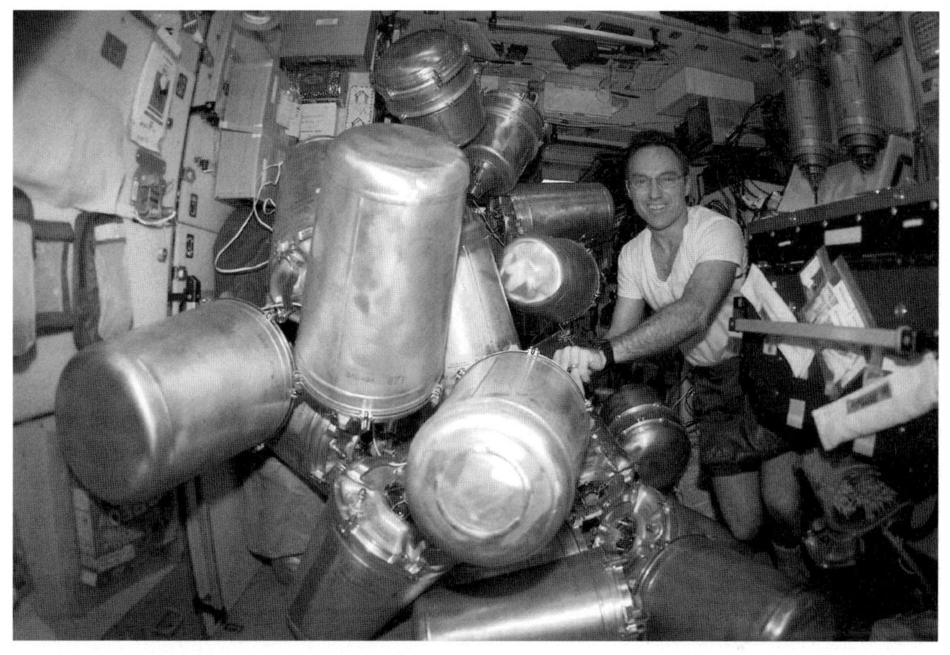

Figure 13. Expedition-4: Carl Walz works with containers of potable water inside Zvezda.

> and involving many different systems. Certainly the pace will be much slower . . . if the Shuttle isn't there . . . So the pace is probably going to be all over the place."

Onufreinko, Walz, and Bursch settled into their own routines in the third week of December. They activated science experiments and unloaded Progress M1-7 as well as the equipment carried to ISS on STS-108. The Expedition's cellular experiments, begun on December 15, were completed after 12 days. On December 19, the crew noticed that despite the installation offered by its insulation blankets the electric motor driving the Beta Gimbal Assembly (BGA) that rotated the Port-6 SAW had experienced a strain and stalled. The motor was restarted and continued to run normally. Engineers at Houston continued to monitor both the Alpha and Beta drive motors.

The crew took a rest day on December 25, and January 1, 2002 to celebrate Christmas and New Year. Although they had to complete their daily exercise regime, they also took time to relax and to talk to their families and friends. On the subject of holidays, Bursch would write,

> "The holidays were a nice break from the rapid pace of a Shuttle mission. I kept thinking about what several experienced Expedition crew members had told me; the Shuttle mission is a sprint, and the Station mission is a marathon. Of course, being away from family during the holidays is always tough. It was very hard for me to be away from my family, but I couldn't help but think of all of the service

men and women that were away from their families as well. And I also couldn't help but think about the tens of thousands of people that were missing friends and family over the holidays because of the terrorist acts of September 11th. And for them there would be no future reunion. I suddenly felt very fortunate to have a healthy family on Earth, knowing that they were sharing the holidays with loved ones."

With the holiday period over, the crew returned to their experiments. Both the Active Rack Isolation System ISS Characterisation Experiment (ARIS-ICE) and the Experiment on Physics of Colloids in Space (EXPPCS) were halted on December 21, and resumed on January 2. The Payload Operation Centre (POC) in Huntsville sent commands to activate the experiments and monitored them as the crew went about their daily tasks. All three men completed their Crew Interaction Questionnaires on December 26–27.

Walz and Bursch both participated in the H-Reflex and the Pulmonary Function Facility (PuFF) experiments. The first experiment studied the spinal cord's adaptation to microgravity. The second studied the effects of spaceflight on lung function. The same pair of astronauts had the opportunity to practise operating the SSRMS on January 3. In moving the SSRMS from one location to another on the exterior of ISS, Walz and Bursch allowed controllers in Houston to record the strains experienced when the arm separated from the fixture holding it to the station. Walz described the crew's daily participation in the station's numerous experiments in the following terms:

"We're like a lab technician: we'll be performing media exchanges for samples, we'll be checking to ensure that the samples are growing as planned, we'll report to the ground if there're any anomalies, we have status checks to perform every day. So, it's just, once we get things started, to make sure everything progresses per the timeline, and then at the end to make sure that we terminate the experiment properly so that when we bring the samples home, the scientists will be able to make the proper evaluations."

On January 7, the EXPPCS began a 120-hour run, the longest yet. On the same date the ARIS-ICE science team began a series of one-minute isolation tests of their equipment using new control software. The crew also checked the individual radiation badges and the monitoring equipment connected with the Extravehicular Activity Radiation Monitoring experiment (EVARM). The radiation badges would be placed inside the cooling garments of pressure suits during future EVAs to monitor radiation reaching the wearer's skin, eyes, and internal organs. Their first use would be on the STS-110 Shuttle flight when it visited ISS later in the year. Walz and Bursch completed the first session of the Renal Stone Experiment, a study of the risk of astronauts developing kidney stones during long-duration flights. Over one 24-hour period the two men monitored their diet and collected samples for return to Earth each time they urinated. They also had to keep logbooks throughout the period of the experiment.

Also on January 7, Bursch removed the hard drive from the Command & Control-2 (C&C-2) computer and replaced it with a new solid-state mass memory card, with three times the memory of the old drive. The computer was located in Avionics Rack-3 in Destiny, from where it processed all commands from Huntsville to the experiments and the flow of telemetry from the experiments back to Huntsville. The new memory card had been delivered to ISS on STS-108.

During the week the crew began preparations for their first EVA, which Walz described in his pre-flight interview published on the NASA Human Spaceflight website,

> "Well, the spacewalk that Yuri and I will do is to move the first Strela cargo boom, which came up during the 2A.2a flight, and we're going to move that from the PMA to the Docking Compartment. And so, we will use the Strela that comes up in the Docking Compartment to drive us over from the Docking Compartment to the PMA, sort of lash the second Strela on, and then bring it over. So it should be a very visually interesting EVA because I'll be hanging at the end of the Strela ... being transferred in free space, with this other large structure. So I think it's going to be very exciting."

That EVA began at 15:59 January 14, when Onufrienko and Walz left Pirs wearing Russian Orlan pressure suits. They assembled and installed an extension to the Strela-1 crane mounted on the exterior of Pirs. Their main task was use the Strela-1 crane to position themselves so that they could detach and reposition the Strela-2 crane from PMA-1 to the base plate on the opposite side of the exterior of Pirs to the Strela-1 crane. Strela-2 was relocated at 19:31. With the relocation complete the two cranes would be able to work in tandem during future EVAs and construction of ISS. The two men also deployed an amateur radio antenna on an EVA handrail at the end of Zarya. They returned to Pirs and prepressurised the airlock at 22:02, after an EVA lasting 6 hours 3 minutes. Immediately after the EVA all three crewmen completed a turn on the PuFF experiment, to study the evenness of gas exchange in their lungs.

On the ground the Active Rack Isolation System (ARIS) control team in Huntsville down-linked data from their experiment and sent new software commands to their equipment in order to maintain good housekeeping on the data storage disk. The EXPPCS completed the 120-hour run begun on January 7. The experiment then began a new 24-hour run, on January 21. This was the last run before a new set of fractal gel tests that would last approximately five weeks and would prevent the crew from examining other samples held within the experiment. All three men completed their Crew Interactions Questionnaire and received new Crew Earth Observations targets, which they would photograph if the opportunity arose.

During the week following the EVA, Walz worked with controllers on the ground to remove the hard drive from the C&C-1 computer and replace it with a solid-state mass memory unit. The task took over four hours, but C&C-1 was brought on-line as the ISS back-up computer on January 23. Meanwhile, Onufrienko and Bursch replenished the two Orlan EVA suits worn on January 14, and prepared the equipment that they would install on the exterior of ISS during their second EVA. Of the 25 experiments planned for Expedition-4, 15 were in progress at this time. During his sleep period on January 24, Bursch was disturbed by a noise. Investigation revealed that one of the push-rods on the ARIS-ICE experiment was broken.

Onufrienko and Bursch began the Expedition's second EVA at 09:19, January 25, when they left Pirs dressed in Orlan pressure suits to mount deflectors behind six of Zvezda's manoeuvring thrusters. They also recovered the Kroma-1-0 experiment package, which had been collecting samples of the thruster effluent deposited on the side of Zvezda when the thrusters fired. They then installed the fresh Kroma-1 collector package in its place. A second ham radio antenna was placed on the exterior of Zvezda along with the Plantan-M package, which was an experiment designed to detect neutral low-energy nuclei both from the Sun and from outside the Solar System. They also installed three materials exposure experiments on the exterior of Zvezda. Finally, they installed fairleads on Zvezda's EVA handrails, guides to prevent an EVA astronaut's safety tether from fouling equipment mounted on the exterior of the module. The EVA ended at 15:18, after 5 hours 59 minutes. On January 26, Onufrienko and Bursch completed their post-EVA PuFF experiments.

January ended with a quiet week during which the crew changed the hard drive in the C&C-3 computer for a solid-state mass memory unit. The crew tested the station's communication systems in an attempt to eradicate an echo that was degrading audio communications. The KURS rendezvous equipment was removed from Progress M1-7 for return to Earth, where it would be refurbished and reused. They also installed a laptop computer in the Quest airlock. On January 29, they withdrew the EVARM radiation badges from the pockets of their liquid-cooled EVA undergarments and recorded the dosages for transmission to the ground. Bursch also logged his dietary data as part of the Renal Stone Experiment. The broken push-rod on the ARIS-ICE experiment was removed and replaced on January 30.

At 08:00, February 4, the Expedition-4 crew's 60th day in space, the main computer in Zvezda crashed, disrupting the station's attitude control. The crew began powering down back-up systems and all experiments in case of a decrease in electrical power caused by the station's SAWs losing their lock on the sun. Controllers in Houston and Korolev worked together to restart the computer, which was achieved at 10:30. One hour later the station's attitude control system was back on line. Power was restored to sensitive experiments within 6 hours and everything was back to normal 24 hours after the computer crash. EXPRESS Rack 4, in Destiny, was the first to be powered on, as it held the Bio-technology Refrigerator and the PCG-STES which both contained biological samples. The MAMS had undergone a period of maintenance prior to the computer crash and was powered on the day after the crash. EXPRESS Racks 1 and 2, and the Space Acceleration Measurement System (SAMS) experiment were powered on during February 5, and the crew began preparations for a 72-hour run of the EXPPCS. EarthKAM was re-instated in Zvezda's window, one week early. The experiment allowed Middle School children to command cameras on ISS to expose photographs of Earth's surface. NASA subsequently published the photographs on the Internet. Additional schools, including one in Germany, had joined the experiment since it was shut down by the Expedition-3 crew in anticipation

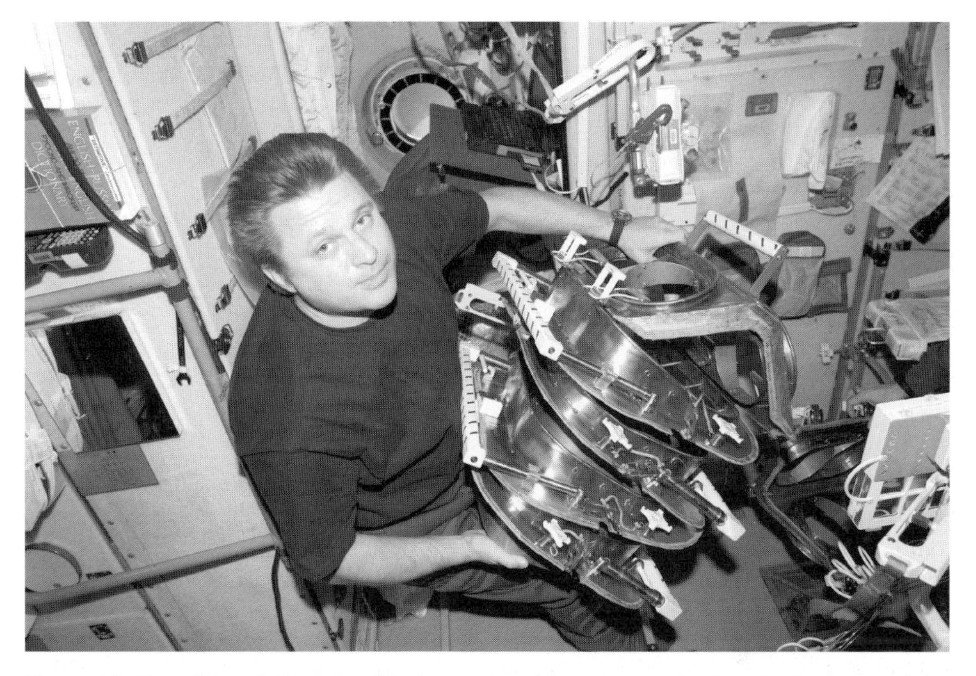

Figure 14. Expedition-4: Yuri Onufrienko works with equipment in Zvezda. In the top right-hand corner of the view is an American–Russian dictionary.

of the school holiday period. Fifty images had been downloaded since the cameras were reinstalled following the school holiday. Activities related to the Education Payload Operation-4 (EPO), which included the crew setting up and video taping a series of simple experiments, was rescheduled from February 4–5.

Onufrienko celebrated his 41st birthday on February 6. Two days later, the crew began to take an inventory of the equipment and food onboard to help planners work towards re-stocking ISS for the Expedition-5 crew. On February 13, they began a Human Research Facility (HRF) workstation test. The following day they carried out testing of the ultrasound life science equipment. The day after that they were familiarising themselves with the Zeolite Crystal Growth furnace before commencing that experiment programme. They also activated one of the cylinders in the PCG-STES Unit 7, which would be used to grow mustard seeds harvested on ISS by the Expedition-2 crew and the Advanced Culture (ADVASC) plant growth experiment.

On the same day a Remote Power Conversion Module (RPCM) failed. The RPCM distributed electricity to the station's systems. The crew worked with engineers in Houston to repair the RPCM over the next few days. Bursch and Walz replaced a circuit breaker box using a spare held on ISS and thereby restored power to the non-essential equipment in Destiny. The repair required Bursch's sleep station in the laboratory module to be removed. Before it was re-established they took the opportunity to install additional high-density radiation protection bricks behind it.

On February 15, the EXPPCS began a 120-hour run. A planned update of the station's computer software, to prepare the computer system for the installation of the Starboard-0 (S-0) ITS during the flight of STS-110, in April 2002, was delayed until after the EVA planned for February 20.

That EVA began at 06:38, when Walz and Bursch made the first egress from Quest without a Shuttle docked to the station. Both men employed a new combination of pure oxygen and exercise to purge the nitrogen from their bloodstream. They wore American EMUs and for the first time the Intravehicular Officer, responsible for supporting the EVA astronauts, Joe Tanner, himself an astronaut, was based in the control room at Houston, rather than in orbit. Throughout the EVA Onufrienko used the video cameras on the SSRMS to view his colleagues' activities. The Americans designated this Stage EVA, "US EVA-1". The two astronauts removed two power cables from their stowage area on the exterior of Destiny and plugged them in to a cable near the base of the Z-1 Truss. Plans to disconnect the cables and return them to storage at the end of the test were cancelled when plugging them in caused unpredicted power readings in the current conversion unit in the circuit that they completed. Working separately, Walz removed four thermal blankets from the Z-1 Truss and stowed them within the truss, while Bursch retrieved tools that would be used during the four EVAs planned during STS-110 and carried them to the Quest airlock. They then joined together to secure the locks holding oxygen tanks to the exterior of Quest that were looser than required following their original installation. They removed two adapters that had been used to hold the Russian Strela cranes on the exterior of the station. One was relocated to the exterior of Zarya, while the other was placed inside the Quest airlock. Finally, they inspected external cable connectors and the Materials on International Space Station Experiment (MISSE), where they found that some of the exposure samples had peeled back from their mounts. Walz and Bursch returned to Quest at the end of their planned activities. The hatch was closed and the EVA ended at 12:25, after 5 hours 47 minutes. Throughout the EVA the MAMS and SAMS equipment recorded the vibrations associated with the external activities. Both EVA astronauts participated in PuFF and EVARM experiments prior to, during, and following their activities.

They were two hours late going to bed at the end of the day. The delay was caused when an unpleasant odour began emanating from the equipment used to clean the air scrubbers in the EMUs that they had used during their EVA. On instruction from Houston the equipment was powered off and the Quest airlock internal hatch was closed to prevent the odour prevailing further into the station. Some of the station's ventilation fans were powered off, while the system used to scrub the station's air was powered on in Destiny. All three men slept in Zvezda that night. On February 21, controllers at Korolev used the rocket motors in the Progress M1-7 spacecraft docked to ISS to raise the station's orbit.

The crew completed their first activities as part of the Education Payload 4 experiment on February 25. During the activities they used a series of small toys to demonstrate basic principles of physics and the microgravity environment. Three days later they activated Cylinder 8 of the PCG-STES experiment. In Destiny, EXPRESS Racks 1, 2, and 4 continued to function normally.

At the beginning of March, STS-109 was launched on a solo flight to repair the Hubble Space Telescope. On ISS the crew spent the first week of the month repairing a shock absorber in the ARIS-ICE and then bringing the experiment back on-line. They also recovered air and water samples from the ADVASC experiment. The samples would be stored before being returned to Earth for study.

Earthkam's cameras were once again made available to schoolchildren. To prepare the Earthkam experiment the astronauts needed to move the SSRMS, which was partially blocking the window in Destiny through which the cameras exposed their images. The primary avionics system failed to release the brakes on the SSRMS, requiring the crew to use the secondary avionics system in order to complete the move. An investigation into the failure began immediately on the ground in Houston. On March 7 the crew began a training run, putting the SSRMS through the series of manoeuvres that it would be required to perform when STS-110 delivered the S-0 Truss in April. During the same day, software tests were sent up to the primary and secondary computer workstations used to control the SSRMS, but the secondary station failed to boot up. When that happened the SSRMS was left parked in a safe position while another investigation began, to identify why the boot-up had failed. During the following week the SSRMS was used to carry out a video review of the station's radiators and SAWs. The secondary software was used to drive the arm. Also at that time, the crew completed an inventory of items on ISS, stowing equipment and preparing items for return to Earth in advance of STS-110's visit, when there would be a total of ten people living and working on the station. On March 19, the rubbish-filled Progress M1-7 was sealed and manoeuvred away from Zvezda's wake, at 12:43. It re-entered the atmosphere and burned up a few hours later, after releasing the short-lived Kolibiri-2000 sub-satellite into orbit.

PROGRESS M1-8

The seventh Progress was launched towards ISS at 15:13 March 21, carrying the usual combination of water, propellant, and dry goods. Following a standard Soyuz rendezvous Progress M1-8 docked to Zvezda's wake port at 15:58 March 24. Pressure and leak checks were completed before the hatches between the two vehicles were opened. The Expedition-4 crew began unloading the new cargo the following day.

Bursch wrote,

> "I was amazed at how much I anticipated the arrival of this Progress. I thought that because we had e-mail and the 'phone,' that I wouldn't think the Progress was such a big deal ... but it was! I underestimated how much I would anticipate the arrival of fresh fruit and care packages ... something from Earth ... something from home ... that my friends and family had touched not too long ago!! ... Everything went well, and after pressure checks we opened the hatch about midnight. I felt as if it was Christmas morning! Everything was tightly packed, but we managed to get to our care packages after about an hour. I honestly forgot

> that we hadn't been visited for the past three months ... and something 'fresh' from home was VERY welcome! We got new books on CD, cards, letters, pictures and some new DVDs."

On March 29, Walz performed hammer tests on the ARIS-ICE.

EUROPEAN COMMERCIALISATION

ESA signed an agreement on March 22 with 11 companies to promote commercialisation of ISS in Europe by making access to the station easier for those companies. ESA agreed to assist the companies in question to promote awareness of the commercial possibilities of ISS within Europe, while supporting projects for such use made by the companies in question. With almost 33% of ESA space on ISS allocated to commercial use rented through the Agency, a spokesman said of the agreement,

> "The co-operation agreement between ESA and our strategic partners in industry forms the foundation of a true partnership between the public and private sectors. With the complementary skills of ESA and our partners we are now in a very good position to optimise the services we can offer to those customers who recognise the unique utilisation opportunities of the International Space Station."

WORDS OF WARNING

In America the Aerospace Safety Advisory Board warned that NASA was too tightly focused on short-term planning for Shuttle flights and was ignoring the long-term safety planning implications of such a narrow focus. The Board warned, "Unless appropriate steps to reduce future risk and increase reliability are taken expeditiously, NASA may eventually be forced to ground the system until time-consuming improvements can be made." With the Shuttle suffering technical difficulties and foam shedding from the ET on practically every flight, it would only be a matter of time before the Board's words would be proved to be prophetic in the extreme.

April began with a hunt for unexplained vibrations detected by the SAMS. The Expedition-4 crew entered and left their sleep stations four times while the experiment's controllers at Huntsville monitored the vibrations picked up by the experiment. Acceleration data was down-linked to Huntsville from one of the MAMS low-frequency sensors. The high-frequency sensor was disabled at that time. The crew continued to take samples from the ADVASC experiment and stored them in the Biotechnology Refrigerator for return to Earth on STS-110. They were also preparing to deactivate the PCG-STES Unit 10.

On a lighter note Bursch described how they entertained themselves on such a long flight,

> "We have some foam balls that we try to throw the length of the station and really haven't been too successful so far. But that's kind of what we do for fun. Of

course, we play with our food, like every good astronaut. Carl's great on the keyboard. He has a keyboard up here that he plays. And we watch movies."

NASA'S COSTINGS CRITICISED

In the run-up to STS-110 the Inspector General's Office issued a report criticising NASA for having no assurance that they had paid the best available price for ISS components. Having spent $334 million on ISS hardware through FY2000, NASA had not negotiated separate prices for each component and had not maintained thorough records of what had been paid. Therefore, the report concluded that NASA could not know if they had paid the best available price, or if they had been overcharged. The report also highlighted the fact that Boeing had not included certain costs in its price estimates and NASA had therefore underestimated its own costs by $39 million over the five years from 1995 to 2000. The criticism came on top of the cost over-runs that had led to the cancellation of the American Habitation module and the X-38 CRV.

NASA's Space Station programme manager Thomas Holloway stated that three teams were studying NASA's ISS finances for the next 3 years. He estimated that the station could be brought to the new "Core Complete" in 2006, for a cost of $8.3 billion. Holloway told the press, "I can assure you that we know where all our money is and where all our money is going. I also do not believe that we are wasting any taxpayers' money."

On the same subject, Joe Mills of Boeing added, "What's disconcerting is that in the face of all this technical accomplishment ... we don't get a more balanced appreciation of the project."

STS-110 DELIVERS THE STARBOARD-0 ITS

STS-110	
COMMANDER	Michael Bloomfield
PILOT	Stephen Frick
MISSION SPECIALISTS	Ellen Ochoa, Steven Smith, Rex Walheim, Jerry Ross, Lee Morin

STS-110 should have been launched on April 4, 2002. On that date the countdown was stopped shortly after the filling of the ET's liquid hydrogen tank had begun. A leak had been found in a weld on a gaseous hydrogen vent line on the Mobile Launch Platform. The line was used to vent hydrogen vapours out of the ET, to prevent a dangerous build-up of pressure. White gas flowed through the leak for 1 minute before it was isolated. Atlantis' crew had not begun suiting up when the launch attempt was scrubbed, so the spacecraft was returned to a safe condition, and the Rotating Service Structure was moved back around the vehicle. Because fuelling had begun the countdown was recycled $T-72$ hours with the next attempt set for April 7, and subsequently moved to April 8. The faulty hydrogen line was vented and purged with nitrogen before a welding team began the 12-hour repair, which was followed by inspection and testing. The STS-110 crew remained at Cape Canaveral throughout the weekend.

On April 4, Bursch wrote,

> "Today is the scheduled launch of STS-110, which will be bringing up the S-0 Truss segment ... I am very much looking forward to the arrival of Atlantis and her crew. They promise to bring new care packages from home, fresh 'smells' of the Earth and old friends. We know that the work pace will once again speed up, but we are ready! We worked many hours together on the ground developing procedures to use the Space Station robotic arm (SSRMS) as a 'cherry picker' as we manoeuvre space walkers 'flying' on the end of the arm. This will be the first time that this new arm will be used in such a capacity ... Later we heard that the launch was delayed. It is disappointing, but I am fairly familiar with launch delays and understand what the crew is feeling right now. There are many things that need to come together before the SRBs ignite. Some we can control, some we can't."

On the new launch day the countdown proceeded to $T-5$ minutes when it was held due to a software difficulty. Atlantis finally climbed away from LC-39B at 16:44, April 8, 2002, just 12 seconds before the end of the 5-minute launch window. Mission Specialist Jerry Ross was beginning his record-breaking seventh spaceflight. The ISS was high over the Atlantic Ocean and approaching the US eastern seaboard as a live video link with MCC Houston allowed the Expedition-4 crew to watch Atlantis climb into space.

In its payload bay Atlantis carried the first segment of the Integrated Truss Structure (ITS). The $600 million, 13.4 m long S-0 (Starboard-0) ITS would be attached to Destiny's upper surface and would form the centre element of the 100 m long ITS. Later Shuttle flights would deliver a further eight ITS elements, to be located four on either side of the S-0 ITS. The elements on the station's port (left) side would be numbered P-1, P-3, P-4, P-5, P-6, with P-6 being re-located from its temporary position on top of the Z-1 Truss. Those segments on the starboard (right) side would be numbered S-1, S-3, S-4, S-5, S-6. P-3, and P-4 would be launched as a single combined unit, as would S-3 and S-4. The P-2 and S-2 ITS elements had been part of a larger Space Station Freedom design, but had been deleted when the ISS design was downsized. The remaining ITS elements were not renumbered at that time and there are no P-2 and S-2 ITS in the current ISS design. STS-110 would also deliver the $190 million Mobile Transporter (MT) for the SSRMS. Built in Canada, the MT will allow the SSRSM to travel along rails on the ITS following the delivery of the Mobile Base System onboard a later flight.

Nine minutes after launch Atlantis was in orbit and Bloomfield set up the orbiter's computers for the rendezvous with ISS. The payload bay doors were opened on schedule allowing the flight to continue. Ochoa described the view of the payload bay through the flight deck's rear windows, including the S-0 ITS, "It takes up pretty much the whole payload bay. You just look out and you think 'Wow, it looks beautiful and we can't wait to get started on operations.' We're just raring to go."

The crew began their sleep period at 21:44, and began their first full day in space at 05:44, April 9. Bloomfield's crew spent the day performing the standard rendezvous routines and preparing their equipment for the docked operations ahead. Onboard ISS, the Expedition-4 crew spent their time preparing for their visitors, having adjusted their sleep period routine to match that of the STS-110 crew. The crew on ISS described Atlantis as a "hot star on our tail". On Atlantis the day ended at 20:44.

Following a textbook rendezvous Bloomfield and Frick gently docked Atlantis to Destiny's ram at 12:05, April 10. On ISS, Bursch rang the ship's bell that had been installed by the Expedition-1 crew to welcome their first visitors since occupying the station in December. After two hours of post-docking checks the hatches between the two spacecraft were opened at 14:07. Bursch reported, "I'm just so happy to see other faces."

Following a Safety Briefing from Onufrienko the two crews began a rehearsal of the procedures required for the installation of the S-0 ITS on the upper surface of Destiny. During the procedures, Ochoa and Bursch put the SSRMS through the manoeuvres required to transfer the S-0 from Atlantis' payload bay to its location on Destiny. Meanwhile, Smith and Walheim transferred the equipment required for their first EVA from Atlantis to the Equipment Lock in Quest. Morin and Ross transferred the Commercial Protein Crystal Growth High Density (CPCG-H) experiment from Atlantis to the station and powered it on. Data received on the ground showed that it had survived the journey into space without damage. The day ended at 20:44.

Both crews were woken up at 04:44, April 11, to begin the assembly work that was the highlight of the flight. Ochoa and Bursch used the SSRMS to grapple the 13.5-tonne S-0 ITS at 05:00. Thirty minutes later S-0 was out of the payload bay. It was then manoeuvred into place above the Lab Cradle Assembly on Destiny and lowered to a semi-rigid fixing just under four hours after the task was begun. Throughout the installation Bloomfield and Frick operated Atlantis' RMS so that its cameras could provide additional views to Ochoa and Bursch.

In his pre-launch interview on the NASA Human Spaceflight website Bursch had described the S-0 ITS and the tasks involved in mounting it on ISS in the following terms:

> "The S-0 is the first segment that comes up that is crossways on the Lab. And so structurally it's the backbone of the truss that will be installed on the Lab. Second of all, the S-0 will bring up some equipment onboard that will allow the American segment ... of the space station to determine its own attitude, and also to determine its rate. Right now we have computers on the American segment that can control the attitude of station, but we get all of our attitude data, meaning how the Space Station is positioned in space and where it's positioned in space,

Figure 15. STS-110 crew (L to R): Steven Smith, Stephen Frick, Rex Walheim, Ellen Ochoa, Jerry Ross, Michael Bloomfield, Lee Morin.

from the Russian segment. It'll be the first time that we'll be able to do that by ourselves on Space Station. It'll have GPS [Global Positioning System satellite] antennas that will be able to not only determine a state vector, or a position and a velocity and acceleration in space, but also the attitude of Space Station while it's flying.

Right now there are four space walks scheduled for, in the installation of the S-0 Truss ... we're going to be using the Big Arm [SSRMS], so myself and Ellen Ochoa will be grappling the S-0 Truss in the Space Shuttle payload bay picking it up out of the payload, and it basically takes up just about the entire payload bay of the Shuttle, pick it up, move it over to the port side, and eventually install it on the top of the Lab. And then there'll be some preliminary mechanical connections that are made just to hold it temporarily on Space Station. [W]e'll actually have the two space walkers, Steve Smith and Rex Walheim, waiting in the airlock while we do that ... and then they go out the hatch. And there's some struts ... to better structurally attach the S-0 Truss to the Laboratory ... There're two in the front and two in the back of the S-0 that go to the Lab, and those have to be attached ... we have mechanical connections that need to be made, to fix the S-0 Truss to the

Lab; there's also umbilicals that need to be connected, and these are for power to some of the equipment that we have on the S-0 Truss, it's also data to some of the computers that we have in the S-0 Truss, there's also video channels that will eventually go to the Mobile Transporter and to the Mobile Base System that comes up on a later mission ... There's a time limit in attaching the S-0 Truss, because there's equipment, once we take it out of the payload bay, we undo a remote umbilical that powers some of this equipment while it's in the Space Shuttle payload bay; once we take it out of the payload bay then there's a clock that starts ticking that says we need to get power to this equipment before, or else we have the chance of losing some of the equipment on S-0."

Bursch had only hinted at the S-0's complexity. It carried over 100 electrical and fluid connectors that would allow current and fluids to flow across the length of the completed ITS. It also contained a number of computers among its 475,000 individual parts. Walz described its function with the following words:

"The addition of the S-0 Truss plus the new command and control software that will come up about a month earlier will allow the U.S. to have a full guidance and navigation capability equivalent to what we have on the Russian segment. So we'll have GPS antennas and also we have the rate gyro assemblies, and so we will have data from both the U.S. and the Russian segment. And it'll just make the station a more reliable station, because if one source of data for some reason cuts out, we have this equivalent data on the other side. So it's just a more robust station."

Smith and Walheim had been inside Quest, preparing for the first of four EVAs on STS-110. They exited the airlock at 10:38, and Walheim became the first person to ride the SSRMS. Smith made his own way around the exterior of ISS. Smith, who was making his sixth EVA remarked, "Beautiful, beautiful," as he viewed the sight in front of him. "This is incredible, unbelievable! Amazing! Amazing!" added Walheim, who was on his first EVA.

They worked together and unfurled two of the four mounting struts on the S-0 ITS and attached them to Destiny, making permanent fixtures. Next, they deployed trays of avionics equipment and connected a series of power, data, and fluid lines between S-0 and Destiny. They also connected an umbilical system between S-0 and the MT. The MT would run along a small rail track mounted on the ram face of the ITS and act as a mobile base for the SSRMS during future assembly tasks. Throughout the multitude of tasks, Ross and Walz directed the EVA from the Shuttle's aft flight deck. When the tasks took more time than expected the EVA was extended by one hour. Even then, the astronauts' limited oxygen supply led to a halt being called before they had installed the final two circuit breakers on S-0. The EVA ended with the astronauts' return to Quest at 18:24, after 7 hours 48 minutes outside. Both crews began their sleep periods at 20:44.

April 12 began at 04:44 for Bloomfield's crew. Onufrienko's crew woke up 30 minutes later. The day was occupied by a series of transfers between the two spacecraft. Morin and Ross moved an experimental plant growth chamber from

Atlantis to a rack in Destiny. It would replace the malfunctioning protein crystal growth experiment, which was moved into Atlantis for return to Earth. Walheim and Ochoa installed a new freezer in Destiny, to store future crystal samples. Oxygen and nitrogen was transferred from Atlantis to ISS to replenish the supply in Quest's tanks. The astronauts completed a number of live interviews in the middle of the day. They also watched a live broadcast by NASA Administrator Sean O'Keefe, in which he voiced his vision for the future of NASA. In the afternoon the transfer work continued. The Biomass Production System-Photosynthesis Experiment and System Testing Operation (BPS-PESTO) and the Protein Crystal Growth-Enhanced Gaseous Nitrogen Dewar (PCG-EGN) were both transferred from Atlantis to the station. Both crews had two hours free time before beginning their sleep periods at 20:44.

Meanwhile, the powering up of the S-0 ITS systems had begun almost as soon as the first EVA had finished and had continued since that time. The four computers on the ITS were performing as planned, as were the new devices for determining the position of the station relative to Earth. The Global Positioning System antennae and the truss' thermal control system were all functioning within normal parameters.

April 13 began at 04:44, with both crews being woken up for another day of EVA. Ross and Morin left Quest at 10:09. Ross, who was on his seventh flight and was making his eighth EVA had been an astronaut for 20 years and had specialised in EVA, participating in the development of many of the tools that his colleagues would use to construct ISS. As he told a pre-flight interview, "I know what to expect. I can't wait to do it again . . . I think it's a religion. Literally, it's the human being in the space element as close as you can get. To do it safely and efficiently is a challenge." Ross and Morin were both grandfathers and had therefore earned themselves the nickname the Silver Team for their EVAs. During the EVA, Smith, working inside the station, asked Morin, "how do you like this fraternity so far?" Morin replied, "It's pretty wild." Once the two men had collected their tools Morin rode the work platform in the end-effector of the SSRMS, while Ross made his own way around the exterior of the station. They lowered the second pair of struts supporting the S-0 ITS and connected them permanently to Destiny before removing the panels and clamps that had supported the truss during launch.

The two men also connected a second cable and reel to the MT; the two systems operated independently, offering redundancy in the case of failure. Ross attempted, unsuccessfully, to remove a restraining bolt from the mechanism designed to guillotine the MT umbilical in an emergency. The attempt was abandoned rather than waste precious time. Both men returned to Quest and the EVA ended at 17:39, after 7 hours 30 minutes. Following the EVA, Frick fired Atlantis' manoeuvring thrusters in the first of three series of burns to raise the station's orbit.

During the day, the first readings were made with the PCG-EGN experiment, which had been placed in Zarya. A muffler installed following transfer to the station had to be removed on April 15, when it caused humidity to rise inside the environment chambers of the experiment. The final samples were also recovered from the ADVASC and the experiment was powered off before being transferred to Atlantis for return to Earth. Both crews ended their day at 20:44.

Smith and Walheim's third EVA was on April 14; they left Quest at 09:48. Smith rode the SSRMS while Walheim relied on the strength in his arms to move himself around. Their first task was to release the claw-like device that had originally held the S-0 ITS in place above Destiny, before its four legs were lowered and secured in place. Next, they reconfigured a number of electrical and data connections on the SSRMS, to allow it to be commanded through the S-0 ITS. The cables had become stiff in the cold of space and Smith reported, "These cables are again sticking together [as they had on the first EVA]. It's a little scary taking them apart because they're fibre optics." They changed and tested the primary series of connectors before changing and testing the back-up system. At one point Smith admired the view of Earth far below him, remarking, "Beautiful place we live."

Moving on, they released the clamps securing the MT in its launch position. This cleared the way for the first tests of the system, planned for April 15. Tests of the new SSRMS connections took longer than planned, so activities using the SSRMS to move an EVA handrail from its launch position on the side of the S-0 ITS were cancelled. Smith and Walheim returned to Quest and ended their EVA at 16:15, after 6 hours 27 minutes. Following the EVA, Frick performed the second series of manoeuvres to raise the station's orbit. Inside the station the crew began preparing the ZCG experiment for a 15-day run, beginning on April 22. They also moved the Commercial Generic Bio-processing Apparatus (CGBA) from Atlantis to the station. The workday ended at 20:44.

After waking up at 04:44, April 15, the crews spent the day transferring equipment from Atlantis to ISS. Meanwhile, controllers in Houston began the first tests of the MT. At 08:22, the MT was commanded to move, at less than 2 cm per minute, from its initial position. The small flatbed car was then commanded to travel 5 m along the track, to its first work position. After approximately 30 minutes of travelling a software problem prevented the car automatically latching itself to the ITS at the work site as it should have done to stabilize itself. Controllers sent a series of commands to achieve the latching down. After moving to a second work site, the car again required the commands to latch it to the ITS to be sent manually, as indeed it did after it returned to the first work site. In all, the car had travelled approximately 22 m at rate of 2.5 cm per second. With the exception of the latching software all MT systems functioned as planned. The work day ended at 20:44.

At a post-EVA press conference at Houston, STS-110 launch package manager Ben Sellari reported, "We did have a very successful activation of the Mobile Transporter ... I think what we're finding out is how the Mobile Transporter works in zero gravity." On the problems encountered with the MT, Sellari stated that he believed that the MT's wheels had become misaligned with the two magnetic strips that they travelled along, causing the system to shut down.

Walz had described the MT's first motion in the following terms:

"It moved from its launch position to the initial work site ... It works on a standard command from the computer and then all of a sudden it started to move, very, very smoothly and, of course, very, very slowly. It got to a position and started to latch and something went wrong with the automatic software. The ground was able to move it manually."

The crew were awake once more at 04:44, April 16. The highlight of the day was the fourth and last EVA, which started at 10:29. Ross and Morin began by pivoting the Airlock Spur away from the S-0 ITS and connected it to Quest, giving future EVA astronauts an easy, direct route between the two. They tested micro-switches on the exterior of the S-0 ITS that would be used to confirm the attachment of later ITS segments and installed two 40-watt halogen lights on the exterior of Unity and Destiny. Next, they erected a work platform on the exterior of the station and installed the two S-0 ITS circuit breakers that had not been installed during EVA-1. Shock absorbers were attached to the MT, to prevent vibrations on the ITS during future EVAs from reaching the SSRMS. During the EVA, Ross joked, "Sure beats the dollar an hour I used to get for baling hay." Struck by the view of Earth he exclaimed, "This is what I call a room with a view." During the EVA they observed a thunderstorm over the Pacific Ocean and the Moon over the Atlantic Ocean.

The two men also returned to the bolt that Ross had not been able to disconnect on the MT umbilical guillotine, during the second EVA. They were still unable to remove the bolt, so the problem was left for a later EVA team to resolve. Finally, they tied down an insulation blanket that had come loose around a navigational antenna on the S-0 ITS. Plans to deploy a gas analyser were abandoned when the equipment proved faulty. The EVA ended at 17:06, after 6 hours 37 minutes. While the EVA went ahead, Onufrienko, Bloomfield, Frick, and Smith continued to transfer equipment in both directions between Atlantis and ISS. The day ended at 20:14, following the crew's evening meal.

One of the final joint activities was to transfer the Arctic-1 refrigerator from Atlantis to ISS. It would be used to store samples awaiting return to Earth. With the new item installed and powered on, the Bio-Technology Refrigerator, which had been used for the same function until that point was powered off. Bloomfield led his crew back to Atlantis and sealed the hatches at 12:04, April 17. Following pressure checks the orbiter was undocked from Destiny at 14:31, and backed away. Frick flew Atlantis on a 1.25 circuit fly-around of ISS before firing her thrusters at 16:15, when his spacecraft was directly above ISS, to complete the separation. The crew spent the remainder of the day preparing their spacecraft for the return to Earth.

The last full day of the flight began at 03:44 April 18. The crew spent it preparing for re-entry. During the end-of-flight press conference they were given the opportunity to sum up their feelings. Bloomfield said, "We made our first step toward our commitment to our International Partners. Once work on the truss and arrays is finished we will actually have enough power that we can add two more laboratories." Those two laboratories would be the European "Columbus" and the Japanese "Kibo" science modules.

Asked about his record-breaking ninth EVA, Ross answered, "I felt the same on this one as I did on the first one, totally enthralled. The spacewalks were incredible."

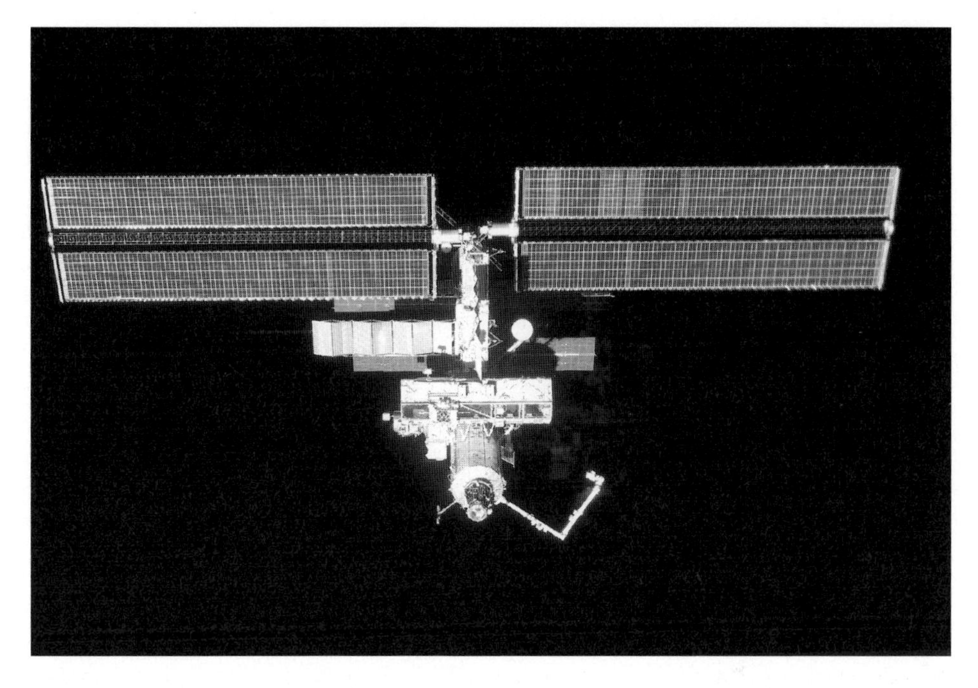

Figure 16. As STS-110 leaves the International Space Station, the Starboard-0 Integrated Truss Structure is visible on Destiny's zenith.

Smith took a question on his feelings when viewing Earth from space, but knowing that there were numerous wars being fought on the planet's surface, he answered,

> "Every time one of us looks out the window, we have really strong feelings about Earth. It looks very peaceful from up here and it really is very hard to believe that there is strife in different places on Earth. I think we are really good ambassadors, when we come back, about being good to the Earth and promoting peace."

Retrofire occurred at 11:20 April 19 and Frick flew Atlantis to a safe landing at KSC at 12:27. Capcom at Houston told the crew, "That was a great landing and a great way to end a mission that has been superb in all respects."

After leaving Atlantis the crew were greeted by the new NASA Administrator, Sean O'Keefe. Bloomfield told the waiting crowd,

> "We just got done with an incredible mission up to the International Space Station. We had an outstanding time while we were up there ... It's great to be back home. Please don't forget those folks that are still up there on the International Space Station. Keep them in your thoughts and prayers."

SOYUZ TM-34 DELIVERS A FRESH CREW RETURN VEHICLE

SOYUZ TM-34

COMMANDER	Yuri Gidzenko
FLIGHT ENGINEER	Roberto Vittori (Italy)
SPACEFLIGHT PARTICPANT	Mark Shuttleworth

Even as STS-110 prepared to return to Earth, Onufrienko, Bursch, and Walz were working hard to place ISS into hibernation mode. On April 20, they sealed themselves inside Soyuz TM-33 and undocked from Zarya's nadir at 05:02. Onufrienko then flew the spacecraft to a docking with Pirs' nadir at 05:35. On returning to ISS they reactivated the systems that they had turned off before the manoeuvre. The change of position had cleared Zarya's nadir for the arrival of Soyuz TM-34, the new CRV due to replace Soyuz TM-33.

The last Soyuz TM spacecraft, Soyuz TM-34, was launched at 02:26, April 25, 2002. Flight Commander Yuri Gidzenko was returning to ISS, having served as a member of the Expedition-1 crew. Interviewed before that flight, Gidzenko described the standard 2-day Soyuz flight as follows:

> "OK, previously it took only one day from taking off 'til docking between the Soyuz vehicle and the space station. After that, our specialists decided to increase this period of time to two days. It's maybe more comfortable for crew and more convenient for ground people to calculate our orbit and to calculate inputs so that we can dock. Then for two days, we are going to alter the orbit so that we gradually approach the station. After two days, in Orbit 32, we'll begin the process of docking. We'll still have a distant approach, then we approach the station, then we have the docking itself. As a rule this is all done automatically, on an automated mode; the astronauts or cosmonauts are monitoring the process, and if something does occur with the automatic docking, well then the crew gets involved and does it manually. Specifically I have two levers that would allow me to control the vehicle to do the approach and docking manually. Then following the docking we check the seal, equalizing pressure between the transport vehicle and the ISS. We open hatches. Then we start working on the station."

Roberto Vittori was a professional Italian astronaut and Mark Shuttleworth, who held joint South African and British citizenship, was the second paying visitor to the station. The launch of Soyuz TM-34 passed without incident and the spacecraft began a standard 2-day rendezvous, docking with Zarya's nadir at 03:56, April 26. Hatches between the two spacecraft were opened at 05:25, and the Soyuz crew floated into Zarya for their safety briefing.

South African President Thabo Mbeki called to congratulate Shuttleworth, who he called "the first African citizen in space". The President told him, "The whole

Figure 17. Expedition-14: The Expedition-4 and Soyuz TM-34 crews pose in Zvezda. (rear row) Carl Walz, Yuri Onufrienko, Daniel Bursch. (front row) Roberto Vittori, Yuri Gidzenko, and South African Spaceflight Participant Mark Shuttleworth.

continent is proud that at last we have one of our own people from Africa up in space ... It's a proud Freedom Day because of what you've done."

In reply, Shuttleworth described his launch in the following terms, "I had moments of terror, moments of sheer upliftment and exhilaration." He also described ISS, "It's amazingly roomy ... Although it's very, very large, we have to move very carefully. As you can see around us, there are tons of very precious and very sophisticated equipment. We hope that we will be good guests." Of his home planet, Shuttleworth said, "I have truly never seen anything as beautiful as the Earth from space. I can't imagine anything that could surpass that."

Following the farce and ill-will surrounding Denis Tito's flight on Soyuz TM-32, NASA and RCS Energia had negotiated a set of rules governing the inclusion of Space Flight Participants (commercial passengers or "space tourists") in the otherwise empty third seat on Soyuz "taxi" flights to replace the Soyuz CRV attached to ISS. Although Shuttleworth would remain the responsibility of Gidzenko and Onufrienko whilst on the station, he was not required to be escorted while in the American ISS modules, as Tito had been. Shuttleworth had also negotiated the use of a NASA laptop computer to send e-mails from the station and some time on the NASA communications link to download audio and images. Like Tito, Shuttleworth had spent one year training in Korolev, and had even received 1-week training in Houston, which NASA had denied to Tito. Unlike his predecessor, Shuttleworth planned to perform experiments whilst on the station. When interviewed by CNN, Tito remarked that he was pleased that negotiations had formalised the position of Space Flight Participants and had allowed Shuttleworth to fly without the acrimony that had surrounded his own flight the previous year.

In orbit, Shuttleworth spent his time performing experiments in crystal growth, stem cell research, and AIDS research. He also spent long periods looking out of the windows at the magnificent view of Earth. During his stay on ISS he became a national hero in South Africa and also received considerable press coverage in Britain. He spoke with Nelson Mandela during a press conference and was embarrassed when a 14-year-old South African schoolgirl asked him to marry her. The short-duration flight came to an end on May 6, when Gidzenko, Vittori, and Shuttleworth sealed themselves inside Soyuz TM-33 and undocked from Pirs at 20:31. They landed in Kazakhstan at 23:51. Shuttleworth described his flight as "the most extraordinary experience".

In 2003, it was announced that female Russian cosmonaut Nadezhda Kutelnaya had been named to the Soyuz TM-34 crew, but had been removed to make way for Shuttleworth who had paid $20 million for his flight. It turned out that Kutelnaya had previously been named in the original crew for Soyuz TM-32, but had also been removed on that occasion, to make way for the fee-paying Denis Tito.

Even as the Soyuz TM-33/Soyuz TM-34 exchange took place in orbit, RCS Energia remained in serious financial difficulty. The failure of the Russian government to pay the company millions of roubles that it was owed meant that Energia was unable to purchase the equipment required to finish manufacturing a number of Progress spacecraft then under construction. The funding problem particularly threatened the Progress vehicle due for launch in January 2003. Energia officials admitted that contingency plans were in place, in full co-operation with NASA, to end permanent occupation of ISS if the problems with funding persisted. The station would then be occupied by visiting crews on self-sufficient, short-duration Shuttle flights, as Skylab and the early Soviet Salyut stations had been. At the same time work on the Soyuz TMA-1 spacecraft, due to replace Soyuz TM-34 in October 2002, and others of its class, was also running behind schedule. If the new spacecraft were not available in time then permanent occupancy of ISS would have to be temporarily suspended.

While demanding that the Russians meet their contractual agreements the Americans continued to move on with their own decision to stop developing ISS after it reached "Core Complete". Few people now believed that they would ever see ISS reach the original "Assembly Complete" configuration. The on-going Russian budgetary difficulties only served to highlight the far-reaching implications of the American decision to cancel the development of the X-38 CRV and the failure to develop relatively inexpensive robotic cargo delivery vehicles. The combination of a new American President (George W. Bush) and a new Administrator (O'Keefe) at NASA, both of whom had vowed to rein in spending on ISS, meant that on this occasion NASA was far less likely to hurry to the Russians' aid with an injection of much needed cash. The time had come for the Russian government to prove that they intended to live up to their contractual agreements and support ISS, or admit that they can no longer afford to do so due to pressing problems at home on Earth and withdraw from the programme. The latter would no doubt have seen the end of the Russian space programme for the foreseeable future and would have been a very sad day indeed.

In the wake of Shuttleworth's successful flight, Energia announced that it was likely that Space Flight Participants would be included in most Soyuz taxi flight crews. The private flights, which cost $20 million each, including a very basic training regime in Korolev, were contracted through the American company Space Adventures. They were a major source of income for the Russians, allowing them to plough the money back into the Soyuz/Progress spacecraft needed to fulfil their contractual commitments to ISS. Lance Bass, a rock band vocalist, was originally in line for the space participant's couch on the October 2002 Soyuz taxi flight, but failed to raise the $20 million to pay for his flight. The Russians removed him from the flight roster.

At the same time, NASA Administrator Sean O'Keefe announced that NASA would select an Educator Astronaut, a teacher, in each subsequent group of new astronauts selected after 2003. Unlike Christa McAuliffe, who was a passenger on STS-51L, the Educator Astronauts would be professional astronauts, serving as Mission Specialists, with the years of training that that would involve before they made their first flight into orbit. The first Educator Astronaut to be selected was Barbara Morgan, Christa McAuliffe's back-up in 1986. The Educator Astronauts would be used to encourage students to study mathematics, engineering, and spaceflight-related science subjects.

Following the departure of Soyuz TM-34 Onufrienko and the Expedition-4 crew had 1.5 days of free time before splitting their time between long-running experiments and preparing to come home. During the second week of May, Onufrienko repaired the Elektron oxygen generator, which had been malfunctioning again and had been off-line. From Korolev, Valeri Lyndin assured the media, "There is enough oxygen in the station, so there is nothing terrible about this. There is enough oxygen to last three or more months." He continued, "Elektron was working on the ISS, taking oxygen from water, but there are other mechanisms, and they are getting oxygen with the help of these systems now ... The system is automated, so you can command it from Earth or from the computers onboard. We are taking the appropriate measures." In the last days of their stay, Onufrienko's crew performed their final experiments.

STS-111 DELIVERS EXPEDITION-5 AND REPAIRS THE SSRMS

STS-111	
COMMANDER	Kenneth Cockrell
PILOT	Paul Lockhart
MISSION SPECIALISTS	Franklin Chang-Diaz, Phillippe Perrin (France)
EXPEDITION-5 (up)	Valeri Korzun (Russia), Sergei Treschev (Russia), Peggy Whitson
EXPEDITION-4 (down)	Yuri Onufrienko (Russia), Daniel Bursch, Carl Walz

STS-111 would take the Expedition-5 crew, Valeri Korzun, Sergei Treschev, and Peggy Whitson, up to ISS and return the Expedition-4 crew, Yuri Onufrienko, Daniel Bursch, Carl Walz, to Earth. It also carried the MPLM Leonardo, with experiment racks and three stowage and re-supply racks for Destiny. The flight would deliver the Mobile Base System (MBS) which, when mounted on the MT, would complete the Canadian Mobile Servicing System (MSS). When the assembly of the MBS was complete the SSRMS would be moved end over end, from its position on the exterior of Destiny, to mount itself on the MBS. Once in place, the MBS would be able to run along rails on the ram face of the ITS. This combination offered new mobility that would be used extensively in the remainder of the ISS construction programme, as well as in support of numerous EVAs and experiments. There would be three EVAs during the flight of STS-111, two in support of the MBS, and one to replace a faulty wrist joint on the SSRMS. The latter repair had resulted in a 1-month launch delay, while the crew learnt the new repair procedures.

The flight was due to be launched at 19:22, May 30, 2002, but on May 28 a problem was found in one of Endeavour's Auxiliary Power Units (APUs). It was repaired on the launch pad, without delaying the launch. On May 30, the launch attempt was cancelled at 19:21, due to the fact that there were thunderstorms within the area around the Kennedy Space Centre.

The launch was rescheduled for 24 hours later, with the countdown resuming at $T-11$ hours and propellant loading commencing at 10:00. However, at the Tanking Meeting, prior to commencing propellant loading, mission managers decided that there was little chance of the weather improving in time for the day's launch. This time the launch was re-set for June 4, and the Rotating Service Structure was moved back around the Shuttle on the launch pad, to protect it from the weather. On June 2, the launch was moved back a further 24 hours, to June 5, due to work on a malfunctioning gaseous nitrogen pressure regulator in Endeavour's left-hand Orbital Attitude and Manoeuvring System (OAMS) pod. The regulator had malfunctioned earlier in the countdown and the decision had been taken to change it.

As the new countdown reached its final stages, Launch Director Mike Leinbach told Cockrell, "Sorry we had to keep you here an extra 6 days. Good luck, have a good flight." Cockrell replied, "We'll do a good job for you."

STS-111 finally lifted off at 17:23, June 5, 2002, while ISS was west of Perth, Australia. At launch Frank Culbertson, who had been Commander of the Expedition-3 crew, and who was working the Capcom's position in Houston, informed the Expedition-4 crew, "It's on the way. I know you will be happy to hear that." After 182 days in space Bursch yelled, "All right!" Walz was more reserved saying, "Good. We look forward to seeing you."

Following a perfect launch, Endeavour entered the correct orbit and was configured for orbital flight. As Endeavour left Earth, Chang-Diaz equalled Jerry Ross' record set on STS-110, becoming the second person to fly in space seven times. Asked about his new record in a pre-flight interview he had replied, "I'm hoping that these kinds of records will be easily broken and many times over. And I'm hoping that there will be many, many people who will fly not 7 or 8 times, but 10, 15 times." The crew's day ended just 3 hours after launch when they settled down to their first night in space.

June 6 was spent doing all of the routine things that a Shuttle crew does during their first full day in space. While Cockrell and Lockhart concentrated on the rendezvous, Perrin and Chang-Diaz fitted the centreline camera and deployed the docking ring. They also checked out the EMUs that they would wear during their three EVAs. On ISS, Onufrienko, Walz, and Bursch prepared the station for the arrival of the Expedition-5 crew and the visitors that would deliver them.

Day 3, June 7, began at 05:30. Endeavour followed a standard rendezvous path. During the final approach Dan Bursch had remarked, "It's a great day, a great day." He then rang the Station's bell to mark Endeavour's arrival. Soft-docking, on Destiny's ram, occurred at 12:25. After waiting 1 hour for oscillations to damp down, hard-docking took place at 13:27. Following pressure checks the hatches between the two spacecraft were opened at 14:08, and the Shuttle and Expedition-5 crews entered Destiny, where the Expedition-4 crew greeted them. One of the Expedition-4 crew noted that the Expedition-5 crew consisted of older individuals and told Cockrell, "We are so glad that you brought our fathers to the International Space Station." Turning to the newcomers he added, "We wish you good luck. Have a good time."

Before launch Korzun had described his views on the Expedition crew hand-over:

> "What is [the] purpose of the hand-over? We have [a] special book; we will use this book and there is each system described in this book, and there are some empty space[s] which we need to use to write changes between [the] condition of systems which we study on the ground and [the] real condition of this system onboard Station. And maybe . . . we will study real situation with stowage of the equipment in station . . . [W]e need to have some times to adapt, maybe first time, I can [watch] Yuri's activity or Peggy will see Carl's activity or, Sergei will follow Dan's example . . . [N]ow we needn't use a lot of time for hand-over because each Shuttle crew bring down some video and, we have [the] opportunity to watch video that crew made in space [to] show us [the] situation. And sometimes I think this is a good idea . . . to watch video and to recognize [the] configuration inside of the module. [It is] very important for us to use INV—this is Inventory Management System."

Following a safety brief, the ten astronauts began transferring equipment and experimental results between the two spacecraft. The Expedition-4 crew removed their couch liners from Soyuz TM-34 and stored them in Endeavour, while the Expedition-5 crew installed their own couch liners in the Soyuz and then checked their Russian Sokol pressure suits. Perrin and Chang-Diaz checked the communication link between their EVA suits and the Quest Airlock. With the latter task completed, they officially took up residence on ISS at 18:55. At the same time Onufrienko, Bursch, and Walz completed their 182-day stay on ISS and became part

Figure 18. STS-111 crew (L to R): Phillippe Perrin, Paul Lockhart, Ken Cockrell, Franklin Chang-Diaz. These four were joined by the Expedition-5 crew during launch and the Expedition-4 crew during recovery.

Figure 19. STS-111: Endeavour delivers the Multi-Purpose Logistics Module Leonardo to the International Space Station.

of Endeavour's crew. The day ended with the failure of the Flash Evaporator System Primary controller. It was one of three such controllers and had no effect on the station's operations. An investigation was begun in Houston.

On June 8, the two Expedition crews continued their hand-over. Onboard Endeavour, Cockrell used the RMS to un-berth Leonardo from the payload bay and dock it to Unity's nadir, at 10:28. Following pressure checks the hatches between Leonardo and Unity were opened at 17:30. All ten astronauts entered the MPLM and began unloading the logistics and equipment that the Expedition-5 crew would use during their stay on ISS. Early in the day the crew reported, "We're hearing a pretty loud audible noise, kind of a growling noise." At the same time, one of the station's four CMGs mounted in the Z-1 Truss seized. It was commanded to spin down (lose speed) and was then shut down. An investigation began, but the remaining three CMGs and a number of alternative systems for controlling the ISS's attitude meant that the failure had no immediate effect on the mission. Flight director Paul Hill told a press conference, "Big picture wise, losing a CMG is a big deal. From a risk perspective right now, we're in good shape. But this is a major component that's failed and we're going to do the best we can to get the next CMG ready to fly." There was a spare CMG in storage in America, but the full Shuttle manifest meant that it could not be launched before 2003.

As the day continued Perrin and Chang-Diaz checked out their EMUs, as well as the tools and procedures for their first EVA, which would take place the following day. During a briefing from the ground the astronauts were updated on the attempt by Lance Bass to purchase a Space Flight Participant's position on the up-coming October Soyuz taxi flight to ISS. Expedition-5 crew Commander Valeri Korzun replied, "We would be happy to see one of the Supermodels." He added, "But this is a joke and we will be very happy to receive any space tourist. They're very welcome here ... Probably someone with certain professional qualities would be better. But it would not make any difference to our greetings." That was still a very Russian point of view. On the subject of his own flight Korzun was equally positive saying, "Physically and psychologically, we are prepared to fly at least a year and a half."

Regarding the CMG failure, during the day Flight Director Rick LaBrode explained:

> "All indications at this point do appear to be a mechanical failure. We saw increases in vibration. We saw increases in bearing temperatures, increased currents and decrease in speed. It looks obviously like a bearing seized."

Perrin and Chang-Diaz left the Quest airlock at 11:27, June 9, at the beginning of a 7-hour 14-minute EVA. Lockhart choreographed the EVA from Endeavour's aft flight deck, while Korzun and Whitson operated the SSRMS. Throughout the EVA Cockrell used the cameras mounted on Endeavour's RMS to view the astronauts' activities. Korzun described the EVA tasks in his pre-launch interview:

> "[E]verybody on our crew has personal tasks during this activity. I will support [the] EVA crewmember on the Shuttle who is [riding the] robot arm during EVA 1, and Peggy will grapple [the] MBS and translate [the] MBS and connect [the] MBS to the MT, and ... Sergei will check [the] station systems during this activity and help us with [a] video view during MBS installation ... I need to transfer Franklin from [the] Airlock to [the] cargo bay of the Shuttle; he will ungrapple [a] spare PDGF and then I will translate him to the P-6 ... He will install this PDGF and then move back to the cargo bay, and unfasten pack of MMOD shield—this is special protection for the Service Module [Zvezda]. And then I will transfer him, with MMOD shield, to the PMA-1. They will, Franklin and Philippe will temporary stowage of this MMOD shield, and [then], during our EVA, we will take this MMOD shield and we will transfer it to the Service Module, and we will install this MMOD shield around the corner of Service Module. They will protect Service Module from meteors during flight."

The first task was to install a PDGA on the P-6 ITS. The PDGA would be used to move the P-6 from its position on the Z-1 Truss to its final location, on the end of the Port ITS, during the flight of STS-119. That move would be one of the final actions in the construction of the ISS "Core Complete" configuration. Chang-Diaz mounted a foot restraint in the end of the SSRMS and climbed on to it. He collected six thermal shields from their storage place within Endeavour's payload bay and was then moved alongside PMA-1, between Unity and Zarya. Perrin made his own way to the same location and helped Chang-Diaz to attach the thermal shields temporarily to PMA-1. In an EVA planned for July, Korzun and Whitson would install the shields in their final location, on the exterior of Zvezda, where they would provide additional micrometeoroid protection, bringing the module up to American standards in that capacity.

Admiring the view from their unique location, Chang-Diaz exclaimed, "This is an amazing experience, I tell you."

Frenchman Perrin agreed, "Unbelievable."

Chang Diaz then conducted a visual and photographic inspection of the failed CMG in the Z-1 Truss, a task added to the EVA only a few days earlier. Their final tasks called for the removal of the thermal blankets covering the Mobile Base System (MBS). With the blankets removed Cockrell released the latches that held the MBS securely in Endeavour's payload bay. Whitson and Walz then grabbed the MBS with the SSRMS and lifted it to a position 1 metre above the MT, on the S-0 ITS. The SSRMS was then powered off, leaving the MBS to become thermally conditioned. Perrin and Chang-Diaz returned to the airlock and the EVA ended at 18:41.

Onufrienko had intended to carry out a short official hand-over ceremony during the day, but a smoke alarm in one of the Russian modules caused that ceremony to be abandoned. After inspecting the relevant module Onufrienko reported, "Everything is OK. Everything is under control. It was a false alarm. Too much dust; that is probably what triggered it."

At 09:30, June 10, Whitson and Walz used the SSRMS to transfer the MBS up to the MT. Controllers in Houston then ordered the latches to close, securing the MBS in place. In the future the SSRMS would be walked end over end until it could attach itself to the MBS. In that position it would be possible to ride the MT up and down the track mounted on the ITS. In that way the SSRMS would be able to assist in attaching future ITS elements to the truss. During the afternoon, Onufrienko's crew held a ceremony to pass command of ISS to Korzun's crew. Following the hand-over ceremony, Endeavour's thrusters were used to raise the station's orbit.

Chang-Diaz and Perrin commenced their second EVA at 11:20, June 11. Lockhart acted as Intravehicular Officer, guiding them through the items in the flight plan. Their first task was to connect a dozen primary and back-up power, data, and video cables to the MBS and the primary power cable between the MBS and the MT. With the latter task completed Houston sent commands to the MT to plug its umbilicals into the S-0 ITS. The two EVA astronauts then connected the Payload Orbital Replacement Unit Accommodation (POA) on the MBS. This was a copy of the end-effector on either end of the SSRMS and would be used in the future to hold cargo while the MT traversed the ITS. Next, they secured four bolts between the MBS and the MT, thereby completing the installation of the MBS. The penultimate task was to move a TV camera to its final position on the MBS, where it would be used to view future assembly work. Finally, the two men added an electrical extension cable to the MBS and photographed all of the connections that they had made. The EVA had been planned to last 6.5 hours but ended at 16:20, after only 5 hours. In Houston, Canadian astronaut Bob Thirsk told them, "We consider it a wrap. Your professionalism and skill really showed through."

Inside ISS, work continued on the Expedition crew hand-over and the transfer of supplies. With Leonardo already unloaded the crews spent much of the day loading equipment and results from the Expedition-4 crew's experiments into the MPLM for return to Earth. At 22:19, Walz and Bursch set a new American endurance record, exceeding Shannon Lucid's 188 consecutive days spent in space.

At 02:55, June 12, Walz also surpassed Lucid's overall record of 223 days in space accrued over numerous flights. The day was spent loading Leonardo and continuing the hand-over of the Expedition crews. Chang-Diaz and Perrin also went over plans for their third EVA, planned for the following day. During the afternoon Endeavour performed a second re-boost manoeuvre. The astronauts also downlinked their pictures of forest fires burning in Colorado.

Chang-Diaz and Perrin left the Quest airlock for the third time at 11:16, June 13. Lockhart controlled the EVA while Cockrell operated Endeavour's RMS, to allow its cameras to return pictures of the astronauts' activities. Their task for the day was to change out the primary wrist roll joint in the SSRMS, which had malfunctioned in March 2002. Since that time all SSRMS functions had been performed using the back-up wrist mechanism. Bill Gerstenmaier, Space Station programme deputy manager had described the SSRMS at a STS-111 launch day press conference, saying, "Without the arm, we could not continue to build the station. So the arm needs to be there, and needs to be functioning."

The repair began with the removal of the Latch End Effector (LEE) which was secured to an EVA handrail on the exterior of Destiny. This exposed the wrist joint that required replacement. The two astronauts disconnected six bolts holding the wrist roll joint in place and a seventh bolt holding power, data, and video umbilicals, before Perrin removed the joint and carried it down to its stowage position in Endeavour's payload bay. He then released the six bolts holding the new joint in place in the payload bay and carried it back up to where Chang-Diaz was waiting. Having positioned the joint the two men tightened the six securing bolts and the umbilical bolt, before recovering and replacing the LEE. While Chang-Diaz and Perrin moved the old joint to its final position for its return to Earth, power was applied to the SSRMS so that Bursch and Korzun could run it through as series of test manoeuvres. The SSRMS was returned to full operation at 16:43. Having collected and stored their tools the two astronauts returned to Quest, and the EVA ended at 18:33, after 7 hours 17 minutes. Throughout the EVA Whitson and Treschev had continued to transfer items from ISS to Endeavour and Leonardo. In the evening an unsuccessful attempt was made to apply electrical power from the MBS to the SSRMS. Initial investigation suggested that a software glitch had prevented the commands travelling between the two units.

The highlight of June 14 was the closing and removal of Leonardo from Unity's nadir. Just after 12:00 Perrin used Endeavour's RMS to manoeuvre the MPLM back into the Shuttle's payload bay, where it was secured at 16:11. Cockrell then performed a third re-boost manoeuvre using Endeavour's thrusters. Back on ISS all ten astronauts brought their group activities to a close.

June 15 began with final farewells and the withdrawal of the STS-111 crew to Endeavour before the sealing of the hatches between the two vehicles at 08:23. Following pressure checks, Endeavour undocked at 10:32, and moved clear of Destiny's ram under Lockhart's control. Onboard ISS, Whitson rang the ship's bell and announced, "Expedition-4 departing, Endeavour departing." This Naval tradition, initiated by Shepherd, commander of the Expedition-1 crew, had become standard practice on the station whenever spacecraft arrived or departed. Onufrienko, Walz, and Bursch had spent 181 days on the station. Once clear of ISS, Endeavour made 1.25 circuits of the station while the crew exposed video and photographs. Finally, Lockhart manoeuvred the Shuttle away from the station and the crew enjoyed some free time. On ISS, the Expedition-5 crew began their sleep period at 16:00, adjusting their daily routine to begin at 02:00, June 26. When they awoke they began the task of unpacking and stowing the items Endeavour had brought to the station.

Preparations for re-entry filled the whole of June 16. Plans to return to Earth were cancelled on both June 17 and 18, when controllers in Houston instructed the Shuttle's crew to back out of preparation for re-entry due to rain and thunderstorms in the region of the Kennedy Space Centre. In the meantime, Edwards Air Force Base, in California, was activated as a back-up landing site. Following a third wave-off to Florida, Endeavour finally returned to Edwards, landing on the dry lakebed at 13:58. The Shuttle flight had lasted 7 days 2 hours 26 minutes. Onufrienko, Walz, and Bursch had completed 196 days in space. In the weeks that followed they underwent the full range of medical and re-acclimatisation studies that previous long-duration space station crews had undergone before them. They were all found to be in good health.

EXPEDITION-5

Following the departure of Endeavour, Korzun, Whitson, and Treschev settled down to the start of their 4.5-month stay aboard ISS. On June 18, they observed and photographed wildfires in Arizona and Colorado. Whitson activated the StelSys Liver Cell Research experiment in the Biotechnology Specimen Temperature Controller (BSTC) on June 21. The experiment compared liver cell function in microgravity with that of similar cells grown on Earth. Processed samples were stored in the Arctic-1 freezer for return to Earth on STS-112. When the experiments were complete the BSTC was powered off.

Having installed a new hard drive in the Zeolite Crystal Growth experiment during their second week on orbit, Huntsville powered on the EXPRESS Rack 2 containing the experiment. The first sample runs were commenced on June 27. All three astronauts completed their first Crew Interactions Questionnaire on June 25, while Korzun and Treschev also completed a running experiment on the station treadmill that required them to take close-up high-definition video of their facial features while running.

The crew also filled the docked Progress with rubbish, in preparation for its departure. The undocking of Progress M1-8, from Zvezda's wake, took place at 04:23, June 25, and the spacecraft re-entered the atmosphere and burned up as planned.

Figure 20. Expedition-5: Valeri Korzun trims Peggy Whitson's hair. Whitson holds a vacuum cleaner hose to take away the loose hair. Routine activities such as this take place regularly on ISS.

SHUTTLE FLEET GROUNDED

NASA suspended all Shuttle flights on June 25, 2002 as a result of small cracks, between 2.5 mm and 7.62 mm in length, being found in the metal liners used to direct liquid hydrogen flow inside the Space Shuttle Main Engine (SSME) propellant lines on Atlantis, on July 11. Three cracks were subsequently found on Columbia's No. 2 SSME and Atlantis' and Discovery's No. 1 SSMEs. One crack was found in the No. 1 and No. 2 SSMEs on Endeavour. Investigation showed that the cracks were most likely caused by bad welding, rather than age, or wear and tear. The entire Shuttle fleet was grounded as a safety measure while an investigation and replacement work was undertaken. The principal concern was that small pieces of metal might be ingested into an engine during a launch. Inspection of the five orbiters was expected to take the remainder of July. With no spare pipe liners in stock, repair would take at least 7 weeks. Replacement would require new liners to be manufactured, a task requiring several months.

In this same period, a mechanical fault was discovered in the bearings inside the hydraulic jacks that maintained the Mobile Launch Platform in the horizontal position during the Shuttle's rollout to the launch pad. Fifteen cracks were discovered in Crawler-1 and a further 13 in Crawler-2. All 16 jacks on the two Crawler Transporters were replaced.

When the Expedition-5 crew were informed that their occupation of ISS had been extended for one month, Bursch wrote,

> "We just got news that our Shuttle flight home has been delayed ... That should send us over the six-month mark and we should break Shannon Lucid's U.S. record of 188 continuous days in space. That feels nice to be able to share in a record ... but I sure do miss my family."

Meanwhile, NASA's own inspectors criticised the Administration for not thoroughly checking on United Space Alliance, the private contractor that serviced the Shuttles and prepared them for launch. NASA had announced that the estimated risk of a catastrophic failure of a Shuttle had risen from 1:78 in 1986, the year STS-51L was lost, to 1:55 in July 2002.

With all of this going on, STS-107, the "Freestar" solo science flight planned for launch on July 19, 2002 was grounded indefinitely, but was expected to be launched at the end of the year. This 16-day flight, with a crew of 7, would carry the new SpaceHab Research Double Module, loaded with more than 80 experiments. The mission was designed to answer the criticism that ISS was not performing sufficient science. However, on the resumption of flights it was decided that STS-107 would not now be launched until after STS-112 and STS-113 had flown to ISS, delivering the Starboard-1 (S-1) and Port-1 (P-1) ITS segments, respectively, which would slip STS-107 to January 2003.

THE INTERNATIONAL PARTNERS HAVE PROBLEMS

In 1997, Brazil had signed a $200 million contract with NASA to provide an EXPRESS Pallet for launch on a Shuttle flight now planned for 2005. Brazil's contract had included the selection of a Brazilian astronaut who would fly on a Shuttle flight to ISS. In mid-2002 Brazil informed NASA that it would be unable to produce the Pallet, citing budget difficulties as the reason for their failure to deliver.

Shortly after Brazil's decision, Japan announced that the Japanese Experiment Module (JEM) Kibo would not now be delivered to NASA's Space Station Processing Facility at KSC until 2006, one year later than scheduled, due to budget difficulties. The JEM Centrifuge Accommodation Module (JEM CAM) would be delivered to NASA in 2006. The delay meant that the ESA Columbus Laboratory Module might be launched earlier than planned, as might the Canadian Special Purpose Dexterous Manipulator, Dextere.

At the same time the speculation over who would fill the third couch on the October Soyuz TMA-1 taxi flight came to an end. Lance Bass' training came to an abrupt end when he failed to make the relevant $20 million payments to the Russians. Russian cosmonaut Yuri Lonchakov would now be the third crew member.

PROGRESS M-46

Progress M-46 was launched from Tyuratam at 13:37, June 26, 2002. While the spacecraft carried out a standard Soyuz rendezvous Korzun and Treschev rehearsed the back-up manual docking procedures using the TORU equipment set up in Zvezda. An automated docking occurred at 02:23, June 29, and the hatches between the two vehicles were opened at 05:30. The Expedition-5 crew began the long job of emptying the Progress and logging all of its contents on to the ISS' computerised monitoring system. They also performed standard maintenance inside Zarya.

The first week of July was one of light duties with lots of free time for the three astronauts. On July 3, Whitson repaired the MCOR and bought it back on-line following a 3-week outage.

In a pre-flight interview Whitson had described the future tasks planned for the MT and the SSRMS:

> "So, currently our arm is sitting on the Laboratory module. The shoulder is sitting on the Laboratory module, and we'll use the arm off the Laboratory, grab the Mobile Base System out of the payload bay, and attach it to the Mobile Transporter. And then once the Shuttle's gone ... one of the things we'll do is we'll check out the Mobile Base, make sure it's working correctly, and then we're going to do the step-off procedure, which means we'll grab one of the Payload and Data Grapple Fixtures with the arm and then release [it] from the Laboratory, so our new shoulder becomes on this Mobile Base System. And that allows us the capability of moving the arm along the truss. And that's important for the next

> phase, when [STS-112] arrives with the next piece of truss, because from that Mobile Base on the end of the truss of S-0, we will reach down into the payload bay and grab the S-1 Truss and pick it up and attach it to S-0. And then during [STS-113] we'll do the same from the other side, except because of the configuration ... the Shuttle arm will pick it up out of the payload bay and then we'll grab it from the Shuttle and attach it to the station. So it's going to be an interesting assembly complex, and the Mobile Base is key in positioning the arm in the appropriate place and it is a platform for the arm from which to work."

Whitson walked the SSRMS off Destiny for the first time on July 10, when she commanded the free end of the arm to attach itself to a fixture on the S-0 ITS. Following the walk-off, Korzun and Whitson put the SSRMS through the manoeuvres that would be required to support the installation of the S-1 ITS, during the flight of STS-112. Two days later the SSRMS was moved to a series of alternative PDGFs, in order to ensure the power and data flows required for the installation of the S-1 and P-1 ITS elements were functioning correctly.

On July 15, Korzun and Whitson worked together to replace the Desiccant/Sorbent Bed Assembly in the Carbon Dioxide Removal Assembly (CDRA), in Destiny. While one bed had been performing normally, the bed being replaced had been malfunctioning since Destiny's launch, in February 2001. A valve between the desiccant and sorbent sides of the bed was stuck in the open position. The replacement took 4 hours to complete, but when the unit was activated on July 20, the new bed showed a similar leak to the original bed, but at a lower rate.

Two days later, the entire crew performed a medical operations drill, to maintain their training in that vital area of crew performance. On the same day, Whitson worked with engineers on the ground to work out a repair procedure for a spacesuit battery that had failed to discharge prior to being recharged. During the week the crew continued with their science programme, working with the Micro-encapsulation Electrostatic Processing Experiment (MEPE), the ADVASC, and the Microgravity Science Glovebox (MSG). Whitson described the importance of the glove box as follows:

> "Well, I think science advances a lot slower than any of us would like it to; but specifically during Expedition 5 we're getting the Microgravity Sciences Glovebox up ... this is a facility payload that is going to allow various different investigators to do materials science inside of a confined environment. In the environment of the Space Station, if we do things that involve toxic materials, we need to have several layers of containment, because obviously we can't just open the window if we have a little toxic fluid escape. So, the Microgravity Sciences Glovebox provides us a level of containment. It allows us to work inside with the rubber gloves up on our arms, and we can manipulate and set up experiments inside a contained environment. And it would be experiments that we couldn't possibly do without that additional level of containment ... We've had other smaller gloveboxes flying, which have flown before either on the Shuttle, in Spacelab, and even one on Mir. So there have been previous ones; this is a kind of a facility-class

payload, very large, and I think it's going to really enhance our capabilities in the materials science world."

They also participated in an educational broadcast called "Toys in Space", whereby they used a number of simple toys to explain the basic principles of physics involved in spaceflight and present on ISS. The scientific work continued throughout the following week, with the crew working on the Solidification Using A Baffle in Sealed Ampules (UABSA) experiment, which was designed to grow semiconductor crystals in microgravity. Whitson activated the MSG and televised the heating and cooling processes involved in heating a semiconductor to melting point and then allowing it to cool. Whitson also monitored the ADVASC, where soybean plants had started growing. All three astronauts performed PuFF experiments in advance of the EVAs planned for mid-August. There was also the Renal Stone Experiment. Whitson said:

"Our experiment is based on some previous data that we've collected on the Shuttle and on the NASA/Mir science program, and there we found that crew-members are at a greater risk of forming renal or kidney stones . . . And that's a big deal in spaceflight because, if you've ever known anybody who's formed a kidney stone, it is excruciatingly painful if that stone begins to move, and in essence it will incapacitate a crewmember, and you would probably have to abort the entire mission. So we are interested in trying to reduce that risk of stone formation. We've had crewmembers form stones after flight, and there's one case where they aborted a Russian mission because of a crewmember who formed a stone during flight . . . that moved. And so . . . we're looking at a countermeasure to try and alleviate some of those effects. We're using a drug that's commonly used on the ground to inhibit calcium-containing stones, and based on the results of our previous research we're going to be using potassium citrate in the crewmembers on a daily basis to see if that actually reduces the risk of forming renal stones, and collecting the same data that we collected . . . before and see if the risk is actually decreased . . . Our research shows that there really is a higher risk, and it has to do with the fact that the crewmembers tend to be somewhat more dehydrated, as well as the fact that their bones are demineralizing, so there's a greater level of calcium and phosphate in the urine, which can form crystals and form the nucleus of the stone that could occur."

Meanwhile, they continued their repair and maintenance work, replacing remote power converter modules in the Quest Airlock after they had shown the initial signs of malfunction. On July 22, the crew's treadmill began making "clanking noises" when they ran on it. Investigation revealed that the problem lay in one of the rollers that the belt ran over, where a ball bearing had seized. The crew also worked on the Elektron oxygen-generating system, but failed to improve its performance.

Figure 21. Expedition-5: Sergei Treschev displays one of the station's many tool kits.

MORE CRITICISM

Following a review of the ISS programme, the director of the US government's Office of Management and Budget described ISS as one of the Bush government's "most inefficient and wasteful programmes." The programme was further described as one of the "biggest [budget] over-runs ever in the federal government."

The NASA Advisory Council, which had also been tasked to review the ISS programme agreed with the conclusions of the Young Committee. Its report stated that the huge budget over-runs in the ISS programme "cannot be excused and must not be ignored." The Council also agreed that NASA must complete the ISS programme without further budget over-runs for at least two years, during which NASA could not hope to expand the Expedition crew beyond three people. Beyond the criticism, the Advisory Council suggested that NASA begin assigning a modest budget to revive the American Habitation Module and an American CRV.

In March Sean O'Keefe had established a task force to review the station's ability to support science of merit. On July 10, the force recommended that 15 of the 35 areas of research reviewed be pursued as "first priority". The task force also recommended that NASA stop referring to ISS as a "science-driven programme", until the size of the Expedition crew was raised to six people. Meanwhile, a Rosaviakosmos spokesman stated that the international agreements on which the ISS programme were based were "deteriorating seriously". He suggested that Russia should demand those agreements be renegotiated, and suggested that, if it wished, Russia could build and launch a "European" space station as an alternative to ISS.

REVIVAL OF THE CREW RETURN VEHICLE

The suggestion that NASA might reconsider developing ISS beyond Core Complete, including the CRV, came as no surprise to many inside the Administration. In 2002, an internal JSC report suggested that NASA could not expect to increase the Expedition crew beyond three people before 2008. The report included a statement that a seven-person CRV (X-38?) should be included in the plans for increased operations. The report stated, "Succeeding with the CRV is key to our long term vision for NASA ... Maintaining our ability to design, build, test, and fly a spacecraft like the CRV is key. This is recognised by every senior manager at JSC, and elsewhere in NASA." The X-38 CRV had been officially subject to cancellation since the Bush Senior Administration's attempt to bring the ISS budget under control.

Despite the X-38 programme's alleged cancellation in mid-2001, Congress had instructed that funding for the programme should be reinstated in November of the same year. In 2002, Aerojet delivered the De-orbit Propulsion Stage (DPS) for the X-38, for use on CRV-201, which had been built for the "ironbird" flight, the return from orbit after delivery into space in the payload bay of the Shuttle Columbia. The eight-thruster DPS would fire to slow the X-38 down, allowing gravitational attraction to pull it back into the atmosphere.

In June 2002, NASA Administrator Sean O'Keefe cancelled the rejuvenated X-38 programme, in advance of a new change in direction: the Integrated Space Transportation Plan (ISTP) and the Orbital Space Plane (OSP) Crew Transfer Vehicle (CTV). The term Crew Return Vehicle was slowly being dropped from NASA's vocabulary, just as the X-38 was being dropped from ISS. Even as the X-38 was finally cancelled, NASA was criticised for not including the budgetary effects of that cancellation in their reports. Rather, some politicians felt that the cancellation announcement had been made in such a way as to suggest that the only reason for the cancellation was so that NASA could present their plans for the new OSP in a better light, in that OSP could perform both the CTV and CRV roles, while X-38 was more narrowly focused on the CRV role.

IN ORBIT, LIFE GOES ON

On August 1, Korzun and Whitson moved the SSRMS so that the camera on its end-effector could view the MBS and the attached POA grapple fixture that was mounted there. The POA was commanded to go through the motions of grasping a payload while being filmed. The controlling computer placed the POA in "Safe Mode" when data suggested that the motors were running too fast. The equipment was powered off and then powered on once more before re-running the test with the POA's motor running more slowly. The test was completed without further problems. During the day Progress M-46's rocket motors were used to boost the station's orbit and place ISS in the best orbit to support the arrival of the next Progress and the Soyuz TMA-1 exchange flight. Whitson also repositioned the SSRMS to a location where its cameras could be used to view the EVA planned for August 16.

All three astronauts spent the following week preparing for their first Stage EVA. They also found the time to follow a set of instructions read up from the ground that allowed them to complete a temporary repair on their malfunctioning treadmill. The repair allowed them to use the treadmill in an un-powered mode until the part for a repair could be delivered on the next Progress, to be launched on September 20. During the week, the Russians announced that Soyuz TMA-1 would be launched on October 28.

The first of two Expedition-5 Stage EVAs began when Korzun and Whitson left Pirs at 05:23, August 16. Their exit was delayed for 1 hour 43 minutes due to a valve controlling the flow of gas to the oxygen tank in their Russian Orlan EVA suits being set in the wrong position. Korzun had made the EVA sound simple:

> "The first EVA ... we will conduct with Peggy, we will take MMOD shield from PMA-1, transfer it with Russian Strela equipment to the Service Module, install this MMOD shield on the corner of the Service Module, and then we will install two antennas which they will use for ham radio; we will install them on the Service Module. And then, we will take old Kromka and install new Kromka. And, we will come back. Approximate time of the EVA will be six hours."

Once outside, they collected their tools together and prepared the Strela crane attached to the Pirs docking module. The Strela was used to move six micrometeoroid shields from their temporary position on PMA-1 to various positions around the exterior of Zvezda where Korzun and Whitson secured them in place. An additional 17 shields would be carried up to ISS on future Shuttle flights and would be installed on Zvezda on future EVAs. Korzun and Whitson were told not to refurbish the Kromka experiment, which was designed to capture residue from Zvezda's thrusters. Plans to collect thruster residue by swabbing the exterior of Zvezda were also abandoned as a result of the late start to the EVA. The EVA ended with the astronauts storing their tools and stowing the Strela crane in its parked position. Pirs' hatch was closed at 09:48, after an EVA lasting 4 hours 25 minutes.

Preparations for their second EVA occupied most of the next week and Korzun and Treschev left Pirs at 01:27, August 26. This time the hatch opening was delayed by the search for a pressure leak in one of the airlock hatches. Once again, Korzun described the activities for the NASA website:

> "Second EVA. We will use ... Docking Compartment—Russian airlock—and we will use Russian spacesuit Orlan. And second EVA we will conduct with Sergei Treschev ... We will install, we will replace flow regulator of the thermal control system of the FGB. And we will remove one of the panel of Japanese experiments, and then we will install special equipment for cable outside of the station. EVA time, it's about six hours."

The crew began work by installing a frame on Zarya to act as a temporary stowage area for equipment on future EVAs. They also installed holders on the exterior of Zvezda, to guide EVA astronauts' umbilicals around the exterior of the

Figure 22. Expedition-5: Peggy Whitson works with the Microgravity Science Glovebox.

Russian modules. Their next task was to exchange the trays of samples in the Japanese materials exposure experiment on the exterior of Zvezda. Having completed their own tasks the two cosmonauts picked up the tasks left from the shortened first EVA. They replaced the Kromka experiment and confirmed that the deflectors installed to reduce the thruster residue build-up on the exterior of the module were performing their task properly. Finally, they installed two ham radio antennae on the exterior of Zvezda. The EVA ended at 06:48, after 5 hours 21 minutes.

Inside ISS work continued on the experiment programme. Whitson repaired and cleaned the Microgravity Science Glovebox before preparing it for the sixth experiment run. She also down-linked a video tour of Destiny, pointing out a number of the principal experiments. On August 30, she serviced the American EMUs in the Quest Airlock Module. She also partially removed the EXPRESS-2 rack in Destiny to replace a smoke detector. The crew had three days off over the American Labor Day weekend.

On the first day after the holiday, the crew began their planning for the arrival of STS-112. In preparation for the three EVAs associated with that flight they processed the batteries in the EMUs held in the Quest airlock. During the remainder of the week they participated in an emergency procedures exercise and continued the experiments in the Microgravity Science Glovebox. On September 5, Whitson and Korzun took turns to control the SSRMS, including allowing the cameras on the end-effector and the POA to view the snare wires on each other as they were operated. They also grappled and un-grappled the fourth and final PDGF on the MBS, ensuring that it was working correctly. The following day much of the equipment in Destiny was powered down to allow the crew to replace an RPCM. Huntsville began applying power to the module's equipment as the crew began their sleep period.

Whitson removed the final sample from the SUBSA experiment on September 11, thereby completing the first experiment in the Microgravity Science glovebox. The following day Korzun and Whitson used the SSRMS to view the nadir CBM on Unity. The move was prompted by the discovery of debris on Leonardo's CBM following the flight of STS-111. Unity's CBM was inspected with its protective petals open and closed, and the images down-linked to Houston for further investigation of the problem.

On September 13 the crew reached their 100th day in orbit. During the day Whitson set up and activated the ultrasound equipment in the HRF rack in Destiny. She then spent the next four hours using the equipment to capture videos of herself. The ultrasound equipment was designed for use in experiments, but it also had the future capability to be used in diagnosing an illness among the crew. That night Korolev commanded the rockets on Progress M-46 to fire, raising the station's orbit in preparation for the launch of Progress M1-9.

Sean O'Keefe spoke to the Expedition-5 crew on September 15. He named Peggy Whitson as NASA's first ISS Science Officer and told the three astronauts that it was time to increase the station's main mission: science. Whitson and Korzun repaired the CDRA in Destiny the following day, and it was subsequently able to function as planned for the first time since its launch in February 2002. Whitson also prepared the Microgravity Science Glovebox for the first in a run of new experiments called Pore

Formation and Mobility Investigation (PFMI), whereby a transparent material was melted in the glovebox to see how bubbles formed and moved in molten materials. Throughout the week, Korzun and Treschev continued to load rubbish into Progress M-46, in preparation for its undocking. The also began packing items for their own return to Earth on STS-113, scheduled for launch on October 2. Progress M-46 was undocked at 09:58, September 24. The Russians kept the spacecraft in orbit for two weeks, employing its cameras to film smog over northeastern Russia, before de-orbiting it on October 14.

PROGRESS M1-9

On September 6, 2002, the Russians announced a delay in the launch of Progress M1-9, from September 20 to September 25, caused by problems with the spacecraft's computer and its KURS docking system antenna. The relevant hardware was returned to RCS Energia for repair on September 3.

Even as Progress M1-9 was being prepared for flight, officials at Energia released details of the company's dire financial position. Valeri Ryumin announced that the company had ten Soyuz and Progress vehicles in manufacture but contractors were reluctant to release vitally needed parts to complete the spacecraft to a company that already owed millions of roubles to the bank and its contractors. He stated that the $79 million that Energia would receive from the Russian government for FY2003 would be only half that required to finance the company's commitment to ISS for the year. This consisted of two Soyuz TMA spacecraft and four Progress M1 vehicles. Ryumin blamed the Russian government for not meeting its commitment to ISS and even went so far as to suggest that the station be powered down and left unmanned between Shuttle flights. American officials once again ignored their own unilateral disregard for the legal contract between the ISS partners and replied to the Russian news by saying that they expected Energia to meet their contractual commitments to ISS.

Progress M1-9 was launched at 13:37, June 26, 2002, and was station-keeping 1 km from ISS on September 28. At that time the spacecraft was used to test the two-way flow of data between the KURS automatic docking system in the two spacecraft. Progress docked automatically to Zvezda's wake, at 13:01, September 29. The internal hatches were opened that afternoon, but unloading did not begin until the following day.

STS-112 DELIVERS THE STARBOARD-1 ITS

STS-112	
COMMANDER	Jeffrey Ashby
PILOT	Pamela Melroy
MISSION SPECIALISTS	David Wolf, Piers Sellers, Sandra Magnus, Fyodor Yurchikhin

STS-112 Atlantis had originally been planned for launch on August 22, 2002, but this was cancelled when cracks were found in the cryogenic propellant duct liners in the SSMEs of each of the Shuttle orbiters, leading to the grounding of the entire Shuttle fleet. Following an engineering review of the cracks, the launch was set for not earlier than September 26. Over August 10–12 the cracks were repaired, by welding and polishing. By August 18, the three SSMEs had been installed in the rear of Atlantis.

On August 22, the launch was set for October 2, 2002, but during the final preparations for launch Hurricane Lilli threatened to make landfall and threaten Houston, the location of NASA's MCC. As there was no way of telling how the hurricane might affect control of the flight, it was decided to delay for 24 hours. On October 2 that was extended for a further four days as the control centre in Houston was shut down and control of ISS handed over to the Russians for the duration of the storm. The hand-over meant a restriction in communications as the Russian network could not handle the station's Ku-band and S-band channels. As Korolev was also unable to monitor the ability of the American P-6 ITS photo-voltaic arrays to track the Sun, the arrays were locked in position. Houston resumed control of ISS on October 4, and it was returned to full operation. The problems caused the Russians to cancel a test of Progress M1-9's thrusters on October 4, and a station re-boost manoeuvre on October 5. The new STS-112 launch date was set for October 7.

On that date Atlantis lifted off at 15:46, to deliver the Starboard-1 (S-1) truss to the station. Throughout the launch a small camera mounted on the ET showed the external view looking back past Atlantis. Similar videos captured on throw-away rockets had proved very popular with the public. As the Shuttle lifted off, ISS was over the Pacific Ocean and the Expedition-5 crew were on their 122nd day onboard and their 124th day in space.

Post-launch inspection showed that all ten back-up pyrotechnics in the SRB hold-down bolts had failed to ignite. Each bolt had a primary charge to sever the bolt at lift-off. The back-up charges were then fired a fraction of a second later, to ensure all of the bolts were separated. On this occasion no serious damage was sustained because all ten primary charges had fired. NASA engineers acknowledged that the problem was most likely in the transmission, or receipt of the firing signal for the secondary charges, an echo of the problems with the wiring in the Shuttle fleet, which had grounded the entire fleet in 2000.

Having achieved orbit, Atlantis followed the standard rendezvous pattern while her crew performed all of the usual activities in advance of docking with ISS. They also performed their own solo experiment programme. As Atlantis performed the rendezvous Korzun, Whitson, and Treschev were unloading Progress M1-9 and packing items for return to Earth on STS-112.

Ashby performed a manual docking with PMA-2 on Destiny's ram at 11:17, October 9. Following pressure checks Whitson asked Ashby if he had brought the salsa that she had asked for. When Ashby replied that he had, Whitson conceded, "OK, we'll let you in." The hatches between the two spacecraft were opened and the Shuttle's crew left Atlantis and moved into Destiny, the first visitors to ISS in four

Figure 23. STS-112 crew (L to Right) Sandra Magnus, David Wolf, Pamela Melroy, Jeffrey Ashby, Piers Sellers, Fyodor Yurchikhin.

Figure 24. STS-112: Atlantis delivers the Starboard-1 Integrated Truss Structure.

months. The two crews immediately began work preparing for the installation of the S-1 ITS, and the first of the three EVAs associated with it.

On October 10, Whitson and Magnus used the SSRMS to lift the 15-metre-long, 14-tonne S-1 ITS out of Atlantis' payload bay and manoeuvre it into a position where it could be soft-docked with the Starboard end of the S-0 ITS, which was mounted on Destiny's zenith CBM. Once soft-docking had been achieved motorised bolts were driven into place to secure the two units together at 09:36.

Wolf and Sellers exited the Quest airlock at 11:21. Wolf rode a foot restraint mounted on the SSRMS while Sellers used his hands to move around. Wolf connected power, fluid, and data umbilicals between the 2 ITS elements while Sellers used a motorised tool to undo the 18 launch locks on the 3 space radiators mounted on S-1, allowing them to be oriented for maximum cooling when they were deployed. The two then worked together to deploy an S-band antenna on the S-1 ITS. Wolf placed the antenna in the end-effector of the SSRMS and it was then moved into place near the join between the S-1 and S-0 ITS elements. Sellers held the antenna in place while Wolf secured the bolts that would hold it there.

As they passed over the Pacific Ocean Sellers looked at Earth and asked, "Where am I? Wow, its too beautiful for words, unbelievable!" After a short break to take in the view he remarked, "That's it, back to work."

They then worked together to release the bolts that held the Crew and Equipment Transition Aid (CETA) to the S-1 truss and configured its brakes. The CETA would

Figure 25. STS-112: Piers Sellers wears an American Extravehicular Mobility Unit near the open hatch of the Quest airlock.

Figure 26. STS-112: The S-1 Integrated Truss Structure. The SSRMS is mounted on the MBS which runs along a track on the ram face of the Starboard-0 Integrated Truss.

be used to transport future EVA astronauts and their equipment along rails travelling the length of the ITS when it was complete. Their final task was to fix the S-1 outboard nadir exterior camera in position. At that point the SSRMS suffered a mechanical failure and the two astronauts completed their tasks without it. This was the first of two cameras to be fixed on the S-1 truss for use in future EVAs. The first STS-112 EVA ended at 18:22, after 7 hours 1 minute. With the two astronauts back in the airlock Houston told them, "You guys did an awesome job. You saved us."

Both crews took time out to relax on October 11, before beginning the task of transferring equipment from Atlantis to Destiny. Wolf and Sellers prepared the equipment for their second EVA, planned for October 12. During the evening both crews answered questions from the Russian and American media. Questioned about their first EVA Wolf replied, "I tell you what, it's pretty tough. We got plenty tired, and I believe our heart rates got up over 170 during that task." On the subject of EVA-2 he added, "We're ready to go again."

Sellers, who had been on his first EVA, described his first view through the open airlock hatch:

> "I was completely knocked out of my socks, which were luckily in my suit. I could see a landscape with clouds and a river, and it was just huge. It was fantastic. So the first five minutes, I was pretty much non-functional. My little brain was overloaded."

The principal task of the second EVA had been identified by engineers working with the ITS elements on the ground. They found that in some of the older ISS plumbing pressure could build up, and the pipes might not twist and disconnect as planned, thereby leading to a destructive failure. After many hours of work, a work-around was designed for the problem: fit Spool Positioning Devices (SPDs) on the plumbing connection to release the pressure.

After breakfast and prior to the second EVA, Ashby and Melroy fired Atlantis' thrusters to raise the station's orbit. EVA-2 then began at 10:31, October 12. Following preparations, Wolf and Sellers worked to further prepare the CETA for use during future EVAs. They also installed 22 Spool Positioning Devices on fluid lines in an attempt to prevent pressure build-up in the pipes preventing the correct use of the quick-disconnect fittings if required in the future. When he fitted the first device Wolf reported that there had already been a pressure build-up in the pipe and he had released it when he fitted the SPD. As Wolf put it, "I heard it burp." Two other SPDs could not be fitted because the fluid lines "were of a different configuration" from that on which the SPD was designed to fit. What the euphemism in the NASA STS-112 Status Report actually meant was that Wolf had discovered that the two pipes in question had been launched with certain parts not installed. That discovery came as a surprise to everyone. Some of the joints were in positions that were hard to reach, but Sellers achieved the task and was congratulated from inside ISS by Melroy, "It is quite possible you're the only person in the astronaut office who could have done that task." The two astronauts also connected the ammonia cooling system to the S-1 ITS radiators and fitted a camera on the exterior of Destiny. The EVA ended approximately 30 minutes early, at 16:35, after 6 hours 4 minutes.

During the end-of-day press conference, mission planners explained why the SSRMS had failed during the first EVA. The SSRMS and the MBS both worked on separate software programmes, which have to work together. If they were not correctly synchronised, then the entire unit stopped working. To ensure that two software programmes were synchronised required both the SSRMS and the MBS to be powered off and then powered back on in order to reset them, like re-booting a computer. There had not been enough time to do that before EVA-2, but Houston expected the MBS/SSRMS combination to be available for use during EVA-3.

October 13 was spent on equipment transfers and repairs. The Expedition-5 crew made a temporary repair to the TVIS, which was returned to a usable condition, despite the fact that they found a broken cable associated with the gyroscope, which would require a new one to be delivered on a future Shuttle, or Progress. The radiators on the S-1 ITS were rotated into position, but their deployment was cancelled after adjustments were required to the tolerance levels of protective circuits used to monitor the initial stages of deployment. The time required to make those adjustments meant that Houston could no longer watch the deployment live, so it was delayed until the following day. Wolf and Sellers made their preparations for EVA-3, scheduled for October 14. During the afternoon press conference Whitson told how she had shared the new salsa delivered by STS-112 with her Expedition-5 crewmates. She also explained how her taste for certain foods that she enjoyed eating on Earth had changed in space so that she no longer enjoyed them. It was a common complaint among long-duration crews that food tasted extremely bland in space. Therefore, strong-flavoured food, or sauces such as Whitson's salsa, became popular among these individuals.

That day began with the de-orbiting of Progress M-46 at 04:34. Ashby and Melroy also performed a second re-boost manoeuvre commencing at 07:20. The two series of manoeuvres had placed the station in the correct orbit to receive Soyuz TMA-1 later in the month, but this second manoeuvre used sufficient propellant to prompt Houston to reduce the planned 360° post-docking fly-around of ISS to just 180°, before Melroy performed the standard separation manoeuvre.

Houston then commanded the middle of three radiator panels on the S-1 ITS to deploy. As it reached its full length the crew played a recording of Handel's *Hallelujah Chorus* and Houston commented, "That's very appropriate music." The remaining two radiators were scheduled to be opened in 2003, at which time their heat-shedding function would be activated.

Wolf and Sellers began EVA-3 at 10:11. Their first task was to remove the bolt that had prevented the cable cutter on the Mobile Transporter, mounted on the S-0 Truss, from activating at the end of the STS-111 flight. Next, they fitted ammonia lines between the S-0 and S-1 ITS elements and removed structural support clamps that had held the S-1 ITS in place during launch. At one point 46-year-old Wolf remarked, "We're over the hill." He quickly added, "I mean over the hill on the station." Sellers, 47, replied, "No comment." They also fitted two more SPDs to the pump motor assembly that circulated the ammonia throughout the cooling system. This was a "get-ahead" task, carried out because the two men had completed their primary tasks well ahead of schedule. As the EVA drew to a close Houston told them, "You guys are doing a great job. Our only concern is that you're making it look too easy for us." The EVA ended at 10:11, after 6 hours 36 minutes.

Next day, both crews were given some free time to spend together before the STS-112 crew began preparing for their departure. While the final items were transferred between the two spacecraft, Ashby and Whitson worked together to replace a humidity separator in the Quest airlock. That evening the two crews said their farewells. Whitson and Magnus hugged each other as they said goodbye. When it came to saying goodbye to her good friend and remaining in orbit for a further month, Whitson commented, "I didn't know it was going to be so hard."

Figure 27. STS-112: As Atlantis departed the International Space Station, the Starboard-1 Integrated Truss Structure was clearly visible.

Melroy thanked Whitson and her colleagues for making the STS-112 crew welcome onboard the station and then concluded with, "You look wonderful. You look great. We miss you. Come home soon." Ashby led his crew back to Atlantis and the hatches between the two vehicles were sealed. Both crews spent the night in their individual spacecraft. In Houston, mission manager Robert Castle told a press conference, "Overall, things are going very, very well, and I don't think they could have done better."

As Atlantis passed over the Russia/Ukraine border Melroy told controllers, "We want to stay." But they could not stay, and she undocked Atlantis from Destiny's ram at 09:13, October 16, and completed the reduced 180° fly-around manoeuvre to allow the other members of the crew to photograph ISS with the S-1 Truss in place. Melroy then performed the separation manoeuvre to allow the two spacecraft to drift apart under the influence of orbital mechanics.

October 17 was spent packing everything away in preparation for retrofire. Houston told the astronauts, "Just to make you jealous, it's [the temperature] in the 50s here in Houston, and the weather is absolutely gorgeous, the air is dry."

As Atlantis continued to fall through the vacuum of space, Ashby replied, "It's pretty dry up here too." He added, "Sounds like a real good day to come home tomorrow." Atlantis performed retrofire on October 18, and Ashby turned his spacecraft for re-entry. As Atlantis fell out of the Florida sky it was struck by some of the strongest crosswinds experienced by any landing Shuttle. Ashby, the most experienced Naval aviator in the Astronaut Office flew his spacecraft to a perfect landing at KSC, touching down on the runway centreline at 11:44. His only comment at the time was "It's great to be back in Florida." Melroy commented on Ashby's smooth landing after the flight saying, "We're so proud of him we could burst our buttons." Ashby saved his comments for the flight in general, commenting at the post-landing ceremony to welcome them home, "What an incredible adventure we've been on. As I stand here. I can't help but think about all the people that helped us take the [S-1] Truss up there. It's been an amazing team effort." British-born Sellers, who had just completed his first flight in space, called ISS an "Island in the sky that is a completely different place that has different rules. It was an experience like nothing I've seen or even dreamed of before. Things float. You're climbing underneath structures like a spider underneath a gutter. It's a magical place."

At a post-flight press conference James Wetherbee, Commander of STS-113, which would fly a similar flight to ISS in November, commented, "It was great to see them pull off the mission so successfully. That makes us feel a lot better and we're that much more prepared."

With Atlantis gone, the Expedition-5 crew had returned to their daily routines. On October 24, Whitson and Korzun put the SSRMS through the manoeuvres that would be required to install the P-1 ITS, which would be delivered by STS-113 in November. That flight would also carry the Expedition-6 crew to ISS. In the same week Whitson brought her experiment programme to an end, in preparation for the Expedition-5 crew's return to Earth. During the same period flight controllers in Houston up-linked new software to the three systems computers housed in Destiny. This was the first major update to the software since the laboratory module had been docked to ISS in February 2001.

SOYUZ TMA-1, FIRST OF A NEW CLASS

SOYUZ TMA-1	
COMMANDER	Sergei Zalyotin
FLIGHT ENGINEER	Frank de Winne (ESA, Belgium)
FLIGHT ENGINEER	Yuri Lonchakov

Soyuz TMA-1 was originally scheduled for launch on October 28, 2002, but it was not to be. On October 15, 2002, a new version of the Soyuz-U launch vehicle, carrying a "Foton" satellite, was launched out of Plesetsk. The launch failed and the vehicle fell in a nearby forest killing 1 soldier and injuring 20 other people. The failure was later identified as being caused by contamination in the hydrogen peroxide system. As a result of the failed satellite launch the Soyuz TMA-1 launch was delayed on October 18, and rescheduled for October 29.

Soyuz TMA-1 was the first of a new class of Soyuz spacecraft, with the new Descent Module interior arrangement to facilitate couch frames designed to be adjustable to allow them to carry taller and heavier, or shorter and lighter than average crew members. The new arrangement had been developed after NASA recognised that the restrictions demanded by the standard Soyuz TM spacecraft meant that many American astronauts would not be able to serve on ISS Expedition crews as they would not be able to squeeze into the Soyuz TM CRV in the event of an emergency return to Earth. The new Descent Module also had improved instrumentation and avionics.

The new spacecraft was launched from Baikonur at 22:11, October 29, 2002. Ten minutes later Soyuz TMA-1 was in orbit with its antennae and photovoltaic arrays deployed. Unlike on previous occasions, the Expedition-5 crew did not transfer Soyuz TM-34 from Zarya's nadir to Pirs' nadir before the launch of the replacement Soyuz. After following the standard 2-day rendezvous, the Soyuz docked to Pirs at 00:01, November 1. Docking took place over central Russia and was monitored by the Expedition-5 crew inside Zvezda. Whitson described the rendezvous:

> "Although the timing for the Soyuz arrival and docking was based on lighting and comm[unication] coverage, it seemed to be choreographed for aesthetic purposes. I was using our new camera at the end of the S-1 Truss to film the docking. I was trying to find a tiny speck of light (we were in eclipse) in the general direction of the approach. I saw a brighter than normal 'star' and zoomed the camera for maximal magnification ... As the Soyuz capsule began to fill my video monitor, the sun began to peek around the edge of the planet, making that incredible royal blue curvilinear entrance. Alpha [ISS] and the new Soyuz capsule were soon bathed in brilliant white light from the sun. While the Earth below was still dark, the Soyuz made contact and became our new rescue vehicle. Valeri and Sergei had

Figure 28. Expedition-5: The Expedition-5 and Soyuz TMA-1 crews pose together in Zvezda. (rear row) Peggy Whitson, Yuri Lonchakov, Sergei Treschev. (centre) Sergei Zalyotin, Frank De Winne. (front) Valeri Korzun.

a close-up view of the docking from the Service Module (SM). From the nadir windows in the SM it is possible to see the docking compartment, which extends below from the forward end of this module. In other words, the new Soyuz docked about 2 meters before their eyes."

Following pressure checks the hatches between the two spacecraft were opened at 01:26. After a safety briefing the three cosmonauts were welcomed aboard ISS. Over the next week, de Winne conducted his own experiment programme related to genetic engineering and the effects of microgravity on genes. The cosmonauts also performed protein crystal growth and materials processing experiments. Having swapped their couch liners with those of the Expedition-5 crew the Soyuz TMA-1 crew undocked Soyuz TM-34 Zarya's nadir at 15:44, November 9, and manoeuvred clear of the station. Following retrofire Soyuz TM-34 landed in Kazakhstan at 07:04, November 10. Soyuz TMA-1 was left docked to Pirs. It would serve as the ISS CRV for the next 6-months.

Whitson wrote:

"After the Soyuz undocked, we were able to watch as it re-entered the Earth's atmosphere, about 2 orbits later. The ground control team had provided instructions of where to look in order to see the spacecraft, and since it was during the eclipse, I shut off all the lights in the [Destiny] lab to watch from the window there. The thing I noticed first was what appeared to be a milky white contrail in the darkness. It brightened and the Soyuz became visible as it began to glow from the heat of re-entry. The Soyuz consists of three parts, the engine section, the 'living compartment', which is not any larger than a subcompact car volume, and the cramped descent module, sandwiched in between them. I was surprised to actually see 'razdalenea' (separation) of these three modules. The three glowing pieces separated, and the engine compartment and the living compartment trailed behind the descent module and began a fiery disintegration, looking much like a bright orange 4th of July sparkler. The central portion, the descent module, has a heat shield to protect the vehicle from the high temperatures (on the order of 3,000°F) generated during re-entry. We were able to see the descent module for a few minutes after separation, before it seemed to be swallowed up in the cloudy darkness below. About 4 hours after separation from the station, the taxi crew had landed in the cold desert of Kazakstan."

On November 10, the Expedition-5 crew waited in vain for the launch of STS-113. Following the cancellation of the launch attempt, Whitson pointed out to flight controllers in Houston that the crew had been rationing their drinks to make them last until Endeavour's arrival at ISS. With the Shuttle's launch delayed for one week she pointed out that the crew would run out of drinks before the Shuttle arrived. She asked for, and was given permission to take additional drinks from the equipment already delivered to the station for the Expedition-6 crew. She also sought permission to commence the Expedition-6 experiment programme, as the

Expedition-5 crew had completed their own science programme and packed away the equipment they had used.

A report entitled *Assessment of Directions in Microgavity and Physical Sciences Research at NASA* by the National Research Council's (NRC) Space Study Board was made public on November 6. The report praised the advance in NASA's microgravity research programme since its beginning during Project Skylab in the 1970s. It said that the present programme consisted of five areas

- biotechnology
- combustion
- fluid physics
- fundamental physics, and
- material sciences

all of which were threatened by the budgetary restrictions placed on ISS by mismanagement and the Bush Administration's restrictions, including the limiting of the Expedition crews to just three people. The report advised NASA to maximise microgravity research both on ISS and in laboratories on Earth.

On November 8, NASA announced plans to alter their FY2003 budget request to allow for the implementation of a new Integrated Space Transportation Plan (ISTP), including plans for the Orbital Space Plane CTV/CRV. In order to fund the ISTP NASA had entered an amendment to its $15 billion FY2003 budget. Of this $6.6 billion would be assigned to the "completion" (Core Complete) of the construction of ISS by 2006. Over the next four years, the OSP would consume $2.4 billion, with the first flight taking place in 2008, carrying up to ten crew members to ISS. A further $1.6 billion would be spent on Space Shuttle enhancements allowing it to continue flying through 2012 and possibly up to 2020. In order to bring ISS to *Core Complete* in 2006, NASA would spend $15.2 billion by adding a fifth Shuttle flight to the annual launch manifest.

As if to prove that there was an urgent requirement for the proposed OSP, Yuri Koptev, head of Rosaviakosmos, reported that the Russian space budget for FY2003 would not increase over that for FY2002, with no allowance for inflation. Koptev stated, "The problem is that our legislators wonder why they need to set aside the same amount or more for space if our partner countries have cut their ISS budgets."

The Russians asked the International Partners to assist with the funding for the new Soyuz TMA spacecraft, stating that production might be delayed, or even brought to an end if no assistance was forthcoming. One suggestion put forward by Rosaviakosmos was to stop flying Expedition crews to the station and evacuate ISS. NASA made it clear that they would not consider that option, saying that, even if Russia stopped flying cosmonauts to the station on Soyuz spacecraft, NASA would continue to fly Shuttle construction flights, involving occupation of the station for a few days at a time. (The original Space Station Freedom, without the Soviets/Russians, would have been constructed with crews only visiting the station on construction flights and not permanently occupying it until the final element, the Habitation Module, had been installed.)

One anonymous Russian source even suggested that if Russia alone was to maintain its ISS budget at the original level, in order to keep its contractual agreements to supply Soyuz and Progress spacecraft, then control of the programme should be passed from America to Russia. This naive view overlooked the millions of US dollars that America had given Russia to keep them in the programme from the beginning and the vast sums that America paid to support ISS operations. The same Russian source also naively suggested that Japan's delaying completion of the Kibo science module threatened that country's political relations with America and the ESA member states.

Meanwhile, the Russian Channel-1 television station had paid an original $20 million to Rosaviokosmos to begin funding a competition to place a journalist on ISS during a Soyuz taxi flight in October 2003. International journalists meeting Rosaviokosmos' strict health and fitness programme and passing the standard Space Flight Participant training programme would be applicable to take part in the competition. The entire selection and training competition would be filmed by Channel-1.

STS-113 INSTALLS THE PORT-1 ITS

STS-113	
COMMANDER	James Wetherbee
PILOT	Paul Lockhart
MISSION SPECIALIST	Michael López-Alegría, John Herrington
EXPEDITION-6 (up)	Ken Bowersox, Donald Pettit, Nikolai Budarin (Russia)
EXPEDITION-5 (down)	Valeri Korzun (Russia), Sergei Treschev (Russia), Peggy Whitson

The Pilot on STS-113 was originally to have been Christopher "Gus" Loria, but on August 14, 2002 Loria requested to be removed from the crew due to an unspecified injury at his home that had caused him to fall behind in training. NASA quoted their privacy rules as a reason for not giving further details at the time. Paul Lockhart, who had flown as Commander on STS-111, a similar flight involving ITS assembly work and an Expedition crew rotation took Loria's position.

Likewise, the Expedition-6 crew had changed just 4 months before launch. It had originally consisted of Ken Bowersox, Nikolai Budarin, and Don Thomas. In June, Thomas was grounded because his combined radiation exposure over the long-duration Expedition-6 flight would take him beyond his allowed lifetime exposure limit. He was replaced by Donald Pettit, who would be making his first spaceflight. Pettit had been training as Thomas' back-up since January 2001. Bowersox paid tribute to Thomas, "Thomas really, really wanted to fly long duration. We know this has been very, very hard for him. But he is a big part of our mission. Everywhere we go we see reminders of him." Despite this compliment, Thomas did not contact the crew on launch day to wish them well.

STS-113 would continue the construction of the ITS with the delivery and installation of the Port-1 (P-1) ITS element to ISS. The P-1 ITS would be mounted on the opposite end of the S-0 ITS to the S-1 ITS, of which it was practically a mirror image. The flight would also deliver the CETA Cart-B, for mounting on the ITS in support of future EVA astronauts.

All preparations proceeded towards a launch on November 10, 2002. On that date, propellant loading had been completed when a problem arose with Endeavour's oxygen system, which had allowed higher than acceptable amounts of oxygen to build up in the mid-body of the orbiter. The launch was cancelled and rescheduled for no earlier than November 18, 2002. De-tanking the propellants in the ET began the following morning and was completed before engineers entered Endeavour to rectify the oxygen system. That work required the payload bay doors to be opened while the STS-113 stack stood on LC-39. A leak was discovered in an oxygen hose located in Endeavour's mid-section, where fatigue from normal use, coupled with a weak design had caused the problem. The affected section of hose was cut out and replaced. While engineers were carrying out that repair a work platform struck the RMS in its parked

Figure 29. STS-113 crew (L to R): John Lockhart, Michael López-Alegría, John Herrington, James Wetherbee. These four were joined by the Expedition-6 crew on launch and the Expedition-5 crew during recovery.

position, alongside the payload bay door hinge. The thermal cover was ripped and the RMS' laminated protective cover was scratched. The RMS was subjected to X-ray and ultrasonic inspections, which revealed an area of de-lamination on the arm. Tests were carried out in Toronto to see if the de-lamination would affect the RMS' performance. Two sets of repair plans were established:

- For a repair to be carried out *in situ* at LC-39, resulting in a new launch date of "no earlier than November 22".
- For a repair if the RMS needed to be removed from Endeavour, causing the launch to be delayed until December.

At the same time the nitrogen flex hose located in the mid-deck, next to the failed oxygen flex hose, was also replaced.

The Expedition-5 crew spent their extra time in space packing and labelling experiment racks and ran through the SSRMS manoeuvres required to fit the P-1 Truss when STS-113 finally arrived at the station.

STS-113 was finally launched at 1950, November 23, 2002, and successfully climbed into orbit. NASA Administrator Sean O'Keefe was in Florida for the launch. At the post-launch press conference he made the comment, "The maximum period of time [in space] we've hit as Americans is 196 days. That's less than half the time needed for a one-way trip to Mars. And we believe in round trips at NASA."

Figure 30. STS-113: Endeavour delivers the Port-1 Integrated Truss Structure.

After following the standard two-day rendezvous, Whitson told the approaching Shuttle crew, "You guys look pretty good out there." As the Shuttle made a slow approach to ISS she told them, "You guys fly that like you stole it." Wetherbee docked Endeavour to Destiny's ram at 16:59, November 25. Following pressure checks the hatches between the two vehicles were opened at 18:31.

The Expedition-5 crew welcomed their visitors to ISS before Korzun gave them their safety briefing. Endeavour's crew began transferring equipment immediately, including the Expedition-6 crew's Soyuz seat liners and Sokol pressure suits. Bowersox, Pettit, and Budarin installed their seat liners in Soyuz TMA-1 and carried out pressure and leak tests on their suits before taking over command of ISS.

Asked to describe the crew exchange, Bowersox explained:

> "Well, it just depends on who you talk to. On paper we're supposed to go across, we're supposed to put our seat liners in the Soyuz, we're supposed to put on our Russian ... Sokol entry suit, try them on, make sure that they're leak-free, and then we're supposed to do a test, all three of us together, to make sure that we're ready to take over the Soyuz. Once we've done that, then officially we can be left on orbit. But if you talk to the guys who are there now, as soon as we show up, they're going home ... and our job is to figure out how we're going to get back."

Korzun, Whitson, and Treschev ended their occupation as the Expedition-5 crew after 171 days 3 hours 33 minutes. They now became part of Endeavour's STS-113 crew.

On November 26, Wetherbee secured Endeavour's RMS to the P-1 ITS, secured in the Shuttle's payload bay. At 10:22, the bolts securing the truss in place were commanded to release and Wetherbee lifted the huge structure out of the payload bay and handed it over to the SSRMS operated by Bowersox and Whitson. The latter pair transferred the P-1 ITS up to the port side of the S-0 ITS and commanded the motorised bolts to secure it in place.

At 14:49, 30 minutes earlier than planned, Herrington and López-Alegría exited the Quest airlock to begin the first of three EVAs to connect the P-1 ITS to the S-0 ITS and the ISS systems. "How do you like the view?" asked López-Alegría, as they made their way outside. "The view is phenomenal, just fabulous. Life is good!" Herrington, the first Native American astronaut, replied.

Copying the procedures used to install the S-1 Truss in October, the two astronauts connected electrical cables and installed SPDs to allow for the quick disconnection of pipes in the case of a future emergency. They also released the locks on the second CETA cart, before removing the two large metal rods, called drag links, that had supported the P-1 Truss during launch. The drag links were secured to the P-1 framework. Herrington then returned to Quest to top off his oxygen supply before rejoining López-Alegría to install the Wireless video system External Tranceiver Assembly (WETA) antenna on Unity. The WETA would allow the pictures from an EVA astronaut's helmet cameras to be received in the control centre without the need for a Shuttle to be present with the necessary antennae. The EVA ended at

21:35, after 6 hours 45 minutes. Mission control told them, "Great work. You've got a happy control team down here."

The two Expedition Crews spent the remainder of the day conducting hand-over briefings. Bowersox had described what he expected these to be like:

> "There's a lot of things that we just can't cover in training because the actual configuration of the station is too fluid and too complex to track on the ground and to reflect in our ground simulators. So there's a lot that we'll pick up during the docked time frame that will help us early in the mission. It's things that we just can't train for: where the cameras really are, where items are located, where cables have been arranged, where people are sleeping, a lot of small things that you just don't have time to cover in training."

November 27 was Wetherbee's 50th birthday. The STS-113 crew spent the day transferring equipment between Endeavour and ISS, while the hand-over briefings continued between the two Expedition crews. Wetherbee, Lockhart, Herrington, and López-Alegría also spent time in the afternoon preparing for the second EVA, while Whitson and Bowersox worked together to clear debris from the vent lines on the Carbon Dioxide Removal system in Destiny. Wetherbee and Lockhart also performed the first of three orbital re-boost manoeuvres using Endeavour's thrusters.

Figure 31. STS-113: John Herrington and Michael López-Alegría install the Port-1 Integrated Truss Structure.

November 28 was American Thanksgiving Day, but there was to be no holiday on ISS. NASA's Bob Castle told a press conference, "It will be a very busy day for them. I suspect they will celebrate Thanksgiving some other day."

The second EVA began at 13:36, November 28, 45 minutes ahead of the time in the flight plan. Herrington and López-Alegría connected two fluid jumpers to connect the P-1 ITS ammonia cooling system to the system in the S-0 ITS and the rest of ISS. Next they removed the starboard keel pin, a launch support. They used the CETA handcart to manoeuvre it to its permanent storage location on the P-1 ITS and secured it in place before installing a second WETA antenna on the P-1 ITS. Having removed and stowed the port keel pin they carried out the "get-ahead" task of releasing the launch locks on the P-1 radiator beams.

Herrington used a foot restraint mounted in the SSRMS to leave both hands free to allow him to pick up the P-1 CETA cart. Whitson and Pettit then swung the SSRMS so that Herrington moved across the front of ISS, passing across Endeavour's payload bay, to secure the P-1 CETA cart on the S-1 ITS, next to the S-1 CETA cart. This cleared the P-1 CETA tracks for the SSRMS to move along them at a later date on its MT in order to extend that side of the ITS. Their final task was to reconnect a cable from the WETA mounted on Unity before stowing their equipment and returning to Quest. The EVA ended at 19:46, after 6 hours 10 minutes. Space Station programme manager Bill Gerstenmaier told the media:

> "It's really been a textbook mission so far. It looks easy. It looks like it comes together without much trouble. That is totally counter to what really happens. It comes out so smoothly because of all the hard work we put in place."

During the end-of-day press conference Herrington described the EVAs, "The work is very difficult. Your hands get very tired. When the Sun goes down, you have this beautiful station illuminated in front of you. It gets incredibly dark, pitch-black, except the little spot your headlamp is aimed at. So you lose the perspective of what is around you." He also admitted, "I was amazed at how massive the Earth is."

On the station, the hand-over briefings continued and the CDRA was working properly after Bowersox and Whitson's repairs. The two crews shared a traditional holiday dinner courtesy of NASA's unique home delivery service: Endeavour.

November 29 was a day of equipment transfers. Whitson transferred the PCG-STES Unit 7 to Endeavour while Bowersox transferred its replacement, the PCG-STES Unit 10, to Destiny. López-Alegría and Pettit transferred the Plant Generic Bio-processing Apparatus (PGBA) from Endeavour to Destiny. The new equipment would allow researchers on the ground to observe plants being grown on ISS. During the morning Wetherbee and Lockhart used Endeavour's thrusters to make a second boost to the station's orbit. Whitson and Pettit also carried out troubleshooting tasks on the Microgravity Science Glovebox, following its failure on November 20.

During the afternoon Korzun, Whitson, and Treschev held a small ceremony to officially hand ISS over to Bowersox, Pettit, and Budarin. During the changeover

Korzun told the Expedition-6 crew, "We were so happy to live here, to work here ... We will miss our space house." He then told Bowersox, "I am ready to be relieved." Bowersox replied, "And I relieve you."

This was followed by a press conference, in which all ten astronauts took part. Bowersox told the conference that the Expedition-5 crew had set a very high standard of work that would be difficult to live up to. He added, "I only hope that my crew, Don, Nikolai, and I, will be able to work as well over the four, or however many months, we end up living on the station; hopefully more than four."

Wetherbee, Commander of STS-113, told the new crew, "Expedition-6, it is your duty to sail on and disappear over the horizon, but return after discovering new land and make the world a better place." It was a tall order.

Whitson admitted, "I do think I'm ready to go, but it's been a kind of a gradual process. A month ago, when I started to pack, I was definitely not ready to go. My husband reminded me it's much better to leave while you still want to stay, rather than the other way round. I'm happy to go, while I still wouldn't mind staying here." She explained how she had asked a NASA food specialist for a special meal on her return, "I asked if they would cook up a nice steak with a caesar salad with lots of garlic on it. I'm looking forward to getting some food that doesn't come in a bag." To go with her meal she wanted a cold drink, "We don't have any carbonated drinks up here so I'm looking forward to that, and anything with ice in it would be nice as well."

Planning towards the third EVA began at 11:21, November 30, when Whitson and Bowersox prepared to command the SSRMS to attach its free end to the MT and then release its hold on the fixture on the exterior of Destiny. The transfer was delayed when the MT stopped 3 metres short of the intended transfer location. As a result, Herrington and López-Alegría exited Quest at 14:25 and began their EVA by searching for anything that may have caused the MT to stop. Harrington found a UHF communications antenna that had failed to deploy and had snagged one of the MT's trailing umbilicals. He cleared the umbilical and deployed the antenna, which allowed the MT to continue on its journey, arriving at Work Point-7 (WP-7) at 17:11. The MT was latched in place, and prepared to receive the SSRMS by 18:00. The delay caused the astronaut's EVA tasks to be re-prioritised and when they voiced the opinion that they could complete all of their tasks without using the SSRMS the walk-off from Destiny to the MT was cancelled. The two men completed all of their tasks, including the connecting of 33 SPDs at various locations around the exterior of the station. They also connected the Ammonia Tank Assembly umbilicals and reconfigured a circuit breaker on the Main Bus Switching Unit. Finally, they reconfigured the Squib Firing Unit on the P-1 radiator unit in preparation for their deployment in 2003. The EVA ended at 21:25 after exactly 7 hours.

December 1 was the final full day of joint operations. Herrington and López-Alegría cleaned and stowed their EMUs and the tools they had used during their three EVAs. Wetherbee and Lockhart completed the third series of re-boost manoeuvres and the two Expedition Crews continued their hand-over briefings. With most of the two-way equipment transfers complete the two crews enjoyed some free time during the day to recover from their hectic earlier schedule.

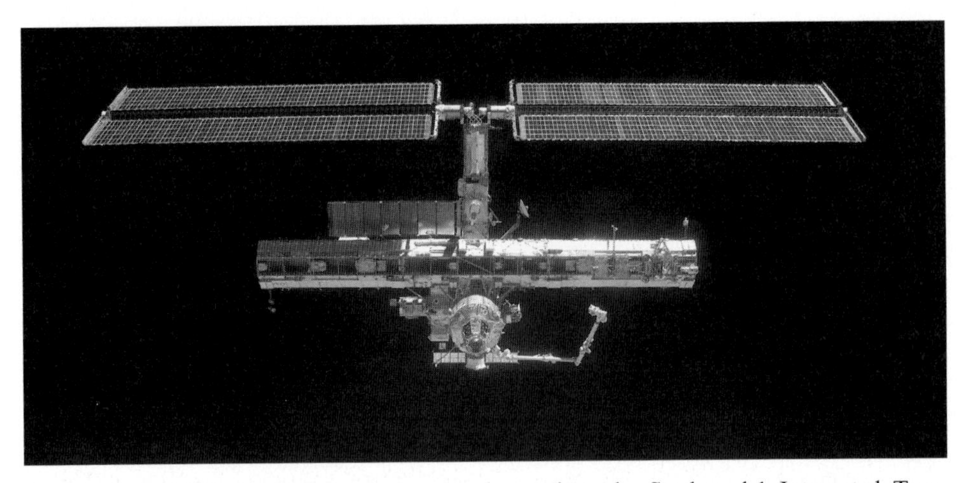

Figure 32. STS-113: As Endeavour departs the station, the Starboard-1 Integrated Truss Structure is mirrored by the Port-1 Integrated Truss Structure. Following the loss of the solo Shuttle flight STS-107 and the grounding of the Shuttle fleet, the station would remain in this configuration for three years.

Wetherbee led his crew back to Endeavour on December 2, with the hatches between the two spacecraft being closed at 12:57. Before that happened, the two Expedition crews embraced each other. Bowersox told the three people he was relieving, "This is a big moment for us." Pettit added, "We promise to take good care of the Space Station." Korzun, the outgoing Station Commander, assured them, "Each crew does better than the last."

With bad weather threatening Florida and a delayed landing likely, the Shuttle crew were told not to leave any of their remaining food with the Expedition-6 crew, as most crews did. Endeavour undocked at 15:05, the same day and made a 90° fly-around of the station before manoeuvring clear. The Expedition-5 crew had been in residence for 178 days. Two Defence Advanced Research Projects Agency (DARPA) mini-satellites were released from Endeavour's payload bay. The two satellites were tethered together and designed to test micro-technologies and nano-technologies during their three days of free flight.

December 3 was spent preparing for landing the following day. That landing attempt was cancelled due to heavy, low cloud and stormy weather over the Kennedy Space Centre. Likewise, the landing attempts on December 5 and 6 were cancelled due to rain and windy conditions in Florida. Endeavour finally came home to KSC, landing under Wetherbee's control at 14:37, December 7, after a flight lasting 13 days 18 hours 25 seconds. As Endeavour came to wheel-stop Houston radioed, "Welcome home to Valeri, Peggy, and Sergei after your half-year off planet. Great job."

Peggy Whitson was just 3 days short of taking the world endurance record for a female astronaut from Shannon Lucid. Even so, she did hold the new endurance record for an American astronaut on a single flight into space. The American record for total flight time accumulated over a number of flights stood at 196 days and was held jointly by Bursch and Walz following the Expedition-5 occupation of ISS. The Expedition-5 crew, who had been in space for 185 days, underwent the usual 45-day long battery of medical and physical re-acclimatisation tests that awaited all returning Expedition crews.

No one knew it on December 7, but this would be the last ISS construction flight for the next three years.

ISS, THE FUTURE

On December 5, 2002, representatives of all the national space programmes participating in ISS met in Tokyo, Japan to discuss the future of the programme. Despite earlier words of warning to the contrary, a Rosaviakosmos representative said that his company was not only prepared to meet its Soyuz and Progress commitments to ISS, but was ready to begin work on a living module designed to raise the Expedition crew to six people. The representative pointed out that NASA was responsible for the station's advanced living quarters and Rosaviakosmos could only construct their new module if NASA compensated them, paid for it. The Russian representative stated that ISS member states had reached an accord to begin six-person operations in 2006.

EXPEDITION-6

Following Endeavour's departure on December 2, Bowersox, Petitt, and Budarin spent the afternoon unpacking items that the Shuttle had delivered to ISS. Asked what his priorities were for the Expedition-6 occupation, Bowersox had told a pre-launch interview:

> "Our number one goal is to have a positive experience for the crew and for the people on the ground. There's a lot of things that come under that sub-goal: [N]umber one being come back alive, for it to be a safe flight; number two, to show people that Americans and Russians can work together in space and accomplish something meaningful; and then, the last thing is to accomplish some of our science objectives for the mission."

He also tried to describe a normal day on ISS:

> "It just depends on who you talk to how normal a day would be. But, the typical days without a Soyuz visit or without [an] EVA on board, or without a Progress arrival, you wake up, you have an hour-and-a-half or so to do your morning cleanup and have some breakfast. Then there's a half-hour conference, or fifteen-minute conference, with the ground, the daily planning conference. Then you start into the work time. There's about eight hours booked for work but part of that is also booked for exercise, so we only consider about six-and-a-half hours as work time. And then in the end of the day it's the same sort of wind down—we have

Figure 33. Expedition-6: Nikolai Budarin wears his Sokol launch and re-entry suit in the Soyuz re-entry module. Kenneth Bowersox is visible in the lower right corner. The view illustrates the cramped conditions inside Soyuz.

> another conference, then a couple of hours to put things away and get ready for bed, relax a little bit. And then an eight-and-a-half-hour sleep period, and the whole day starts again."

Their first day alone on the station began with some free time to get over the hectic hand-over period of joint activities. Thereafter, they began their daily regime of maintenance, experiments, and personal exercise by reconfiguring the station's computer network and loading it with new software. They also checked the HRF rack in Destiny and the station's defribillator, as well as performing maintenance tasks around the station. By the end of their third week in space they had completed their first PuFF and Renal Stone Experiment runs. Pettit also completed a monthly check of the GASMAP experiment.

As Christmas approached, the Expedition-6 crew were finishing their first month on the station. They spent the week ending December 20 working on their experiments, including the Zeolite Crystal Growth (ZCG) experiment, designed to produce zeolite crystals in space that are larger than those produced on Earth. Bowersox completed a practice run of the Foot/Ground Forces experiment (FOOT), which he had described before launch:

> "The way it works is there's a suit that the subject wears, and it's got sensors on it that measure the angles of the ankle and the knee and the hip, in addition to

electrodes on different muscles on the leg and some on the arm. And that senses the electrical activity in the muscles and all that's being recorded as you do normal daily tasks. There's also some sensors that are on the bottom of shoes so that if you're running on a treadmill or standing on the platform doing resistive exercise, those pads will measure the amount of force on your feet. And we'll collect seven or eight hours of data in that suit three, four times during the mission, about one month apart."

Pettit set up the EXPRESS Rack-3 laptop computer prior to activating and checking out the rack itself. The crew also tested the station's KURS automatic docking system, working with Russian controllers. On December 19, they completed more than 3 hours of SSRMS operations, including a series of grapples on an MBS fixture, to collect Force Movement Sensor (FMS) data. Their final task on December 20 was to install the High Rate Communications Outage Recorder (HCOR) on Destiny. This recorder would store data for later transmission to Earth when the station was not in contact with an American TDRS satellite. It replaced a medium-rate recorder, thereby offering a greater storage capacity. Just before 01:00, December 21, the crew manoeuvred ISS so that the starboard side was facing the direction of travel. This was called the YVV attitude. The manoeuvre was carried out because some areas of the station had been overheating.

Christmas Day, December 25, marked the beginning of the crew's second month on ISS. They had a day off, although they had to perform some housekeeping chores and their usual two hours of physical exercise each. NASA Administrator Sean O'Keefe spoke to the crew and each man had a 15-minute private conversation with his family and opened presents that had been delivered to the station by STS-113. Later in the day they returned ISS to its standard attitude, with Destiny at the ram and the P-6 SAWs locked to the Sun once more, this was the so-called XPOP attitude. During the next week, Bowersox completed the FOOT experiment, recording data on changes in microgravity of his leg joints and muscles. Throughout the same week Budarin worked on the Russian plant growth experiment and Pettit continued working inside Destiny. Prior to launch Pettit had described the basic principles behind the microgravity experiments on the station:

"The microgravity science experiments are generally physical science experiments: crystal growth, combustion, things like that. And, they are utilizing an environment where there are small sedimentation forces, no buoyancy forces or reduced buoyancy forces, things that will allow you to do containerless processing, where you can have something floating around without touching the walls of a container, or a high vacuum, high pumping rate environment like an experiment done outside of the pressurized modules on an exposure platform ... many observations in science are key around the balance of forces, measuring one force in the absence of another. And many of the phenomenon that we see on Earth are governed by the balance of these forces. So if you remove, say, gravitational force, now all of a sudden you can see surface tension force. And so, experiments done on Space Station are designed around the reduction in the gravitational force so that you can see other forces manifest themselves and you can make new observations that are very difficult, if not impossible, to make any other way."

The crew celebrated New Year 2003 at midnight GMT, December 31. January 2 saw them carry out a fire drill and setting up the ultrasound equipment in the HRF, which would be used to "image" the crew's body organs for both research and medical use. The following day, they recorded sound levels in the different ISS modules for health and safety at work monitoring, and continued their work with the Zeolite Crystal Growth experiment. Budarin continued to work with experiments in the Russian sector of the station and also checked the wake docking port on Zvezda in preparation for the arrival of Progress M-47, in February.

All three men spent the week preparing Quest for the upcoming Stage EVA and completed their monthly lung function test. Bowersox and Pettit operated the SSRMS on January 9, to complete a video survey of the thermal control equipment associated with the growing ITS. NASA made an announcement the following day that the 50th EVA dedicated to the construction of ISS was delayed until January 15, with Pettit accompanying Bowersox outside in place of Budarin. During a December 5, pre-EVA session on the station's stationary bicycle, Budarin had registered a rate of oxygen consumption that was too low to meet American protocols for an EVA that used the EMU and the American-controlled Quest airlock. For NASA, Rob Navias announced that the delay had "no mission impact whatsoever ... There is no mission impact to anything else that this crew is doing on orbit or to the objectives." He added, "There was no rush to conduct this spacewalk and we decided to delay it."

While America refused to release the reason behind Budarin's replacement, quoting the medical privacy of the individual concerned, the Russians did release the information, stating that if it had been a Russian EVA using the Russian Orlan suit and the Pirs airlock Budarin's oxygen consumption rate would not have barred him from making the EVA. Bowersox had discussed the idea behind the EVA:

"[I]t's proof of concept as much as anything. We're trying to show that a station crew, with just three people, really can get suited up, go outside, and do simultaneous EVA arm ops with the Canadian robot arm, and a mixed-nationality EVA. It's a lot to take on, if you think about it, and there's only three of us there when we do these things during the docked time frame, with a Shuttle crew there, there's a whole lot more support, there's more cameras from the orbiter, there's an extra airlock, an extra door to go in and out and a lot more people to help you get things done. It's quite the challenge to do it with just three people, and so what we're going to be doing mostly is proving that it is possible."

Ironically, it was Budarin who had described the EVA during his own pre-launch interview:

"Well, speaking about EVA, I very much hope that we'll have this EVA. There will be two crewmembers going outside, stepping outside the station; one will stay behind, supporting their activity in space. Jim Wetherbee's crew will have installed the P-1 segment on the S-0 Truss; we will pick up with installing equipment on this truss segment. We will install a UHF antenna, we will install a radiator, we will have to deploy it. It is stowed and latched. In order to deploy this radiator we will need to open the latches, open the locks ... there are eighteen of them, so there will be a lot of tedious work. I'm doing these locks. We will also have to install some struts with lights on the CETA cart ... and we will also be transferring tools from one truss segment to another; we will be using the robotic arm. I'm hoping that I will get a chance to participate in this EVA. I have eight spacewalks under my belt from the Mir experience, and I'm hoping to get EVA experience on the International Space Station. Maybe there will be other objectives, but for now, this is the program of our EVA. But, we are ready to do whatever comes our way."

On the day, the EVA was delayed by problems opening Quest's outer hatch, which the Americans said was caused by dirt in the lock, but the Russians quickly blamed on the inexperience of the two astronauts, who were both making their first EVA. With the hatch finally open, the two men switched their EMUs to internal battery power at 07:50. Bowersox reported a loss of digital data in his EMU, but the problem cleared up when he cycled the internal power switch. Having collected their tools they made their way to the P-1 ITS, where they released ten launch restraint straps on the P-1 radiators. A further eight restraints had been previously released during the STS-113 EVAs in November 2002, when the P-1 ITS was installed on ISS. Controllers in Houston then commanded the central radiator to extend to its full 15 metres. The deployment took 9 minutes to complete. The two men then inspected equipment on the P-1 ITS before making their way to the exterior of Unity, where Pettit used sticky tape to remove grit from the CBM docking seals in preparation for the arrival of the MPLM Raffaello on the next Shuttle flight, STS-114, then planned for March.

Moving to the S-1 ITS, the two men failed to remove a stanchion from its stowed location for installation on one of the two handcarts that future EVA astronauts would use to move themselves along the completed ITS. A pin was interfering with the stanchion's movement and its installation was deferred to a later EVA. Pettit retrieved tools from a storage box on the Z-1 Truss prior to checking the ammonia system on the P-6 ITS. This task was performed in advance of an ammonia cooling system test on a Shuttle flight planned for later in the year. Returning to Quest, the astronauts used a pair of scissors to cut the strap that had delayed the hatch opening at the start of the EVA. The EVA was completed at 14:41, after 6 hours 51 minutes. All three men had a rest day on January 16, performing only routine exercise and maintenance.

4

Triumph and tragedy

"COLUMBIA IS LOST—THERE ARE NO SURVIVORS"

STS-107	
COMMANDER	Rick Husband
PILOT	William McCool
PAYLOAD COMMANDER	Michael Anderson
MISSION SPECIALISTS	Kalapana Chawla, David Brown, Laurel Clark
PAYLOAD SPECIALIST	Ilan Ramon (Israel)

STS-107 lifted off at the beginning of a solo Shuttle flight on January 16, 2003. The "Freestar" flight carried a seven-person crew and was designed to appease the critics of the ISS science programme. Columbia, the oldest of NASA's Shuttle orbiter fleet, had just completed an extensive refit programme and this was her first flight after being returned to operations. The orbiter now carried the Extended Duration Orbiter fuel cell system, which would allow Columbia to remain in orbit for 16 days. STS-107 did not visit ISS, which was in a completely different orbit, and did not carry sufficient propellant to make the manoeuvres to do so, but it would have a major impact on the ISS programme.

One minute after lift-off, one of two bipod ramps, made of shaped insulation foam, was seen to detach itself from the ET, at a point near the bipod holding Columbia's nose. A white cloud suggested that the foam impacted the underside of the left wing and broke up. With no RMS fitted and no EVA planned, it was not possible for the crew to inspect the underside of the left wing. Although the crew could have made a contingency EVA, they were not cleared to leave the payload bay. As Columbia's orbit meant that she could not reach ISS, there was no opportunity for the ISS crew to inspect the orbiter for damage. The incident was investigated on the ground in Houston, and it was decided that it would not affect the mission, which

Figure 34. The STS-107 solo Shuttle flight crew pose alongside a NASA T-38 jet trainer. They are (L to R): Rick Husband, William McCool, David Brown, Laurel Clark, Ilan Ramon, Michael Anderson, and Kalpana Chawla. Their tragic deaths in the Shuttle orbiter Columbia would ground the remaining three Shuttles and stall the construction of ISS for the next three years.

continued as per the flight plan. Despite requests from engineers connected with the flight, NASA managers failed to ask the US Air Force to use its satellites to image the orbiter in space in an attempt to identify any damage. On the subject of the observed impact, Flight Director Steve Stich e-mailed the crew saying, "We have seen this phenomenon on several other flights and there is absolutely no concern for re-entry."

This was the first flight of the new SpaceHab Double Science Module, which was carried in Columbia's payload bay. Among the 80 experiments that the crew performed 30 were sponsored by ESA. STS-107 was dedicated to science, with the crew splitting into two groups and working rotating 12-hour shifts in order to maximise their time in space. For 16 days the 7 astronauts performed their experiments largely ignored by the press and media which had been so critical of the lack of such work on ISS.

On January 28, both the STS-107 crew and the ISS Expedition crew joined the rest of NASA in remembering the astronauts lost in the January 27, 1967 Apollo-1 fire and the January 28, 1986 STS-51L Challenger explosion. At Cape Canaveral flags flew at half mast at LC-34 and LC-39 for the two days of remembrance.

Columbia's Commander Rick Husband radioed, before a moment of silence to mark the exact time that STS-51L exploded:

> "It is today that we remember and honour the crews of Apollo 1 and Challenger. They made the ultimate sacrifice, giving their lives and service to their country and

for all mankind . . . Their dedication and devotion to the exploration of space was an inspiration to each of us and still motivates people around the world to achieve great things and service to others."

On February 1, their mission behind them, the STS-107 crew strapped into their seats and prepared for re-entry. As they crossed the Indian Ocean Husband turned his spacecraft so that the closed payload bay doors were facing Earth and the three SSMEs were facing the direction of travel. Retrofire occurred at 09:17, and the hour-long descent to KSC began. Husband turned Columbia through 180°, with the nose forward and pitched up, to present the orbiter's flat, high-drag underside to the atmosphere for re-entry. Columbia entered the upper atmosphere and crossed the Californian coast at 09:51. As it did so, the telemetry from the hydraulic system on the inboard and outboard elevons on the left wing dropped out. Three minutes later telemetry was lost from the brake lines and tyres on the main undercarriage on the left side. At that time the spacecraft's computer alerted the crew to the loss of data. Charles Hobaugh, Capcom in Houston radioed, "Columbia, Houston. We see your tyre pressure message." Scanning his instruments Husband acknowledged with a simple, "Roger."

It was 09:59. All communications and data transmissions from Columbia abruptly stopped. Streaking across North America, 61,000 km high and travelling at Mach 18.3 Columbia broke up. STS-107 was 15 minutes away from landing in Florida, where the crew's families were waiting to watch them return home. Beneath the Shuttle's flight path people reported a loud sound, like an explosion, or an aircraft breaking the sound barrier. Looking up, they saw trails in the sky as the remains of Columbia passed through the thick lower atmosphere. Debris fell to ground in a long swath covering parts of Texas and Louisiana. Among the debris were human remains. DNA testing would be required to identify the remains of individual crew members.

In Houston, all data relating to the flight was secured, in preparation for the investigation in to the cause of Columbia's loss. The doors to the control room were locked and all communication with the outside world was stopped. In the VIP viewing area at KSC, NASA personnel escorted the astronauts' families away from the empty runway where they had been expecting to see Columbia land at 10:15. The remaining Shuttle fleet was immediately grounded indefinitely. At 10:30, the Expedition-6 crew were informed that Columbia had been lost during re-entry. NASA organised a press conference to inform the media of what had happened. This was in total contrast to the loss of STS-51L, Challenger, in January 1986 when NASA had adopted a "no comment" policy.

President Bush was informed of the tragedy at Camp David. He returned to Washington DC by motorcade because the weather was too bad to allow him to be flown there by helicopter. When he arrived he made a speech to the American People, telling them:

"My fellow Americans, this day has brought terrible news and great sadness to our country. At 9 AM this morning mission control in Houston lost contact with our Space Shuttle Columbia. A short time later, debris was seen falling from the skies over Texas. The Columbia is lost; there are no survivors.

On board was a crew of seven: Colonel Rick Husband; Lt Colonel Michael Anderson; Commander Laurel Clark; Captain David Brown; Commander William McCool; Dr. Kalpana Chawla; and Ilan Ramon, a Colonel in the Israeli Air Force. These men and women assumed great risk in the service to all humanity.

In an age when space flight has come to seem almost routine; it is easy to overlook the dangers of travel by rocket, and the difficulties of navigating the fierce outer atmosphere of the Earth. These astronauts knew the dangers, and they faced them willingly, knowing they had a high and noble purpose in life. Because of their courage and daring and idealism, we will miss them all the more.

All Americans today are thinking, as well, of the families of these men and women who have been given this sudden shock and grief. You're not alone. Our entire nation grieves with you. And those you loved will always have the respect and gratitude of this country. The cause for which they died will continue. Mankind is led into the darkness beyond our world by the inspiration of discovery and longing to understand. Our journey into space will go on."

NASA immediately established both internal and external investigation teams while members of the public kept their local police services busy reporting the location of debris and human remains. While public broadcasts were made making clear how every piece of debris was potential evidence and should be handed over to the authorities, some individuals still tried to sell Columbia debris on the Internet. Those people fell foul of the authorities, who were monitoring the Internet for just such attempts. Those arrested potentially faced ten years in gaol and a huge fine if found guilty.

In the seven days following the disaster the press and media had a feeding frenzy. Journalists that had ignored the space programme for years suddenly became experts overnight. Despite President Bush's speech, there were the usual, predictable, demands for the Shuttle to be scrapped and the human space programme to be cancelled. All of the old complaints about ISS were aired and demands were made for the Expedition-6 crew to be bought home and the station abandoned. Many claimed that the STS-107 crew had been sent to their deaths by a NASA that had cut spending on safety checks to the point that Columbia was a creaking wreck just waiting to fall apart around them. Such charges ignored the 100 plus modifications that had just been made to Columbia during a major re-fit. Much was made of the impact incident during launch, and of Columbia's age. There was speculation about a computer malfunction that positioned the spacecraft incorrectly for re-entry, impact with a micrometeorite, or a piece of junk from an earlier space launch. With America preparing for a war in Iraq, some Muslim extremists publicly claimed that Allah had struck Columbia down because of the combination of Americans, an Indian, and an Israeli in the crew. They pointed to the fact that some of the debris fell in Palestine, Texas as proof of their claim. Some television channels showed film of Arabs parading in the streets and cheering at America's loss in several Middle Eastern

Figure 35. Following the loss of the crew of STS-107, members of the public placed personal memorials alongside the Johnson Space Centre sign in Houston, Texas. The NASA facility is where the Astronaut Office is located and is where crews do much of their training before flying in space.

cities. The greater majority of Muslims disassociated themselves from these extremist ideas and expressed their sympathy for the loss of the Shuttle's crew. Although the flight had been largely ignored by the American media, it had been the focus of much attention in Israel and India, where the public were deeply shocked at the loss of their citizens.

One American television presenter publicly aired his total lack of knowledge of his nation's space programme when he told his audience in all seriousness, "This is not like Tom Hanks' film *Apollo-13*, where they fixed the spacecraft and got home safely. This is real life." Just, for his information, the film *Apollo-13* was based on an actual spaceflight. The Apollo-13 spacecraft exploded on the way to the Moon, in April 1970. The crew did not "fix the spacecraft" and only "got home safely" because of the professional expertise and personal tenacity of everyone concerned with the flight. The crew consisted of James Lovell, John Swigert, and Fredrick Haise. At the time, Tom Hanks was a schoolboy sitting at home, watching it all on television and pretending to land his plastic model of an Apollo Lunar Module on the lounge carpet. Hanks is an excellent actor, but he has never been selected as an astronaut, at least, not in "real life".

As time progressed the hysteria subsided and Columbia left the headlines, temporarily. The search for debris and human remains continued and the official investigation into what had happened got underway. Meanwhile, there were still three people on ISS.

PROGRESS M-47

While the search for debris from Columbia began on the ground, life went on onboard ISS, and NASA had to continue to operate the station despite their recent loss. At 10:59, February 1, Progress M1-9 was undocked, manoeuvred clear, and commanded to re-enter Earth's atmosphere and burn up. The undocking cleared Zvezda's wake for Progress M-47. Following the undocking, the Expedition-6 crew had expected to have an easy day in front of them. Bowersox, Budarin, and Pettit were informed of the loss of Columbia approximately one hour after it occurred, at which point the mood onboard turned very sombre.

The following day NASA Administrator Sean O'Keefe spoke on the NASA Television channel about the three men still on ISS:

> "We're focused on making sure they are fully supported and as early as we can possibly get back up there to rotate that crew and bring them back home and send a new crew, that's exactly what we're going to do ... All three of them are apprised of the facts of what's going on. They've been given all the information and they're

Figure 36. Expedition-6: Kenneth Bowersox inside Zvezda, surrounded by items unloaded from Progress M1-9.

prepared to do what's necessary to stiff this out throughout the course of the investigation ... We've got time. We've got an opportunity, I think, to sustain them with what they need, but we're always focused on making sure that we recognise that people are depending on us."

Progress M-47 was launched from Baikonur at 07:59, February 2, and docked at Zvezda's wake at 09:49 February 4. Following pressure and leak checks the hatches between the two vehicles were opened at 14:00, but unloading did not begin until the following morning. The cargo on Progress M-47 would ensure that the crew had sufficient essential supplies to continue the occupation of ISS through the end of June and early July 2003. During the day the Expedition-6 crew paid their private tributes to the STS-107 crew and listened to an audio feed from the Memorial Service held at JSC attended by President Bush and NASA Administrator Sean O'Keefe.

Following unloading of the Progress, Pettit replaced the two power cells in the MSG with two brought up by the Progress M-47. When the change-out was complete he applied power to the MSG and two circuit breakers in Destiny tripped. This was exactly what had happened when the MSG malfunctioned in November 2002. To avoid damage to the new components Pettit powered the MSG off and the investigation into why it had failed continued on the ground.

The Expedition-6 crew's schedule had changed due to the loss of Columbia. With all Shuttle flights suspended indefinitely until the cause of the loss could be identified and corrected, the crew knew that they would have to return to Earth in a Soyuz at the end of their occupation, rather than on STS-114 in March, as originally planned. The most likely scenario was that the Expedition-7 crew would bump the Soyuz TMA-2 taxi crew due for launch in April, and the Expedition-6 crew would return to Earth in Soyuz TMA-1, with Budarin serving as Soyuz Commander. This was exactly the reason a Soyuz spacecraft was attached to ISS at all times. The Expedition-6 occupation could be extended for one month, to maintain the Soyuz TMA-2 launch schedule. Meanwhile, they were instructed to put together a full inventory of all supplies onboard, while flight controllers in Houston and Korolev carefully monitored and recorded everything that the crew used. Put together, the two audits would allow mission planners to ensure that future Progress flights were loaded with essential supplies to maintain the mission.

On February 10, it was announced that the rocket motors on Progress M-47 would be used to raise the station's orbit in advance of the Soyuz TMA-2 launch. The first of three re-boosts took place at 06:34 the following day. At the same time it was announced that NASA had requested the Russians to provide an additional Progress for 2003. Following a Russian request, it was announced on February 12 that, until the Shuttle began flying again, all future Expedition crews would consist of one American astronaut and one Russian cosmonaut. All crew deliveries and returns would be made using Soyuz TMA spacecraft, with the two-man Expedition-7 crew flying to ISS in Soyuz TMA-2 no earlier than April 26, 2003. Following the usual hand-over in orbit the Expedition-6 crew would return to Earth in Soyuz TMA-1.

At the time of the reduction to two crew members the schedule for the next four Expedition crews had been

Expedition-7	
ISS COMMANDER	Yuri Malenchenko
FLIGHT ENGINEER	Edward Lu
FLIGHT ENGINEER	Aleksandr Kaleri
Delivery	STS-114
Recovery	STS-116

Expedition-8	
ISS COMMANDER	Michael Foale
FLIGHT ENGINEER	Valeri Tokarev
FLIGHT ENGINEER	William McArthur
Delivery	STS-116
Recovery	STS-119

Expedition-9	
ISS COMMANDER	Gennady Padalka
FLIGHT ENGINEER	Michael Fincke
FLIGHT ENGINEER	Oleg Kononenko
Delivery	STS-119
Recovery	STS-121

Expedition-10	
ISS COMMANDER	Leroy Chiao
FLIGHT ENGINEER	Salizhan Sharipov
FLIGHT ENGINEER	John Phillips
Delivery	STS-121
Recovery	STS-123

On February 10, Sean O'keefe was able to tell the press and media:

> "The International Space Station is doing very well. My last conversation with Captain Bowersox was, 'Don't worry about us. You know where to find us, and we're not going anywhere; everything's fine.' So they're in good spirits and moving along."

During the same day the Expedition-6 crew spoke of how they had learnt of the loss of STS-107. Bowersox explained:

> "We were scheduled for a normal planning meeting on Saturday [February 1]. General Howell, the Director of the Johnson Space Centre, came in and told us we lost the vehicle on entry. My first reaction was sheer shock, I was numb and it was

hard to believe that what we were experiencing was really happening. Then, as the reality wore on, we were able to feel some sadness. It's the classic grieving responses our psychologists had warned us about, you feel sad, you feel angry, all those things. And now, as time goes on, we're able to put those aside and focus a lot better on our work ... The folks on the ground have been real good about reducing our schedule and we've had some time to grieve our friends; and that was very important. When you're up here this long, you can't just bottle up your emotions and focus all the time ...

It's important for us to acknowledge that the people on STS-107 were our friends, that we had a connection with them and that we feel their loss. Each of us had a chance to shed some tears. But now, it's time to move forward and we're doing that, slowly ... At the conclusion of the memorial service, after the bells had rung on the ground and the T-38s had flown by and it was very quiet onboard, we rang our ship bell seven times. At that point it was very, very quiet on board the International Space Station. We spent 15–20 minutes in silence and then we moved on. We had work to do. We had to unload our Progress. At that point, we started thinking about good things, we pulled out the fresh fruit, the oranges, the mail we got from home and it gave us quite a lift after the memorial service.

We are enjoying our mission up here, we enjoy the environment of the Space Station and we're going to enjoy the next two and a half, three months here. So the extra stay is not something that we consider negative. In fact, for us it is positive. We actually volunteered to stay longer. We told our management if they need us to stay a year that's fine, they've got blanket approval for that. If they want us to stay longer than a year, please give us a couple of months notice. So we like living on Space Station, and we feel comfortable that we have a way home, we have complete confidence in our Soyuz vehicle and the ability of our Russian partners to operate that vehicle and get us home safely. Because it's been a few months since we've been in a simulator, we'll do additional training here on board if it should be required for us to come home in the Soyuz."

Pettit remarked that he felt that grieving was a very personal thing and he felt comfortable grieving in the privacy and quiet offered by the various individual modules of ISS. He said that a two-man crew would be

"real busy just maintaining the systems on the station ... However, there would be time to do some level of research and by virtue of having people here, you're always doing research on your body itself, looking at the effects of long duration weightlessness on human physiology. So it's important to keep people on station."

By February 13, NASA was able to announce that DNA testing had positively identified the remains of all seven STS-107 astronauts and all searches for additional human remains had been halted. Ramon's remains had been returned to Israel earlier in the week and preparations were underway for the other astronauts' remains to be handed over to their families for private burial arrangements.

The following day, Bowersox and Pettit operated the SSRMS, to perform routine checkout procedures and maintain their proficiency in using the system. Other maintenance during the week included removing and cleaning several of the fans responsible for driving the airflow throughout the station. Pettit continued to support the ground-based work on the continuing MSG malfunction investigation.

On February 24, Rosaviakosmos named two caretaker crews.

Soyuz TMA-2	
Prime crew	Malenchenko and Lu
Back-up crew	Foale and Kaleri
Launch	May 2003
Recovery	October 2003

Soyuz TMA-3	
Prime crew	Foale and Kaleri
Back-up crew	McArthur and Tokarev
Launch	October 2003
Recovery	March 2004

Although no announcement was made at the time, it seemed likely that McArthur and Tokarev would fly Soyuz TMA-4 to ISS and serve as the third caretaker crew if the Shuttle was not flying again by the end of the TMA-3 tour. It was also announced that the Spanish ESA astronaut Pedro Duque, who had originally been scheduled to fly on Soyuz TMA-2, would now fly on Soyuz TMA-3 under commercial contract with the Russians. He would return to Earth with the Expedition-7 Crew.

The first three-person Expedition crew to occupy ISS when the Shuttle returned to flight was named at the same time. They were Krikalov, Volkov, and Phillips. All other crews that were in place when Columbia was lost were disbanded.

CONTINUED OPERATIONS

The third week in February saw the crew sample their potable water and disinfect parts of the water supply system in the Russian sector of ISS in order to ensure its continued functioning to the highest standard. Pettit replaced the remote power control module in Destiny due to a bad power switch in the unit causing a video recorder to power off unexpectedly. The crew also completed regular maintenance of some of their fitness equipment in Destiny. They conducted an inventory of all articles in Quest. At the same time, new software was loaded into the Command and Control (C&C) computers, and the Guidance, Navigation, and Computers (GNC) on the station.

Figure 37. Expedition-6: Donald Pettit works with the PuFF experiment inside Destiny.

On February 24, Bowersox and Pettit donned American EMUs without assistance from Budarin. This was a test in advance of launching two-man Expedition crews to continue station occupation while the Shuttle remained grounded. The two men went as far as setting up the equipment for pre-breathing oxygen before discontinuing the experiment and removing the suits. In Houston, Carl Walz told them "Bravo, great job." He joked with Pettit, "We think you are losing too much weight. You make it look altogether too easy." Three days later, O'Keefe told Congress that if at any time the crew of ISS were at serious risk then they would "dim the lights, get into the Soyuz and head for home."

As March began the Expedition-6 crew celebrated their 100th day in space. All three men were continuing to perform human life sciences experiments, Earth observations, and other onboard scientific experiments. Pettit continued to work on repairing the MSG, working with engineers on the ground. This work appeared to be complete by the end of the month when Pettit applied electrical power to the unit and ran a test run of the Pore Formation and Mobility Investigation (PFMI), which had been in the glovebox since it failed in November 2002. Following the test, the PFMI was replaced by another experiment: Investigating the Structure of Paramagnetic Aggregates from Colloidal Emulsions (InSPACE). The crew also continued to prepare for a Stage EVA that was planned for April.

Pettit had put the human life science experiments in their historic context during his pre-launch interview:

"When you do research associated with exploration, you're in a unique environment and you learn new things about people and new things about nature. And these in themselves enrich the knowledge for everybody that doesn't go on the exploration trip. One example I like about human physiology in exploration is transoceanic exploration in the 14th and 15th century and the role of diet and vitamin deficiencies. And, it was this kind of exploration that helped open the can of worms leading to things like vitamin C and its role in scurvy, and this information was prised, so to speak, from the souls of the early explorers. And once you learn this information, then it helps and benefits everybody back on the continent that didn't get a chance to go on these trips. And I see this as the goal of the life science research on Space Station."

In his own interview, Budarin had discussed the experiments in the Russian segment of ISS:

"[The] Russian segment will have its own share of science and medical experiments. We will be performing Earth monitoring and observation for [the] Russian scientific program. One of the experiments is called Diatomeya. It involves observation of ocean surface in order to determine ... regions that are best suitable for fishing; fertile regions of the ocean. Currently, these particular regions of the ocean are well-studied and their location is known, but in nature, everything changes, everything morphs, and these regions are changing as well. So we'll be determining the new characteristics of the regions. Also, we will be monitoring the glaciers. Everybody's talking about global warming, so we will be watching out for glacier dynamics. Medical, well, the goals and objectives are similar across all programs. The equipment may be different, but we will be working towards the same end pretty much."

The third week of March began with Pettit installing a new Pump Package Assembly (PPA) in the Moderate Temperature Cooling Loop (MTCL) of Destiny's Thermal Control System. The original pump had failed the previous day. Due to seating problems in one of the valves the system was not returned to operation until March 20. Meanwhile, Budarin upgraded the Russian computer software.

When the Russian computer system was re-booted on March 19, a Russian Terminal Computer in Zvezda was unable to communicate with the American GNC Computer 2, which was controlling the station at the time. This caused a failure of the routine hand-over of control to the Russian computer. As a result, the American computer began an automatic shutdown of non-critical systems. But attitude control was not lost and, after the communication problem between the two computers had been overcome, all systems were brought back on-line within a few hours.

The last week of March was filled with three primary activities. Work to return the MSG to operational status was finally completed and preparations were made for the first run of InSPACE experiments, which were completed during the first week of April. The crew also began reviewing plans and preparing equipment for their second Stage EVA. Success would reduce the likelihood of the two-man Expedition-7 crew having to make a Stage EVA during their occupation of ISS. Finally, the Expedition-6 crew began computer-based training for their return to Earth in Soyuz TMA-1.

On March 29, representatives from America, Russia, and Europe met to discuss the financing of the two additional Progress vehicles requested for flight in 2004. Russia had hoped that America or Europe might find emergency funds to pay for the two vehicles, but it was not to be. The Americans reminded the Russians that the contract which they had signed when they became an ISS partner committed them to building and launching two Soyuz and five Progress vehicles each year, a target that they had only met in 2001. In 2002, the Russians had quoted financial difficulties as the reason for cutting the number of Progress vehicles to two each year. In March 2003, NASA now demanded that the Russians meet the terms of their original contract and supply the five Progress vehicles required in 2004. In the aftermath of STS-107 no one asked why America continually held the Russians to the letter of their contract and yet they had unilaterally ignored their own legal requirements under the same contract. In related negotiations, Russia offered an Expedition crew position to an ESA astronaut if the Europeans would finance the two additional Progress vehicles. Having failed to extract additional money from their ISS partners, Russian officials were quoted in the press as saying that Russia could not afford to finance the continued crewing of ISS on their own. Once again they suggested that the station may have to be abandoned and mothballed until the Shuttle was flying once more.

On April 4, 2003, Progress M-47 was used to raise the station's orbit for the second time. The new orbit optimised the conditions for the docking for Soyuz TMA-2. Two days later, the SSRMS was positioned so that its lights could provide support for the up-coming Stage EVA.

At 08:40, April 8, Bowersox and Pettit commenced their second EVA, exiting the station through the Quest airlock. After preparing their tools, the two men began work on separate lists of tasks. Bowersox reconfigured electrical connectors between the S-0 and P-1 ITS. The work put in place additional protection to prevent the unintentional separation of the entire truss structure from the S-0 through the Bolt Bus Controller System: they didn't want a malfunctioning circuit to withdraw the bolts that held the segments together. Bowersox also inspected a faulty heater cable on the P-1 ITS Nitrogen Tank Assembly, but found nothing obviously wrong. Meanwhile, Pettit replaced a power relay box in one of the CETA carts. The cart had suffered from electrical problems since it had been installed.

Both men then moved to the Z-1 Truss and re-routed power cables to CMG-2 and CMG-3 at that location. The changes would prevent the two CMGs being disabled if they suffered a power failure. Next, they installed two SPDs on the fluid quick-disconnect lines for Destiny's heat exchanger. Moving on to the S-1 ITS, they worked together to secure a thermal cover on the Radiator Beam Valve Module, which controlled the flow of ammonia to the S-1 radiators. Returning to the CETA carts, Pettit used a hammer to free a stanchion from its stored position and deployed it on the cart, before deploying a light on the stanchion. The two men then stowed their equipment and returned to Quest, bringing the EVA to a close at 15:06, after 6 hours 26 minutes.

On April 11, Progress M-47 was used to complete the third of three re-boost manoeuvres in preparation for the arrival of Soyuz TMA-2. Three days later the three men donned their Sokol pressure suits and climbed into Soyuz TMA-1. The short exercise was part of their preparations for return to Earth. As the week advanced the crew began packing personal items and answering questions from controllers about onboard maintenance.

Soyuz TMA-2 was rolled out and erected on the launch pad at Baikonur on April 24. Everyone was at pains to point out to the media exactly what the flight represented. Sergei Gorbunov said, "Obviously, this mission is very important in terms of the survival of the International Space Station."

Meanwhile, a NASA spokesman insisted, "I think everybody that has been sceptical about the strength of the Russian space program and in a broad sense about the viability and status of the Russian space program—those sceptics have been proven wrong."

SOYUZ TMA-2 DELIVERS EXPEDITION-7, THE FIRST "CARETAKER CREW"

SOYUZ TMA-2	
COMMANDER	Yuri Malenchenko
FLIGHT ENGINEER	Edward Lu

Prior to the loss of STS-107, Soyuz TMA-2 was just another "taxi" flight to replace Soyuz TMA-1, due for launch in April 2003. That changed when the decision was taken to keep ISS occupied by having two-man Expedition crews fly to and from the station in the available Soyuz TMA spacecraft. The crew for this flight were originally members of a three-man Expedition-7 crew with Sergei Moschenko as the third member. They should have been launched on STS-114 with the following Shuttle crew:

COMMANDER:	Eileen Collins
PILOT:	James Kelly
MISSION SPECIALIST:	Soichi Noguchi (Japan), Stephen Robinson

Soyuz TMA-2 was launched at 23:54, April 25, 2003. In Washington, Sean O'Keefe told journalists, "The real testimonial to how strong that partnership is, is tonight's launch of the Soyuz." At the same time O'Keefe announced the names of the Shuttle crew that would fly the "Return to Flight" mission, possibly in December 2003. That launch date would be pushed back to March 2004, and even then it would not be met. The Shuttle crew would deliver a fully loaded MPLM to ISS as well as the

Figure 38. The Expedition-6 and 7 crews pose together during hand-over operations. They were (L to R) Edward Lu, Kenneth Bowersox, Donald Pettit, Nikolai Budarin, and Yuri Malenchenko. Malenchenko and Lu were the first 2-man "caretaker" crew after the loss of STS-107 grounded the American Shuttle fleet.

first three-person Expedition crew since the Expedition-6 crew. The Shuttle would be flown by Eileen Collins' crew (named above).

Following a two-day rendezvous Soyuz TMA-2 docked Zarya's nadir at 01:56, April 28. As the Soyuz approached the station Budarin performed a pitch-up manoeuvre to allow the ISS crew to photograph his spacecraft. The photographs would be studied in America, where plans were under consideration to have all future Shuttle orbiters perform a similar manoeuvre before docking to the station. The photographs taken on those occasions would be sent down to the ground, where experts would review the Shuttle before declaring it safe for re-entry.

The hatches between the two vehicles were opened at 02:27 and the Expedition-7 crew made their way into ISS where the Expedition-6 crew greeted them. Following their official welcome and the standard safety brief the two crews began a five-day hand-over period.

In a press conference held on April 28, Bowersox joked, "I feel a little bit like I'm being kicked out of my apartment for not paying my rent. But when I get back to Earth, the best part is going to be, to be able to hug my wife and hug my kids." At one point Bowersox was wistful about returning to Earth in Soyuz TMA-1, "I've been looking down quite a bit from orbit, looking down on Kazakhstan. It's a beautiful country ... I think it's going to be a very, very interesting life experience." The fickle

American media made much of the fact that Bowersox and Pettit would be the first American astronauts to return from ISS in a Soyuz spacecraft, but Pettit put their minds at rest, "We've had a heap of training for both Soyuz and Shuttle entries and either one is fine with us. I don't think there's any extraordinary angst about the particular entry we're planning to do here." More important to Pettit was the fact that he would soon see his wife and twin 2-year-old sons. On the subject of the loss of STS-107, Bowersox noted, "I think it's going to be hard for Don and I, after being away from it all, to suddenly be confronted with all that emotion ... But at the same time, I think it's going to be very good for us to be back there with our friends and help them work through it and let them help us work through the changes that we'll be going through."

Asked about the fact that there were only two people on the new Expedition crew, Lu said, "I think we'll be able to do just fine." On May 1, the station's computer server went down. Malenchenko and Lu worked with controllers to solve the problem and the server was back on-line the following day. May 2 also saw the Expedition-7 crew carry out familiarisation training on the SSRMS.

The change of command ceremony took place at 13:15, May 3, before Bowersox led his crew into Soyuz TMA-1 and handed over to Budarin, who assumed his role as Soyuz Commander. Bowersox told the new Expedition crew, "You guys have to be the two luckiest guys who come from planet Earth today. Over the next six months you get to live aboard this beautiful ship." Then he turned to Malechenko and told him calmly, "Yuri, I'm ready to be relieved." Of his own crew's occupation of ISS Bowersox said, "We carried out everything we intended to, but most important is that we worked well together as an international crew." Listening in Korolev, O'Keefe joked with the outgoing crew, "Put in your order for how you want your steaks done so we can have them ready for when you arrive."

The hatches between the two spacecraft were closed at 15:38, and Budarin undocked Soyuz TMA-1 from the station at 18:40. Three hours later, at 21:07, the re-entry module landed inside Kazakhstan, some 400 km short of its predicted landing site. The offset centre of mass in the re-entry module enabled the Soyuz to generate lift, and thereby control its passage through the atmosphere, generally extending it by several hundred kilometres and aiming for the assigned target spot. Soyuz TMA-1 had defaulted to a ballistic trajectory and therefore fallen "short" of its target. During their unplanned ballistic re-entry, the crew, who had spent 5.5 months in microgravity, were subjected to more g forces than they had expected. When the main parachute deployed, some of its lines snapped, including one that carried the main communications antenna. As a result, the Soyuz TMA-1 re-entry module completed its final descent and landing in radio silence. Touchdown was also harder than expected and Pettit's shoulder was injured. The wind caught the parachute and dragged the module 13 m across the steppe, before leaving it on its side. Recovery helicopters had to re-fuel before they could reach the off-target site. Contact was made with the crew at 23:30, and aircrew reported that all three men were out of the spacecraft and waving to them as they landed. It would take 2.5 hours for the recovery team to reach them, so the crew lay on the ground to avoid the cardiovascular stress of gravity.

In reply to their questions about returning to Earth, Bowersox told journalists, "We could smell the dirt. We could smell the grass. It was fantastic."

Petitt added:

> "When the hatch was just cracked open, there were real Earth smells because we stirred up a fair amount of dirt when we landed. You had this fresh dirt smell, which was just a beautiful smell. It had a little bit of crushed grass in it because there was all that fresh spring grass coming up in little clumps ... The next thing that hit me were all the birds chirping. It was just music to our ears."

He continued:

> "I was actually relieved to ooze out of the spacecraft and lay on Mother Earth and have a solitude moment in which to get reacquainted."

On the subject on what had caused the switch to a ballistic re-entry trajectory Budarin was non-committal, "It's for the specialists to figure out what was the cause. Let's wait and see, but for now I can say that it was not our own doing."

The crew were recovered by helicopter, with Pettit being placed on a stretcher. The following day they were flown to Baikonur. Both the Russians and the Americans played down the difficulties at the end of the Expedition-6 flight.

Having watched the recovery from TsUP in Korolev, O'Keefe was damning when he returned to America. He told journalists that just 8 hours before the landing he had used new cellphone technology to talk to the crew on ISS, but following their off-target landing there was no communication for 2.5 hours. "First we're talking on a cellphone, and eight hours later we couldn't reach them ... Two tin cans and a string would have been an improvement. It was an absolutely phenomenal contrast."

The Russians began an investigation into the cause of the spacecraft leaving its controlled trajectory and commencing a ballistic re-entry. On June 28, it was announced that the fault had been identified as having been caused by the spacecraft's "yaw gyroscope experiencing gimbal-lock when its angular excursion exceeded its permissible range of 54 degrees." RSC Energia said that corrections would be made to the control systems of all later Soyuz TMA spacecraft.

Prior to flying to ISS, Bowersox had been asked how he viewed the end of the flight, and what, in his mind, would make the Expedition-6 occupation a successful one. Answering, before the loss of STS-107, he replied:

> "I guess the most important thing will have been the unity of the crew at the end of the flight. To be successful we have to come back as a crew that was able to support each other, able to forgive each other when we made mistakes or when we accidentally offended someone, when we didn't mean to, that we were able to get past all those human frailties, and stay united as a supportive crew. And that's not just the three of us on board but also with our team on the ground, because there will be tons of frustrations that will come down upon us as we're going through our mission. We'll be in a high-stress environment, and typically when people are

stressed and they have more stress being dumped on them, their teams can break down. And what we want to do instead is to support each other so that we become stronger with that stress. And if we can do that, we'll be successful; everything else will work out and take care of itself."

Judged against that description, Expedition-6 had been very successful indeed.

ESA MOVES AHEAD

At this time ESA awarded a €3.7 million contract to the German DLR Space Centre to establish the Columbus Control Centre at the German Space Operations Centre, in Oberpfaffenhofen. The new centre would also control Automated Transfer Vehicle (ATV) operations when they commenced. In the contemporary launch manifest, both the Columbus launch to ISS and the first ATV launch were due to take place in 2004.

Meanwhile, Node-2, the first of two Nodes built by Aleno Spazio in Turin, Italy, had completed its Acceptance Review and was due to be delivered to ESA in mid-May. Due to be launched by Shuttle in 2004, Node-2 would serve as a mounting for Columbus and Kibo. Node-2 would be named "Harmony". Following ESA acceptance testing the Harmony and Columbus modules were to be shipped to KSC in Florida, where they were officially handed over to NASA and placed in line for their respective Shuttle launches.

EXPEDITION-7

Malenchenko and Lu began their official increment as the Expedition-7 crew following the undocking of Soyuz TMA-1. The first two days were free days, to allow them to adapt to their new home. The third and fourth days were spent on a familiarisation tour and routine maintenance. May 9 was a Russian holiday, which gave the crew another day off, even so they performed more routine maintenance and two periods of physical fitness training each.

In his pre-flight interview Malenchenko had discussed how the Expedition-7 occupation had changed following the loss of STS-107:

"Our whole program [will] be revised. Some of the things we will not do because ... there won't be any Shuttle flights to deliver consumables and the hardware. And, even the items that were originally planned to be delivered on Progress will not be delivered because Progress will be delivering something else. But, some of the scientific experiments we will do nevertheless ... And, I think we will be pretty busy with science as well."

Asked how he expected history to view the Expedition-7 occupation Malenchenko replied:

"It's hard to say. I think that the tragedy that has occurred, the fact that we lost our comrades, the fact is that they gave their lives for the continued space exploration. The fact that we are together and that it's an international project allows us to continue this effort. We have the capabilities of different countries that we can put together to continue. I think that our Expedition confirms that, shows that we continue working even in such a difficult time period."

In answer to a similar question he had previously replied:

"Of course, we will have fewer resources and fewer capabilities available to us. We won't have any Shuttle flights. Originally, there were three Shuttle flights scheduled for our Expedition, and we had a lot of activities scheduled for the construction of the station. All of this has been postponed. We will use the resources that we have remaining and all our capabilities to continue. We still have our program. It looks different, but we will continue working. We will continue supporting the station. We will continue performing scientific experiments... We will be missing our third crewmember, but we realize that two people are enough to maintain the station in a working state and, additionally, to conduct work on science experiments. That's how I see our future work."

Their second week on ISS began with fire and evacuation training. The crew also performed maintenance of Zvezda's ventilation ducts, and took an inventory of Russian communication equipment on the station. Malenchenko and Lu also harvested the "Red and White" peas planted by the Expedition-6 crew as part of the Russian PLANTS-2 experiment. Lu worked in Destiny, servicing the experiment racks and preparing the InSPACE experiment in the MSG. The experiment studied the behaviour of magnetic particles in a fluid when subjected to a pulsed magnetic field. It was activated on May 20. As the crew completed their first month in orbit they began a series of maintenance tasks to ensure the station remained in good working order. These included monitoring the quality of the station's internal atmosphere and the operation of the station's LSS.

During the week ending May 30, Malenchenko and Lu replaced a faulty battery in Zvezda. They also practised donning American EMUs, inside Quest, although part of the test was cancelled when the water flow failed in the water-cooled undergarment of Lu's EMU. With no Stage EVAs planned during the Expedition-7 occupation, the two men were practicing donning and doffing the EMUs without the assistance of a third crew member in case an unexpected situation developed that required them to complete emergency EVA. On May 30, Progress M-47 raised the station's orbit.

On June 1, American President George W. Bush met with Russian President Vladimir Putin in St Petersburg, Russia. In a joint statement issued after their meeting Bush told journalists:

"The United States is committed to safely returning the Space Shuttle to flight, and the Russian Federation is committed to meeting the ISS crew transport and logistics re-supply requirements."

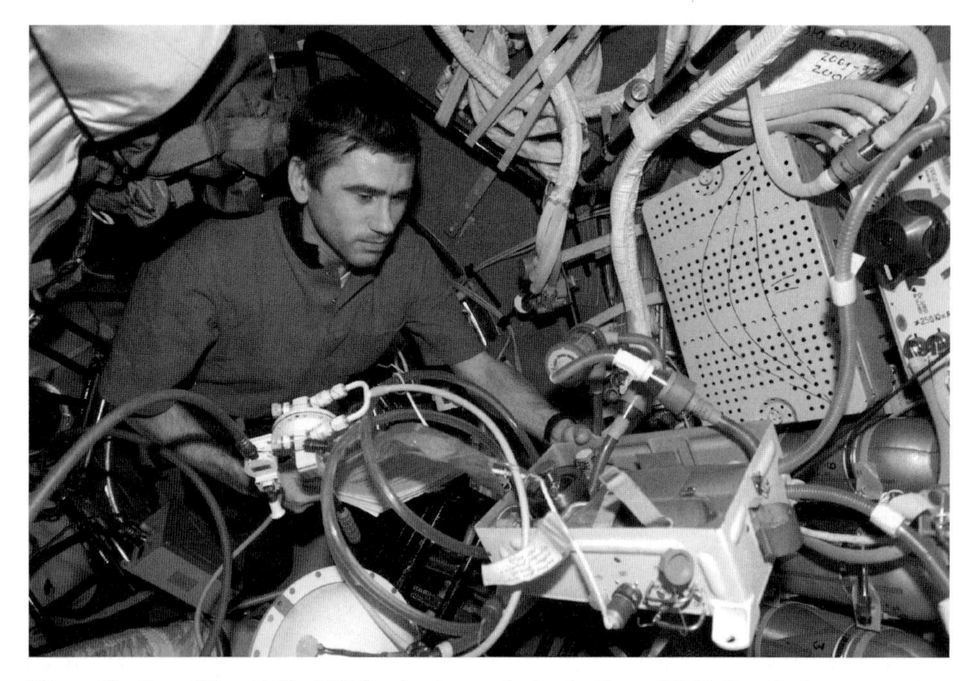

Figure 39. Expedition-7: Yuri Malenchenko works in the Soyuz TMA-2 orbital compartment.

The President thanked his counterpart for Russia's commitment to keeping ISS occupied and stocked with vital consumables. President Putin replied:

> "Space remains a vital part of our cooperation."

Malenchenko and Lu both practised operating the SSRMS on June 4, grappling and releasing a target on the exterior of Destiny. The training session also served to exercise an adjustment made to the arm to improve its grappling procedure. In anticipation of the arrival of Progress M1-10, Malenchenko practised manual docking procedures with the TORU system in Zvezda.

One of the initial recommendations of the STS-107 Investigation Board was that all future Shuttle flights should be imaged while in orbit in an attempt to allow any damage to be identified and assessed. To this end Malenchenko and Lu spent June 7 calibrating and focusing a number of cameras on the exterior of ISS. In future, all Shuttles would be imaged in real time as they approach to dock with ISS. Images would be down-linked to MCC-Houston for assessment before the Shuttle was allowed to return to Earth.

PROGRESS M1-10

Progress M1-10 was launched at 06:34, June 8, 2003 and docked to the station at Pirs' nadir at 07:15, June 11, after a standard approach. It carried food, drinking water, and equipment for the Expedition-7 mission as well as propellants for the thrusters on Zvezda. It also carried two experiments to be carried out by ESA astronaut Pedro Duque, who would spend a week on ISS during the Expedition-7–8 crew changeover. Following pressure checks, the hatches between Pirs and Progress M1-10 were opened allowing Malenchenko and Lu to unload the cargo, a task that they commenced on June 13.

During the week, Lu continued to work on the InSPACE experiment based within the MSG. At one point, Lu was looking out of Destiny's window when he saw a rectangular piece of metal, 5 cm long, drift away from the station. Controllers thought that it was most likely a metal label that had become detached from the exterior of the station. The following week was a busy one, with the two men continuing to unload the new Progress and overseeing the pumping of water and propellant from the cargo vehicle to ISS. Lu continued his work with the InSPACE experiment. A third series of InSPACE experiments also occupied Lu during his eighth week of living on ISS, while Malenchenko began loading rubbish into the now empty Progress M1-10.

Both men worked to replace the flexpacks in the canisters of the Resistive Exercise Device (RED). The flexpacks provided the resistance as the crew used the machine to exercise the major muscle groups of the body. The new flexpacks had been lifted into orbit on Progress M1-10. Lu also calibrated an ultrasound device in the HRF rack in Destiny, while Malenchenko replaced a pump in one of Zarya's cooling loops. Lu also set up and calibrated the Portable Clinical Blood Analyser (PCBA), which they both used during their routine medical checks the following day. In his pre-launch interview Lu had discussed the importance of such exercises:

> "I think that the most exciting results so far we've had on Space Station will be continued. And, that is that it does seem possible to reduce or even eliminate possibly the calcium loss in bones from astronauts. One thing that we're doing differently on ISS than, say, on Mir or on Skylab is that we now [have] the capability to do heavy weight-bearing, weightlifting-type exercises that we did not have before. And, interestingly, on the first six increments thus far, we found that you can very, very much reduce the calcium loss in bones by doing heavy weightlifting things such as squats, dead lifts, exercises like that. We know on the ground, that to build bone and muscle mass, you need to do heavy weight-bearing exercises ... To me this is by far the most interesting scientific thing we've found so far in the early stages of Space Station ... We are changing our exercise protocols, and we will be switching to as many of these sorts of heavy weight-bearing-type exercises as we can ... Because if you want to fly on long, extended missions across the solar system, as we do someday, you have to solve the problem of bone and muscle loss. And, we may have actually essentially solved that. Or, come very much of the way towards solving that."

On June, 17, Space Adventures and the RSA announced completion of an agreement reached on April 30, to secure positions for two spaceflight participants on Soyuz TMA flights to ISS in 2004–2005. Such flights had been stopped in the wake of Russia's request to reduce the ISS Expeditions crews to just two people. With NASA preparing to begin work on bringing the Shuttle back into operations after the CAIB's Final Report was published, probably in July, Space Adventures negotiated for the purchase of two positions on future Soyuz taxi flights. A NASA spokeswoman told the media that Russia had yet to clear the plan with the Americans:

> "We're expecting that the Russian Aviation and Space Agency will discuss with NASA and the other partners how this project can be conducted within the procedures that exist within the International Space Station partnership."

In Florida, six members of the CAIB visited the hangar where 84,000 individual pieces of debris from STS-107 were being investigated on June 17. Retired Admiral Harold Gehman, chairman of the Investigation Board, told the media:

> "At this stage, the Board has not come across any show stoppers that in our mind would prevent the Shuttle from returning to flight ... Now, how high is the stack of return-to-flight items when we get finished? I can't tell you now, but right now, it looks manageable."

He continued,

> "We get briefings continuously on what the debris and the metallurgy tells us. Many of us felt it was our duty to come down to see it for ourselves ... We saw the things today which we believe are compelling pieces of evidence that tell us how the heat got into the vehicle and where the flaw started."

Meanwhile, some Republican politicians were calling for the 20-year-old Shuttle to be scrapped and replaced by the proposed Orbital Space Plane.

On June 24, the two men in ISS spoke to the six people of the Aquarius crew, inside NASA Extreme Environment Mission Operations (NEEMO). Peggy Whitson, Expedition-5 Science Officer, was Commander of the 14-day underwater NEEMO mission.

In advance of the publication of the CAIB Final Report, then expected in July, NASA Administrator Sean O'Keefe told a meeting of the Florida Society of Newspaper Editors and the Florida Press Association that NASA intended to establish an "Engineering and Safety Centre" initiative, to establish and assure high standards of safety on future Shuttle flights. He said:

> "The effort we need to go through, the high bar we need to set for ourselves, ought to be higher than anything anybody else would levy on us ... We've got to not only focus on the CAIB's findings and recommendations, but beyond that, to correct everything we think might stand in the way of flying as safely as humanly possible."

Figure 40. Expedition-7: Edward Lu plays his keyboard during a period of free time. The view shows the conditions inside Destiny when the Expedition crew were running numerous experiments simultaneously.

As July began, the Expedition-7 crew were starting their third month on ISS. On July 1, Lu celebrated his 40th birthday. Hawaii, his home state, marked the occasion with "Edward Tsang Lu Day" and MCC-Houston held an "Aloha Day", with members of the control room staff wearing Hawaiian shirts. NASA Administrator Sean O'Keefe spoke to Lu during the day. As the week continued, Lu completed his work with the InSPACE experiment. The NASA website contained Lu's letters from ISS in a series called "Greetings Earthlings". These included simple explanations of the basic scientific principles governing life on ISS and a number of highly personal essays relating to Lu's experiences on the station. They are all highly readable and informative.

Malenchenko and Lu also completed routine maintenance checks of the Pirs docking module. July 4 was an American holiday and a rest day for the crew, in addition to the weekend that followed it. Exercise and routine maintenance were the only activities that were allowed to interrupt their time off.

As June ended, NASA confirmed plans for the remainder of the Shuttle orbiter fleet. Prior to STS-107, Columbia, as the oldest orbiter, had undergone the first 2.5-year refit to update most of its aging systems and install a new "glass cockpit". With Columbia's refit complete the plan had been for the remaining three orbiters to be removed from flight duties in turn to undergo a similar refit. NASA confirmed that despite the loss of one orbiter the refit schedule would be upheld, with Endeavour being next in line, followed by Discovery, and then Atlantis.

THEORY TESTED, AND CONFIRMED

By this time the favoured theory for the cause of the loss of STS-107 was that the left-hand bipod ramp, a piece of shaped insulation foam the size of a suitcase, had separated from the ET at $T+81.7$ seconds after launch, it struck the left wing of Columbia 0.2 seconds later, striking the leading edge of the wing in the area of Carbon–Carbon Panels 8 and 9.

On July 7, CAIB representatives watched as a gas-powered cannon was used to fire a similar piece of insulation foam at a mock-up of the leading edge of the orbiter's left wing, including all of the relevant carbon–carbon panels. The test was designed so that the foam would strike Panels 8 and 9 at an angle of 22° and travelling at 237 cm per second, as a true representation of the impact event seen in the STS-107 launch day film.

The test resulted in impact forces that were more than 50% above those the carbon–carbon panels were designed to survive and produced a hole 40.5 cm by 43 cm in the underside of Panel 8. The remainder of the panel was severely cracked. Damage to the T-seal lug (the T-seal linked Panel 8 to the neighbouring panel) was similar to that found on Columbia's recovered T-seal lug. The foam test subject disintegrated on impact. Two pieces of the carbon–carbon panel were found inside the U-shaped panel and observers suggested that this was similar to the object picked up by radar drifting away from Columbia one day after launch. Investigators deduced that such a large hole would have led to an earlier break-up of Columbia during re-entry and therefore arrived at a hole approximately 25 cm square for the STS-107 damage. In August, the recovered Columbia debris was crated and stored on Level 16 of the VAB.

During their 11th week of Expedition-7's occupation, Lu installed the EarthKam camera on the Earth-facing window in Destiny for use during the new school term. Malenchenko repaired the Satellite Navigation System in the Russian segment of the station. He also replaced pipe conduits in the condensate separation and pumping unit using items delivered by Progress M1-10. Working together, Malenchenko and Lu upgraded a relay unit in the Russian audio system, and inspected the life support system, smoke detectors, and microbe filters throughout the station. They also re-built laptop computer hard drives and carried out an audit to assist programme managers to decide what to launch on future Progress flights. Despite this workload they also managed to participate in Russian medical experiments on their own bodies and to talk to a number of amateur radio hams as they passed around the planet.

The week ending July 18 included Lu imaging Hurricane Claudette as it approached the Texas coast. Houston had made preparations in case the hurricane threatened their location, but it did not. Lu installed and checked out the Coarsening of Solid–Liquid Mixtures (CSLM) experiment in the MSG. He also swapped software from the six Station Support Computers (SSCs) to next-generation laptops. Together, the two men inspected the windows in Zvezda and Pirs and down-linked digital images of them to Korolev. Lu attempted to repair the cooling loop in his EMU, but was unsuccessful.

On July 21, both men participated in medical experiments before working on an inventory of Russian items in the station. The following day saw more medical experiments and a test of a new Russian satellite navigation antenna. July 23 was occupied with descent procedures training in Soyuz TMA-2 and conditioning the batteries in their EMUs, a process that took all week. During the following day, Malenchenko continued with the Russian medical experiments while Lu prepared for a ground-based test of the SSRMS. The test, which took place on July 25, involved ground-based control of SSRMS activities. Lu was required to complete station-based activities that could not be controlled from the ground.

NASA officially announced the Expedition-8 crew on July 25. Michael Foale would be Commander, with Alexander Kaleri as Flight Engineer. Spanish ESA astronaut Pedro Duque would fly a short mission to the station, under contract to the Russians, performing experiments before returning to Earth in Soyuz TMA-2 with the Expedition-7 crew. Soyuz TMA-3 would be launched on October 18, 2003, with Foale and Kaleri spending almost 200 days on ISS.

The Expedition-7 crew marked the 1,000th day of continuous occupation of ISS on July 29. They down-linked a message and received calls from the heads of the 16 national space agencies involved in the programme. At a meeting held in Monterey, Canada, NASA representatives thanked the heads of all ISS International Partner agencies for their continued support of ISS. Jean-Jacques Dordain, Director of ESA, said that "The Columbia tragedy is not just NASA's tragedy—it's our tragedy." Particular thanks went to the Russians, whose capabilities had allowed ISS to remain occupied and re-stocked throughout the period following the loss of STS-107 and the suspension of Shuttle flights. Yuri Koptev, director of the RSA, took the opportunity to talk about the additional financial burden placed on his cash-strapped Agency but admitted, "Sometimes a partner has to take more responsibility ... When such big projects are involved, there is no other way to do it."

Partners were informed of the preliminary findings of the CAIB and agreed to continue following the ISS Programme Action Plan, adopted in 2002, as the basis of ISS operations. The Plan would be up-dated in October to include operations by the ESA-operated ATV.

On ISS, Malenchenko spent the week working on Russian medical experiments. He also operated the Russian/German Plasma Crystal-3 (PK-3), which examined particles in an evacuated chamber that have been excited by radio frequencies. Lu continued to work with the CSLM experiment in the MSG. He also performed a function test of the Biotechnology Specimen Temperature Controller, part of a fluid dynamics experiment to be used later in the flight. Regular maintenance also occupied a large portion of their time, as did their daily exercise regime.

On August 4 the crew reached their 100th day in space. The following day ISS shifted into "survival mode", when the onboard computers failed to recognise the thermal system loops in the Russian sector of the station. Non-essential items were automatically powered off before controllers on the ground began working with the crew to bring everything back on-line. The event had no major impact on operations, or the science programme.

On August 6, Fred Gregory, NASA's Deputy Administrator, spoke to the media at KSC regarding the CAIB Final Report, which was due to be released later in the month. He told journalists, "My assumption is we will follow to the letter the recommendations. There will be no attempt, whatsoever, to argue or defend recommendations from the Columbia Accident Investigation Board."

During the week on ISS, Malenchenko continued to work on Russian medical and agricultural experiments, while Lu ran the second series of CSLM experiments (CSLM-2) in the MSG. The two astronauts worked together to re-size the spare EMU in Quest so that it fitted Lu. The cooling loop in the suit originally failed but then began working. Engineers in Houston added the second EMU to their workload in an attempt to establish what had happened. Troubleshooting the two malfunctioning EMUs continued into the following week, when the two men inspected valves and filters in the coolant water loop.

Malenchenko married Eketrina Dimitriev, a US citizen of Russian birth, on August 10. For the ceremony he wore a tuxedo and bow tie sent up to him on Progress M1-10. The ceremony was carried out over a secure radio link between a room in MCC-Houston that had been decked out to look like a wedding chapel, where Dimitriev was located. Malenchenko and Lu were on ISS, but were shown on a large video screen in the room where the legal ceremony was taking place. Lu served as best man, he also played the *Wedding March* on a portable keyboard that he had with him on ISS, while Dimitriev walked down the aisle. Both individuals had to place their new wedding rings on their own fingers. Operations on the station were not interrupted as it was a Sunday and therefore a rest day for the crew. The couple had planned their wedding in August, before Malenchenko was assigned to the 6-month Expedition-7 crew on ISS. Texas law allowed the wedding to take place with the groom absent provided there was a sufficiently good reason. Malenchenko had written to his lawyer in Texas and explained the situation. The letter resulted in the marriage licence being granted. Russian officials then forbade the wedding taking place during Malenchenko's time on ISS, stating that Russian government equipment was not available for such private use. Following the wedding, NASA made no comment, but Russian space officials were at pains to point out that, as a member of the Russian military, Malenchenko had to have written authority to marry a foreign national. They stated that he could face criminal charges for holding the wedding ceremony without official permission, but it was made clear that such charges would probably not be brought against the cosmonaut.

In Houston, Michael Foale and Alexander Kaleri, the ISS Expedition-8 crew, talked to the media when they arrived for continued training with equipment that they would use during their occupation of the station. Asked about comments that the station should be abandoned until the Shuttle was flying again Foale said, "For us to not step up and not continue in space on the International Space Station is, for me, not really an option. We need to show perseverance in our goals and dreams by maintaining a human presence in space." Ignoring the fact that American politicians and journalist were suggesting that Russian equipment and the Russian infrastructure were incapable of sustaining a human presence on ISS, Kaleri replied, "If we are able to maintain manned flight on board ... we must do it. That is why the station is up there."

On ISS life went on. A continuation from the previous week was the use of oxygen in the Progress M-47, docked to Zvezda's wake, and Progress M1-10, docked to Pirs, to repressurise the station. The oxygen was used in this manner to refresh the oxygen in the station before the two Progress vehicles were undocked, prior to their re-entry and destruction. Lu removed the CSLM from the MSG and replaced it with the Pore Formation and Mobility Investigation (PFMI) experiment. The PFMI experiment involved the melting of plastic samples to study the formation of bubbles that might weaken metals, crystals, and other materials at high temperatures. Plans to begin working with the experiment had to be delayed until the following week when Lu could not locate a cable. Ultimately, Lu spoke to Don Pettit, whose suggestions allowed him to locate the cable. Work with the new experiment's first sample began on August 20, and was concluded two days later. Both men continued to perform medical experiments for their respective nations and both also worked on filling Progress M-47 with rubbish, as its oxygen supply approached depletion.

During the week ending August 29, Lu ran the PFMI experiment on a second sample. Two more samples would be processed in the following week and three more in the week after that. He also activated the Commercial Bio-processing Apparatus for use in future biology experiments. Lu also installed a new laptop computer to control the repaired SAMS vibration measurement equipment. Russian mission managers reported that the charge/discharge unit on Zvezda's Battery 2 had been declared failed, and would need replacing. The module continued to function normally on seven batteries.

Progress M-47 undocked from Zvezda's wake at 18:48, August 27, as ISS was flying over China. It re-entered and burned up later the same day. The undocking cleared the way for the arrival of Progress M-48 later in the month.

Even in the wake of the loss of STS-107, the Russians announced that they intended to launch a Soyuz TMA flight to ISS with two spaceflight participants and a single Russian cosmonaut acting as Soyuz Commander. The Russians announced that they hoped to raise up to 50% of its annual spaceflight revenue from spaceflight participants flying for a fee of $20 million each. However, no Soyuz flight with two spaceflight participants had taken place up to the close of this manuscript (February 2008).

COLUMBIA ACCIDENT INVESTIGATION REPORT

For several months in advance of its final report on the loss of STS-107 the chairman of the Investigation Board had been briefing NASA and the media on what had been found. Speaking about what the final report might contain NASA Administrator Sean O'Keefe had warned, "It's going to be ugly ... This is not going to be anything that anybody's going to be particularly happy with."

Likewise, Bill Gerstenmaier, Shuttle program manager said:

> "We are well aware of what is coming out of the Columbia Accident Investigation Board and what is coming from the Shuttle Return to Flight discussions ... We are looking at all of the systems on board the station and evaluating whether we need

to do something directly or whether we made decisions in the past ... we ought to go back and look at."

NASA had been accused of excessive use of waivers during the preparation of the STS-107 flight. At the time of launch some 5,800 had been recorded during Columbia's preparation. The Administration was accused of "bureaucratic fumbling and administrative missed signals".

The Final Report of the Columbia Accident Investigation Board (CAIB) was published in August 2003. It established that the left bipod ramp had fallen off of the External Tank during launch and had struck the leading edge of Columbia's left wing in the region of Reinforced Carbon–Carbon Panels 6 through 9, on the internal bend where the wing root moves away from the fuselage. The suitcase-size block of foam caused a large hole in the RCC panels in a region that could not be seen through the flight deck windows. Columbia performed near-flawlessly throughout its mission until re-entry, when super-heated plasma entered the hole in the leading edge of the left wing and caused the destruction of the spacecraft. NASA management was severely criticised. Managers had refused requests from their engineers to have the orbiter photographed by a military reconnaissance satellite when the foam impact was identified on film of the launch. They had also adopted the attitude that if Columbia was fatally damaged then nothing could be done to save the crew.

The report made 15 recommendations for Return to Flight:

1. Initiate an aggressive programme to eliminate all External Tank Thermal Protection System debris shedding at the source with particular emphasis on the region where the bipod struts attach to the external tank.
2. Initiate a programme designed to increase the orbiter's ability to sustain minor debris damage by measures such as improved impact-resistant Reinforced Carbon–Carbon and acreage titles.
3. Develop and implement a comprehensive inspection plan to determine the structural integrity of all Reinforced Carbon–Carbon system components.
4. For missions to the International Space Station, develop a practicable capability to inspect and effect emergency repairs to the widest possible range of damage to the Thermal Protection System, including both tile and Reinforced Carbon–Carbon, taking advantage of the additional capabilities available when near to or docked at the International Space Station ... Accomplish an on-orbit Thermal Protection System inspection using appropriate assets and capabilities, early in all missions ... The ultimate objective should be a fully autonomous capability for all missions to address the possibility that an International Space Station mission fails to achieve the correct orbit, fails to dock successfully, or is damaged during or after undocking.
5. Upgrade the imaging system capable of providing a minimum of three useful views of the Space Shuttle from lift-off to at least Solid Rocket Booster separation ... The operational criteria of these assets should be included in the Launch Commit Criteria for future launches.

6. Provide a capability to obtain and down-link high-resolution images of the External Tank after its separation.
7. Provide a capability to obtain and down-link high-resolution images of the underside of the Orbiter wing leading edge and forward section of both wings' Thermal Protection System.
8. Modify the Memorandum of Agreement with the National Imagery and Mapping Agency to make the imaging of each Shuttle flight while on orbit a standard requirement.
9. Test and qualify the flight hardware bolt catchers.
10. Require that at least two employees attend all final closeouts and intertank area hand-spraying procedures.
11. Kennedy Space Centre Quality Assurance and United Space Alliance must return to the straightforward, industry-standard definition of "Foreign Object Debris" and eliminate any alternate or statistically deceptive definitions like "process debris".
12. Adopt and maintain a Shuttle flight schedule that is consistent with available resources. Although schedule deadlines are an important management tool, those deadlines must be regularly evaluated to ensure that any additional risk incurred to meet the schedule is recognised, understood, and acceptable.
13. Implement an expanded training programme in which the Mission Management Team faces potential crew and vehicle safety contingencies beyond launch and ascent. These contingencies should involve potential loss of Shuttle or crew, contain numerous uncertainties and unknowns, and require the Mission Management Team to assemble and interact support organisations across NASA/Contractor lines and in various locations.
14. Prepare a detailed plan for defining, establishing, transitioning, and implementing an independent Technical Engineering Authority, independent safety programme, and a reorganised Space Shuttle Integration Office.
15. Develop an interim programme of close-out photographs for all critical subsystems that differ from engineering drawings. Digitise the close-out photograph system so that images are immediately available for on-orbit troubleshooting.

Chapter 9 of the Report addressed the future of American human access to space. The report noted:

> "The Board observes that none of the competing long-term visions for space have found support from the nation's leadership, or indeed among the general public. The U.S. civilian space effort has moved forward for more than 30 years without a guiding vision, and none seems imminent. In the past, this absence of a strategic vision in itself has reflected a policy decision, since there have been many opportunities for the national leaders to agree on ambitious goals for space, and none have done so."

The CAIB did note that almost everyone seemed to agree that

> "The United States needs improved access for humans to low-Earth orbit as a foundation for whatever directions the nation's space programme takes in the future."

Board members called for a national debate to define America's future in space.

The report highlighted the short-sightedness of developing the Shuttle in isolation. Lack of funding and an often-hostile Congress meant that for the next 20 years after the decision to build the Shuttle, NASA made little or no attempt to develop parallel "access to space" technologies until the X-33 and X-34 programmes of 1994. As a result, those programmes were begun with a limited technology base, and X-33 proved beyond the technological capabilities of Lockheed-Martin at that time. The Report then recognised NASA's attempts to broaden their technology base with the Space Launch Initiative (SLI) in 2000, and promptly narrowed that vision once more with the decision, in 2002, to redirect SLI to commence the Integrated Space Transportation Plan (ISTP) with its proposed Orbital Space Plane (OSP).

The Board made it clear that they did not study NASA's plans for the ISTP, or the OSP in depth. Even so, they concluded:

> "Because of the risks inherent in the original design of the Space Shuttle, because that design was based in many aspects on now-obsolete technologies, and because the Shuttle is now an ageing system but still developmental in character, it is in the nation's interest to replace the Shuttle as soon as possible as the primary means for transporting humans to and from Earth orbit."

CAIB members recognised that in the mid-term that replacement would, more than likely, be the OSP and demanded:

> "The design of the system should give overriding priority to crew safety, rather than trade safety against other performance criteria, such as low-cost, re-usability, or against advanced space operation capabilities other than crew transfer. This conclusion implies that whatever design NASA chooses should become the primary means for taking people to and from the International Space Station, not just a complement to the Space Shuttle."

This represented a major change in direction for the OSP, which NASA had originally intended to begin flying while the Shuttle continued to operate.

The CAIB members stated that there was considerable urgency in developing the OSP, which would require commitment and financial support from Congress and the American people. They stated that America must be prepared to support the OSP as a long-term commitment, and not shy away from the long-term cost, as ISS was likely to be the primary destination for Americans in space for the next decade, or longer.

The report called the failure to develop a Shuttle replacement vehicle "a failure of national leadership" caused by "... continuing to expect major technological advances in that vehicle". The CAIB recommended that everyone concerned should agree that the overriding design principal of the OSP should be "to move humans safely and reliably in to and out of Earth orbit. To demand more would be to fall into the same trap as previous unsuccessful efforts." The paragraph concluded, "Continued US leadership in space is an important national objective. That leadership depends on a willingness to pay the costs of achieving it."

On August 26, Sean O'Keefe made a speech that was broadcast to all NASA field centres. In that speech he told NASA's employees:

> "We must go forward and follow this blueprint in an effort to make this a much stronger organization. All of us at NASA are part of the solution ... ultimately ... to return to the exploration objectives that they dedicated their lives to."

PROGRESS M-48

Progress M-48 lifted off at 21:48, August 28, 2003, and docked to Zvezda's wake two days later, at 23:40, August 30. The new Progress carried food, water, and propellants as well as replacement parts for ISS, tools, and a new laptop computer. There was also a cellphone and global positioning equipment, for use by the Soyuz TMA-2 crew in the event they land off-target, as the Soyuz TMA-1 crew had. There were a number of experiments for the Expedition-8 crew and others for ESA astronaut Pedro Duque to perform during his 8-day stay on ISS.

The launch was made amidst continuing concerns over funding for the Russian Soyuz and Progress vehicles needed to continue support of ISS occupation in the absence of Shuttle flights. The Russians felt that the details of their legal contract did not cover the unique situation that they now found themselves in and Energia managers were having difficulties making their political leaders and their budget controllers in Russia and America understand their difficulties. The subsequent allocation of 3 billion roubles to Rosaviakosmos would be used to launch 11 Soyuz TMA spacecraft on crew rotation flights and sufficient Progress vehicles to carry 80 tonnes of supplies to the station. None of the money would be used to develop and construct new Russian station modules. Rosaviakosmos managers pointed out that if the Russian government did not fund new station modules separately then Russian participation in ISS would remain limited to cosmonauts serving on Expedition crews, Soyuz TMA taxi crews, and robotic Progress cargo flights.

Meanwhile, Donald Thomas, the Mission Specialist who had been removed from the Expedition-5 crew due to concerns over his total radiation exposure, was named as ISS Programme Scientist, the head of America's science programme on the station. In the same period, in Washington DC, NASA's FY2004 budget had been settled at $15.3 billion, with politicians removing $20 million from the proposed ISS funding for that financial year.

Having delivered its oxygen supply and been loaded with rubbish, Progress M1-10 undocked from Pirs at 15:42, September 4. The undocking cleared Pirs' nadir for the arrival of Soyuz TMA-3 occupied by the Expedition-8 crew. Progress M1-10 would stay in orbit for the next month performing independent Russian scientific experiments involving using the spacecraft's cameras to view sites of ecological interest, before it was commanded to re-enter and burn up on October 3, 2003.

Malenchenko and Lu spent the following week unloading Progress M-48 and using its supply of nitrogen to increase the pressure inside ISS. Repressurisation using the Progress' oxygen would occur at a later date. The Progress' thrusters were also tested in advance of a burn to raise the station's altitude.

On September 10, Lu experienced trouble adjusting the resistance on one of the canisters on the RED exercise apparatus. He removed the canister and repaired it during the following week. Two new canisters for the RED were delivered on Progress M-48, but Lu hoped to leave them untouched so that the Expedition-8 crew could hold them in reserve in case of further failures.

Two days later the crew informed the ground that they could barely hear their transmissions. The fault was traced to equipment at Houston that relayed the audio uplink to ISS from the three control rooms at Houston, Huntsville, and Korolev. The problem was resolved by bypassing the equipment in question while it was repaired. In orbit, Lu continued his PFMI experiment programme. He also performed the first operations of the Hand Posture Analyser, an experiment that required him to wear an instrumented glove while performing a range of tasks. The experiment allowed its investigators to study how astronauts use their hands in microgravity. Lu also performed two educational tasks, making films in Destiny for use in American schools. Malenchenko replaced the failed battery in Zvezda and a computer hard drive. As the week ended, cameras on the exterior of ISS were used to image Hurricane Isobel as it crossed the Atlantic Ocean *en route* to its landfall in North Carolina.

The two astronauts powered up the SSRMS on September 23, for training and mechanical tests. The manoeuvres they performed placed part of the arm in sunlight so that any performance differences of the moment sensor in sunlight and shadow might be recorded. They also performed maintenance on two Russian Orlan EVA suits, ensuring they remained in good condition. Lu completed the Expedition-7 work with the PFMI experiment, while Malenchenko performed Russian medical experiments. He also made the first use of the station's ultrasound equipment, using it to monitor Lu whilst he was exercising on the station's stationary bicycle.

On October 1, the rocket motors on Progress M-48 were used to raise the station's orbit and two days later Progress M1-10, which had spent a month performing Russian experiments in orbit, was de-orbited and burned up in the atmosphere. The first week of October saw Lu install a Protein Crystal Growth experiment in the MSG, for Duque to use later in the month. He also set up a soldering experiment and an automated Earth observation camera. Malenchenko continued to perform Russian biomedical experiments, as well as observing thunderstorms, ocean biology, and studies of human-made disaster prediction. Weekly maintenance included Malenchenko inspecting fire sensors and checking systems in Pirs, prior to the arrival of Soyuz TMA-3. Lu configured the American laptops for the Expedition-8 crew and both men worked together to perform maintenance on the treadmill and the RED.

During the following week, the crew began spending more time preparing for their return to Earth, at the end of the month. They donned their Sokol launch and re-entry suits and measured how well they fitted within their Soyuz couch liners. On

October 10, the two men in orbit had the opportunity to talk with the Expedition-8 crew in Korolev. Lu spent much of his time in Destiny, where he worked with the SAMS and made electrical connections as part of the In Space Soldering Investigation (ISSI) experiment. Later in the week one of the station's RPCMs, which routed electrical data throughout the station, failed. The failure disabled one camera and some onboard redundancy, but caused no problems to the arrival of Soyuz TMA-3. Work to identify the problem began immediately on the ground.

During their final week alone on ISS, Malenchenko and Lu concentrated on their preparation for returning to Earth. They carried out systems checks in Soyuz TMA-2 and began transferring items from ISS to their spacecraft. Lu continued to perform experiments and maintenance. On October 15, he replaced the malfunctioning RPCM in Destiny. Two days later, he spent several hours collecting water samples from the cooling system in Quest, which was used to cool the suits of astronauts making EVAs from the airlock. The samples would be returned to Earth for analysis.

As his occupation of ISS drew to a close, Lu told of earlier plans to abandon ISS during a press conference:

> "The critical things the ground cannot do, of course, is repair and change out things up here ... Luckily, nothing has happened that could cripple the Space Station, while we were up here ... I'm much more comfortable with a crew on board knowing they could take care of something you had not planned for."

Indeed, this was a lesson learned the hard way by the Russians with their Salyut and Mir stations.

CHINA JOINS THE SPACEFARING NATIONS

China launched its first crewed spacecraft on October 15, 2003. The Soyuz-based ShenZhou-5 was launched from the Jiuquan Satellite Launch Centre in Gansu Province, on a Long March II-F launch vehicle. The single taikonaut (astronaut), Yang Liwei, made 14 orbits before returning to Earth and landing in Inner Mongolia on October 16, after a flight lasting 21 hours 23 minutes. China had become only the third nation to develop the ability to launch a crewed spacecraft into orbit.

Following the Chinese launch, Houston had told the Expedition-7 crew, "We have a news item to pass on. The world's spacefaring nations have been joined by a new member tonight. For the next few hours, Russia and the United States will share the heavens with China."

Lu replied, "That is very good news. From one spacefaring nation to another, we wish them congratulations." When discussing the Chinese flight, both of the Expedition-7 crew members were positive. Lu stated, "Personally, I think it's a great thing. The more people in space, the better off we all are."

Malenchenko added, "I'm glad to have somebody else in space (besides) Ed and me. It was great work by thousands and thousands of people from China. I congratulate all of them."

NASA Administrator Sean O'Keefe was equally positive, "They are developing a capability—this can't be understated—to accomplish something that only two other nations on the planet have ever done. That's a rather historic, hallmark achievement." Despite these positive words, in the coming years NASA would reject China's attempts to become part of ISS.

ISS, SAFE HAVEN FOR SHUTTLE CREWS

Following the publication of the Columbia Accident Investigation Board's (CAIB) Final Report, NASA began preparing for the Shuttle's Return to Flight. One idea that was developed was the idea of a safe haven on ISS for the crew of any future Shuttle that was damaged during ascent into orbit.

In keeping with the CAIB report's recommendations, all future Shuttle flights would use cameras mounted on a new Orbiter Boom Sensor System (OBSS) to inspect the previously inaccessible areas of the orbiter's Thermal Protection System (TPS). Additional photographs would be taken by the ISS Expedition crew as the Shuttle performed a nose-over-tail pitch manoeuvre prior to docking. The photographs and videos would then be downloaded to MCC-Houston, where engineers would study them and declare the TPS fit, or unfit, for re-entry. If the TPS was damaged and declared unfit for re-entry, the Shuttle crew would dock with ISS, which would then serve as a safe haven until a new means of recovery could be launched to recover them. This might be a second Shuttle, or in extreme cases, additional Russian Soyuz spacecraft.

With a two-person Expedition crew and a seven-person Shuttle crew onboard, supplies available on ISS would be limited. Under existing conditions a Shuttle crew would be able to utilise the safe haven for up to 86 days, at which point NASA would have to be able to launch the rescue vehicle, a second Shuttle. Excess water and food from the crippled Shuttle would be transferred to the station to support the additional astronauts. In addition, the launch schedule for Progress cargo vehicles would have to be sustained.

SOYUZ TMA-3 DELIVERS THE EXPEDITION-8 CREW

SOYUZ TMA-3	
COMMANDER	Michael Foale
FLIGHT ENGINEER	Alexander Kaleri
ENGINEER	Pedro Duque (Spain)

Prior to the loss of STS-107, the Expedition-8 crew consisting of Michael Foale, William McArthur, and Valeri Tokarev was due to be launched to ISS on STS-116, with a Shuttle crew consisting of:

COMMANDER: Terrence Wilcutt
PILOT: William Oefelein
MISSION SPECIALISTS: Robert Curbeam, Christer Fuglesang

STS-116 was to have returned the original (three-man) Expedition-7 crew to Earth at the end of their mission.

When the two-person caretaker crews were named on February 24, 2003, Foale and Alexander Kaleri (the third member of the original Expedition-7 crew) were named as the new Expedition-8 crew, with McArthur and Tokarev serving as their back-up crew. Spaniard Pedro Duque was flying under the Roscosmos contract that offered the third couch on a Soyuz flight to ESA if it was not filled by a commercial spaceflight participant. Duque would return to Earth in Soyuz TMA-2 with the Expedition-7 crew. Foale was on his sixth spaceflight, Kaleri his fourth, and Duque his second.

Discussing his training for flying the Soyuz spacecraft during his various crew allocations, Foale has said:

> "Well, I've had a lot of experience, compared to other U.S. astronauts training for Soyuz flight. First of all, on Mir we had to train for Soyuz emergency descent, and that was basically as a passenger cosmonaut–astronaut in the right seat. On this last training flow before Columbia I was training for the left seat emergency descent with a cosmonaut in the center seat. After Columbia, Sasha and I were named to be the backups to Ed Lu and Yuri Malenchenko that are on orbit right now, and there I had to train both for launch in the left seat, rendezvous, and then descent. And then at the same time I had to take the same classes that Sasha, the Commander, was taking and do those in the centre seat. So I have seen the whole smorgasbord of crew roles and responsibilities on board the Soyuz."

Asked about the possibility that he might also have to return to Earth in a Soyuz spacecraft at the end of the Expedition-8 crew's occupation, Foale remarked:

> "I think, to be quite honest, I'd like to come home in the Soyuz. And that's mostly because it's well understood right now. Its ballistic entry was demonstrated by Ken Bowersox and Don Pettit and Nikolai Budarin on the last entry. People called that off-nominal—it certainly was not an expected entry—but it demonstrated yet another aspect of the Soyuz, which is its robustness ... [H]owever, I do want the Shuttle to succeed ... But I don't believe ... that the Shuttle's architecture will allow it to be significantly safer without adding crew escape to it."

Soyuz TMA-3 was launched at 01:38, October 18, 2003. Kaleri served as Soyuz Commander, with Foale assuming command only after entering the station. Docking to Pirs nadir occurred at 03:16, October 20, as ISS passed over Central Asia. Meanwhile, in orbit, pressure and leak checks were followed by the opening of the hatches between the two spacecraft at 05:19, and the new crew entered the station, where the

Expedition-7 crew, who were on their 177th day in space, greeted them. The flags of Spain and ESA were displayed on the station alongside those of America and Russia. After eating lunch, the three newcomers received the standard safety brief before Duque's couch liner was transferred to Soyuz TMA-2. Meanwhile, NASA announced details of two minor problems with the flight. The potentially more serious of the two was a small helium leak between the helium pressurisation tank and the propellant tanks of Manifold 2 in the Soyuz propulsion system. Manifold 1 would be used for the remainder of the Soyuz TMA-3 flight.

After Soyuz TMA-3 had safely docked to Pirs, the *Washington Post* released details of how two "mid-level scientists and physicians" had refused to sign the initial approval for the launch due to their concerns over the deterioration of environmental monitors and the medical and exercise equipment on ISS. The report also claimed that ISS was running low of medicines and intravenous fluids, and the noise levels in the Russian sector of the station were greater than desired, and required a noise-deadening programme. NASA Administrator Sean O'Keefe stated publicly that the he did not believe the crew to be in any danger. Both the Expedition-7 and Expedition-8 crews acknowledged the difficulties, but stated that they believed they were in no immediate danger. These reports led to a number of politicians jumping on the bandwagon and using the incident as an excuse to attack NASA and its Administrator. Representative Sheila Jackson Lee (Democrat-Houston) stated, "Safety, above all, has to be the highest priority. After the Columbia Accident Investigation Board's report came out, I asked specifically about the safety of the Space Station and the response coming back was not as strong as I wanted. Now it seems there is not only a problem, there is a crisis." NASA also had its supporters, Dana Rohrabacher, head of the House Space and Aeronautics subcommittee, told the press, "I have faith Mr. O'Keefe is doing his best and shouldn't be second-guessed by politicians."

Meanwhile, the two individuals concerned told the media that they had not identified any immediate threats to crew safety, but were concerned over the deterioration and long-term failure of equipment measuring the environment on the station since the loss of STS-107, and the grounding of the Shuttle fleet. They stated that they had spoken to NASA officials from JSC and were content that NASA was taking their concerns seriously and had begun to address the issues that they had raised.

When the issues were first raised and became public, in the second week of October, O'Keefe had been blunt. He told the media that if the situation reached a point where the crew was at risk then "... the answer is, get aboard the Soyuz, turn down the lights and leave."

Docking was followed by 8 days of joint operations, with Malenchenko and Lu handing over to Foale and Kaleri, while Duque performed his "Cervantes" experiment programme. This consisted of 24 experiments to be completed, in both the Russian and American sectors of the station, over 40 hours. Foale described his feelings on what would happen during the hand-over in the following terms:

> "[T]here are a number of goals to be achieved during the joint operations ... of the oncoming crew and the off-going crew. First and foremost ... Spain, through ESA ... are paying the lion's share of this Soyuz flight, and they have serious

Figure 41. Expedition-8: Michael Foale and Alexander Kaleri pose alongside the mess table in Zvezda.

science objectives to accomplish during the five days that Pedro is planned to be on board the Station with us. And so, to be honest ... I feel obliged, just as Sasha does, to help Pedro get kicked off to a running start as soon as we arrive. There is nothing more important than getting Pedro running. The remaining four days I will spend my time with Ed Lu, and I hope to learn everything he has learned about the Destiny module, about the Airlock, the Node, and all of our stowage there and all of our equipment there, and its operations with the control center here in Houston. However, I must not ignore what's going on in the Russian segment, where Sasha Kaleri will be spending a lot of time with Malenchenko and learning about Russian operations, work in the Service Module and in the FGB, and in their docking module ... By the time four days have gone by, I will know just the bare minimum to be able to find my clothes, wash my body, do my exercise, and work the radio ... but the practical knowledge will be there after four days. And at that point, we will be ready to say, Pedro, you're going to that spacecraft; your seat liner's in that spacecraft—the old one, the returning one—with Ed Lu and Yuri Malenchenko, and then we'll close the hatch and breathe a sigh of relief, because joint ops is a very hard time because everybody has a lot to do in a short time."

The ceremonial change of command ceremony took place on October 24. Three days later Malenchenko and Lu and Duque sealed themselves in Soyuz TMA-2. They

Figure 42. Expedition-8: Launched with the Expedition-8 crew, Spaniard Pedro Duque participated in an ESA experiment programme before returning to Earth with the Expedition-7 crew.

were preparing to undock when the complex rolled unexpectedly. This caused the thrusters on Zvezda to fire, to correct the station's attitude. NASA described the incident:

> "At 2.57 PM, while ISS was in the XPOP momentum management mode, the station experienced a large unexpected roll manoeuvre event, with momentum increasing from 16 percent to 90 percent in four minutes. As 90 percent triggers a de-saturation of the CMGs, with Service Module thrusters, several de-saturation burns followed, using several kilograms of propellant. Proper attitude control was re-established for the undocking ... Troubleshooting by Moscow determined that the cause of the torque was a crewmember, during ingress, contacting the Soyuz hand controllers, which are not supposed to be active at that time. TsUP further determined that the override commands to activate the hand controllers were inadvertently initiated by a Soyuz control panel pushbutton, while the crew was loading return items."

As usual TsUP had managed to put the blame on the crew, despite the fact that no one in the Russian control room had noticed that the Soyuz hand controllers had been inadvertently activated.

Soyuz TMA-2 undocked at 18:17, October 27. The de-orbit burn took place at 20:47, and the re-entry module touched down on target in Kazakhstan at 21:41. The

Expedition-7 crew had spent 184 days 22 hours 46 minutes and 9 seconds in flight. Duque had been away from Earth for 9 days 21 hours 1 minute 58 seconds. Initial medical examinations at the landing site showed all three men to be in good health, but the Expedition-7 crew underwent the usual 45-day rehabilitation programme at Korolev, where Malenchenko's new wife was waiting to welcome him back to Earth.

EXPEDITION-8

Foale and Kaleri spent their first week in space completing experiments and familiarising themselves with ISS. Their experiment programme included 42 Russian experiments, 38 of which had been used on previous Expedition occupations. They quickly established a daily routine of briefings, exercise, experiments, and maintenance. October 28 was a light work day for the new crew, offering them a chance to rest after the intensity of their Soyuz flight and the hand-over week. Kaleri donned a compression cuff on his upper thigh as part of the Russian Braslet-M/Anketa microgravity adaptation experiment. He also donated blood for the Hematokrit experiment. At the start of the new week they performed well in an emergency evacuation drill, before performing maintenance and station configuration tasks. The day after that was a full workday, with both men performing their own countries' medical experiments, and an inspection of the station's exercise devices, and maintenance.

After a quiet weekend the crew began work with the Body Mass Measurements experiments before stowing the EarthKam. Foale spoke to former Skylab astronauts who were meeting at MSFC, in Huntsville, to celebrate the 30th anniversary of the launch of Skylab-4, the final crewed flight in that programme. During the week a series of solar flares erupted from the Sun, sending radiation towards Earth, where the planet's magnetic field directed some of it into the upper atmosphere, causing aurorae. At various times the two astronauts sheltered in Zvezda and the third sleep station in Destiny as the station passed close to areas of high activity within Earth's ionosphere. Korolev made it clear to the public that ISS orbited between 1,000 km and 2,000 km below the areas of highest activity in Earth's atmosphere, and that the station's radiation shielding was sufficient to protect the crew. Despite this and the fact that broken radiation-monitoring equipment had been part of the general deterioration of ISS complained about before the Soyuz TMA-3 launch, radiation levels outside and inside the station were carefully monitored.

Throughout all of this the crew continued to perform their experiments, with Kaleri performing Russian medical and physiological experiments in Zvezda, while Foale worked on American experiments in Destiny. On October 31, the Russian veloergometer, a vital exercise machine mounted in Zvezda, failed. Engineers in Korolev began an investigation into what had happened and how to repair it. The veloergometer was part of the crew's daily exercise routine to ward off the effects of prolonged microgravity, so its speedy repair was important to them. On November 2, ISS passed its third year of permanent occupation.

Both men continued with their national experiment programmes, while also spending time trying to find suitable storage for all of the items on the station. This even included pressurising PMA-2, which was normally kept in a vacuum state, opening the hatch between Destiny and the PMA, and mounting items around its internal walls, before closing the hatch once more. Foale installed the EarthKam in Destiny's nadir window and activated the Protein Crystal Growth experiment. He also used a special glove to measure hand muscle action in microgravity. The Hand Posture Analyser was an Italian experiment flown by ESA in an attempt to "quantify muscle fatigue associated with long-duration spaceflight."

Foale had discussed some of the experiments during his pre-launch interview:

> "we have ... one experiment that looks at the melting and then the re-solidification of metal analogues in the glovebox facility in the Destiny Laboratory Module. We have a very interesting experiment that looks like it's straight out of *Star Wars* called SPHERES ... And this experiment is, the set of spheres—actually polyhedrons—that manoeuvre themselves in relation to each other and fly in formation to each other. We do this inside ... the Node [Unity], and I'll be doing that at different times during the mission. And then we have experiments that are life science-oriented. There's an experiment that measures how I move in space ... for a number of days I'll be wearing some pretty fancy, expensive tights that are fully instrumented with instrumentation that measure how my muscles are moving, how the nerves that are triggering my muscles are firing, and indeed, the actual resulting position of my leg. It's called FOOT. Although, actually it's not only the foot that it's studying, it's studying the whole leg. And so that will then bring data back on basically how a human being adjusts, or just naturally assumes a neutral position in space during a normal workday."

In mid-November Kaleri reconfigured the TORU cabling inside Zvezda. Meanwhile, spikes in the current vibration lasting approximately 30 minutes were noted in CMG-3, on November 8. Three gyroscopes, mounted in the Z-1 Truss, were used to control the Station's attitude, but were taken off-line when large manoeuvres were required. The fourth gyroscope had been taken off-line in 2002, following a malfunction, and a replacement gyroscope was waiting in Florida for the Shuttle to resume flying, because it was too big to fit in Progress.

November 11 saw the crew completing periodic hearing tests, while inspecting the Thermal Vibration Isolation System (TVIS) and checking the batteries for the station's defibrillator. The following day they worked on reorganising equipment on ISS. November 12 was spent changing ten smoke detectors in the station and inspecting equipment at the request of mission control. The crew ended the week by commencing a course of potassium citrate pills, or placebos, in advance of a study of renal (kidney) stone development, an ongoing problem in long-duration spaceflight. Foale prepared the CBOSS experiment and its Fluid Dynamics Investigation experiment for later operations when it would be used to grow three-dimensional cell cultures. At the same time Kaleri worked with the Russian Profilakita experiment to study the effects of long-duration spaceflight on the human body. The week ended

Figure 43. Expedition-8: Alexander Kaleri demonstrated transferring from the Pirs airlock to a docked Soyuz spacecraft while wearing an Orlan suit. This was an emergency procedure to be used if the astronauts could not open the hatch between Pirs and Zvezda at the end of a period of extravehicular activity.

with the crew making preparations for the following week's rehearsal for the Stage EVA that was planned for February 2004.

Both men spent the beginning of the week evaluating emergency procedures to be used if Pirs failed to pressurise at the end of the planned EVA. In that event they would have to move directly from Pirs to Soyuz TMA-3 while still wearing their Russian Orlan suits. On November 18, Kaleri donned an Orlan EVA suit and attempted to make the transfer through the hatch from Pirs to the orbital module of Soyuz TMA-3. Clearances between the suit's Life Support System and the hatch rim were minimal and, even with the unsuited Foale pushing him from behind, the attempt took much longer than expected and was abandoned with Kaleri only partway through the hatch.

November 20 was the fifth anniversary of Zarya's launch. Representatives from all 16 countries participating in the ISS programme celebrated the launch of the first element. Foale completed a computer-guided refresher course on the use of the SSRMS on November 21. He also completed alterations to the instrumented suit that was part of the FOOT experiment.

At one point the crew heard an unusual noise from the treadmill's Vibration Isolation System (VIS). They were told to stop using the treadmill while the noise was investigated. The cause was later established as a possible gyroscope failure within the treadmill. They began running the treadmill with the VIS powered off a few days later, as part of a Russian experiment to see exactly what vibrations were present and how they affected the station's other experiments. Following this, the crew were given permission to use the treadmill, with the VIS powered off, for their daily exercises over the weekend.

At 02:59, November 26, both Foale and Kaleri heard what they initially believed to be an external impact on the aft end of Zvezda. Foale later explained how he had been immediately sure that ISS had not been ruptured because his ears had not popped, due to a drop in internal air pressure, as they had when he was onboard the Russian Mir station and a Progress vehicle had collided with one of its modules. The SSRMS was used to scan the area, but no signs of an impact were found. Internal pressure readings and the coolant system were monitored both on the ground and in space, but no leaks were discovered. The crew ultimately went back to work, while Houston and Korolev continued to investigate the noise. In the wake of the STS-107 investigation and the criticism that they had received on that occasion, NASA asked the Pentagon to turn a reconnaissance satellite's cameras on ISS, to image the area where the impact was thought to have taken place.

Foale set up video cameras in Destiny to document the Fluid Dynamics Investigation, as part of the CBOSS experiment. He also installed equipment in the MSG for the PFMI experiment. Finally, he completed the alterations to the FOOT experiment's instrumented suit. November 27 was American Thanksgiving Day. The crew enjoyed a light workload and some free time to listen to music and watch films. NASA Administrator Sean O'Keefe spoke to both men by telephone.

Foale spent much of December 2 participating in the FOOT experiment. He had to wear the Lower Extremities Monitoring Suit which was fitted with 20 sensors to measure the wearer's daily activities. During the same day Kaleri worked on the veloergometer, stripping it down and reassembling it. Although the device worked normally after reassembly, Kaleri found no cause for the earlier malfunction. The exercise machine malfunctioned again six days later. On December 4, Kaleri completed the first run on the Russian Plasma Crystal-3 experiment, after spending the earlier part of the week setting it up. The automated experiment studied crystallisation of plasma dust subjected to high-frequency radio waves in a vacuum chamber. Both men also inspected the TVIS on the treadmill in Zvezda, before they both ran on the treadmill with its Vibration Isolation Stabilisation turned off. Instruments in Zvezda and Unity measured the resulting vibrations.

On that day NASA announced that, due to the malfunction of one CMG in the Z-1 Truss and the recent vibration experienced by a second, "ISS attitude hold and all attitude manoeuvres will be performed on Russian thrusters only, between now and mid-January." During the week the crew also discharged and re-charged the batteries in the various pressure suits on the station, reloaded the station laptop computers, and participated in the ongoing Renal Stone Experiment.

December 8 saw Michael Foale set a new record for an individual American astronaut's cumulative time in space, passing the previous record of 230 days 13 hours 3 minutes 38 seconds set by Carl Walz. Thirty Russian cosmonauts had spent longer cumulative times in space than Foale, including Kaleri, who had spent 415 days in space, even before commencing his Expedition-8 flight to ISS. The Russian record was held by Sergei Avdeyev's cumulative time of 748 days. Carl Walz, the previous holder of the American record telephoned Foale to pass on his congratulations. The following two days were spent stripping down the treadmill and confirmed that a bearing in one of its gyroscopes was the cause of the problem. A replacement would be carried up to ISS on Progress M1-11, due for launch in January 2004. The following day the Elektron oxygen generator's prime and back-up pumps failed, causing the machine to shut down. Engineers suggested that an air bubble had caused the problem, and Elektron was later powered on for 10 hours to clear it. Foale used the SSRSM on December 9 to continue the inspection of the station's exterior following the crew's report of an external sound earlier in the flight. Its cameras were also used to inspect the exterior of the station to search for any changes, a task usually carried out by Shuttle crews during their post-undocking fly-around of the station.

NASA announced four new ISS Expedition crews on December 17, 2003. The details released at the time were

Expedition-9	
Prime crew	McArthur and Tokarev, with Kuipers as visiting astronaut
Back-up crew	Padalka and Fincke, with Thiele backing up Kuipers
To be launched on	Soyuz TMA-4, on April 19, 2004

Expedition-10	
Prime crew	Padalka and Fincke
Back-up crew	Onufrienko and Tani
To be launched on	Soyuz TMA-5, on October 9, 2004

Expedition-11	
Prime crew	Sharipov, Phillips, and Kononenko
Back-up crew	Pettit and two Russian cosmonauts to be named
To be launched on	STS-121, on November 15, 2004

Expedition-12	
Prime crew	Chiao, S. Williams, and a Russian cosmonaut to be named
Back-up crew	To be named
To be launched on	STS-116, on April 14, 2005

The Shuttle launch dates were subject to the Shuttle returning to flight following the loss of STS-107.

When McArthur was temporarily medically disqualified, Chiao was teamed with Tokarov on Expedition-9. In time McArthur returned to flight status and the original pairings were reinstated with Padalka, Fincke, and Kuipers being teamed up on Expedition-9, Chiao and Sharipov on Expedition-10 and McArthur and Tokarev slipped to Expedition-11, giving McArthur time to catch up on lost training.

Prior to the STS-107 tragedy, the Expedition-9 and Expedition-10 crews had been:

Expedition-9

Prime crew:	Padalka, Fincke, and Kononenko
Back-up crew:	Poleshchuk, Romanenko, and Tani

Expedition-10

Prime crew:	Chiao, Sharipov, and Phillips
Back-up crew:	J. Williams, Kozeyev, and S. Williams

December 17 was the 100th anniversary of powered flight. Foale and Kaleri were given the day off, although they did talk to schoolchildren at the Wright Brothers' Memorial at Kitty Hawk. Some aerospace journalists had expected President Bush to make an announcement regarding the future of the American human spaceflight programme during the day, but he did not do so. The ISS crew ended the week by sending down the results from some of their experiments. Their weekend of light duties was interrupted by the Elektron shutting down a further three times. The unit was turned down to the lowest power setting, reducing atmospheric pressure on the station. The atmosphere was supplemented with oxygen from tanks in Progress M-48. NASA announced:

> "The Elektron has been operating only intermittently, shutting down when air gets into pumps that help separate liquid and gas. The problem is believed to be one that is sometimes experienced as membranes in that unit age. A replacement is onboard, but flight controllers plan to continue operations as they are for as long as possible before using the new equipment. Oxygen also is being provided to the cabin air from tanks aboard the Progress craft that is docked to the complex. The oxygen in those tanks must be used in the next few weeks to prepare for the undocking of that supply craft in January. With the Progress oxygen being used, continuous use of the Elektron is not necessary."

Two days later Foale and Kaleri completed their ninth week in space. Foale began the week by performing a leak check on the seal between Quest and Unity. The connecting hatch was sealed and the airlock's internal pressure reduced and left overnight. No leaks were detected. At the same time Kaleri worked to replace a faulty heat exchanger in Zvezda's back-up air-conditioning system. He successfully removed the old unit, but then experienced difficulty aligning the new one. The work was left while controllers in Korolev diagnosed the problem. The primary system continued to function flawlessly. Troubleshooting also took place on Zvezda's Elektron system.

The two days before the Christmas holidays were spent repairing various experiments and taking sound level measurements in Zvezda. On December 24, Christmas Eve, Kaleri worked on the Elektron system in Zvezda, replacing filters, but the system failed to restart at the end of his efforts. The sensors in the tanks that were supposed to fill with water indicated that they were full when they were in fact empty. Meanwhile, Foale stowed clothing in Unity, including many items that had been used by previous crews and would not be used again. Both men had the holiday period off, although they had to perform their daily exercises, and routine housekeeping and maintenance on the station. They had a smoked turkey dinner, received a call from Sean O'Keefe, and Foale spoke to his family. On December 26, it was back to work, carrying out routine maintenance on the station.

Kaleri spent three hours on December 29, removing no longer needed attitude control equipment from Zarya; the equipment would be discarded in Progress M-48 in late January. He also worked with the Russian Harmful Impurities Removal System, which helped to purify the station's atmosphere. Foale spent the day working with the CBOSS experiment. The following day, Foale completed some soldering and repacked the ISS medical kit with fresh supplies from Progress M-48. On December 31, both men completed 1 hour of emergency medical training. They also turned off the Elektron and activated their first SFOG candle, in an attempt to use up those candles that were approaching their use-by date. NASA explained:

> "Each candle releases 600 litres of O_2, enough for one person per day. There are 142 SFOGs on board, and the certified lifetime of all of them expires today. Russian formalities required for extending their lifetimes are being expedited as much as possible."

The pair exchanged New Year greetings with each of their many control rooms as they passed around the planet. They celebrated New Year at midnight GMT.

As 2003 ended Sean O'Keefe made it clear that STS-114, the Shuttle Return to Flight mission, would not fly in 2004. He stated that the requirement to develop the OBSS, required to inspect the Shuttle's TPS on each flight, and the requirement to develop a TPS repair kit, as demanded by the CAIB, were the principal causes of the delay. O'Keefe said that the cost of returning the Shuttle to flight had risen from the $280 million estimated in November 2003, to the present figure of $400 million.

January 1, 2004 was a day off for the crew. Both men spoke to their families and completed only vital technical maintenance and personal exercise. During the day, controllers noticed that ISS was suffering a slow pressure leak. Systems on the station were able to compensate for the leak, but the two men still began checking all valves that gave access to open space to ensure they were properly closed. They found nothing untoward. Both men had first-hand experience of leaks in space. Foale had been onboard Mir when it was struck by a Progress spacecraft in 1997, and Kaleri had been sent to Mir in 2000 to look for an unidentified slow leak on the station at that time.

The following day Foale returned to work on the CBOSS and Kaleri followed instructions that were radioed up from Korolev to adjust the Elektron oxygen generator in Zvezda. Kaleri also activated two SFOGs to enrich the station's atmosphere. A further two SFOGs were burned the next day. Meanwhile, oxygen was also introduced to the station's atmosphere from tanks in Progress M-48 on January 1 and again on January 3.

Two days later the crew were informed of a drop in internal air pressure registered in Houston. The leak had first been registered on January 1, and had grown steadily worse since then. They were told, "There's no action for you at this time and no immediate concerns ... We'll continue to investigate this on the next shift and we may have some actions for you tomorrow." Even so, the crew carried out some basic checks onboard ISS but they found nothing amiss. On January 6, they returned to the search for the cause of a pressure leak on the station. NASA's Mike Suffredini explained, "We're going to take a very measured and methodical approach to sort through this problem ... If this was in fact a leak, which we're not certain that it is, we have, oh, about a little over half a year's worth of gas on board to feed it and so we're in no particular hurry to overreact."

Foale finally identified the location of the leak on January 11. It was caused by a flexible cable called a vacuum jumper that was used to equalise pressure between the individual panes in Destiny's main window, where it entered a steel harness at the edge of the window. The vacuum jumper would be capped off and replaced later, after relevant equipment had been lifted up to ISS on a future Progress. Both men had checked the window before, but their efforts to listen for the leak had been frustrated by the noise coming from a science experiment being run in the laboratory. NASA told journalists, "Foale reported that as soon as the flex hose was disconnected, the noise stopped. While additional evaluation is needed for confirmation, the pressure in the station appears to have stabilised since the removal of the hose." Having identified the leak, the crew requested, and were given the remainder of the day off. At Houston's request the crew ended the week by closing the internal hatches and spending the weekend in Zvezda. Controllers then monitored the internal pressure in the various isolated modules of the station throughout the weekend. In Korolev, engineers were considering a total replacement of Zvezda's Elektron unit, which continued to function intermittently.

"LET US CONTINUE THE JOURNEY"

After much anticipation, President George W. Bush finally addressed the nation from NASA HQ, Washington DC, on January 14, 2003. The speech that he made was the culmination of months of work by the White House staff and numerous other agencies to give NASA new goals for the future. This speech was thought by many to be the equivalent of President Kennedy's speech made on May 25, 1961. That speech sent Project Apollo to the Moon, but George Bush's presentation was much less dynamic. The highlights, as they related to ISS, were as follows.

The President was introduced by NASA Administrator Sean O'Keefe. The President talked from the auditorium, and Michael Foale via a video link from ISS:

> "America is proud of our space programme. The risk-takers and visionaries of this agency [NASA] have expanded human knowledge, have revolutionised our understanding of the Universe and produced technological advances that have benefited all of humanity. Inspired by all that has come before, and guided by clear objectives, today we set a new course for America's space programme. We will give NASA a new focus and vision for future exploration. We will build new ships to carry Man forward into the universe, to gain a new foothold on the Moon and to prepare for new journeys to the worlds beyond our own ...
>
> Our programmes and vehicles for exploring space have brought us far, and they have served us well. The Space Shuttle has flown more than 100 missions. It has been used to conduct important research and to increase the sum of human knowledge. Shuttle crews and the scientists and engineers who support them have helped build the International Space Station ... Yet, for all these successes, much remains for us to explore and learn.
>
> In the past 30 years, no human being has set foot on another world, or ventured farther up into space than 386 miles [621.1 km], roughly the distance from Washington DC to Boston, Massachusetts. America has not developed a new vehicle to advance human exploration in space in nearly a quarter century. It is time for America to take the next steps ...
>
> Today I announce a new plan to explore space and extend a human presence across our Solar System. We will begin the effort quickly, using existing programmes and personnel. We'll make steady progress, one mission, one voyage, one landing at a time ...
>
> Our first goal is to complete the International Space Station by 2010. We will finish what we have started. We will meet our obligations to our 15 international partners on this project. We will focus our future research aboard this station on the long-term effects of space travel on human biology. The environment of space is hostile to human beings. Radiation and weightlessness [microgravity] pose dangers to human health. And we have much to learn about their long-term effects before human crews can venture through the vast voids of space for months at a time. Research onboard the station and here on Earth will help us better understand and overcome the obstacles that limit exploration. Through these efforts, we will develop the skills and techniques necessary to sustain further space exploration. To meet this goal, we will return the Space Shuttle to flight as soon as possible, consistent with safety concerns and the recommendations of the Columbia Accident Investigation Board. The Shuttle's chief purpose over the next several years will be to help finish assembly of the International Space Station. In 2010, the Space Shuttle, after nearly 30 years of duty, will be retired from service ...
>
> Our second goal is to develop and test a new spacecraft, the Crew Exploration Vehicle, by 2008, and to conduct the first manned mission no later than 2014. The Crew Exploration Vehicle will be capable of ferrying astronauts and scientists to

the Space Station after the Shuttle is retired. But the main purpose of this spacecraft will be to carry astronauts beyond our orbit to other worlds. This will be the first spacecraft of its kind since the Apollo Command Module ...

Our third goal is to return to the Moon by 2020, as the launching point for missions beyond ... Using the Crew Exploration Vehicle, we will undertake extended human missions to the Moon as early as 2015, with the goal of living and working there for increasingly extended periods of time ... With the experience and knowledge gained on the Moon, we will then be ready to take the next steps of space exploration: human missions to Mars and to worlds beyond ...

We do not know where this journey will end. Yet we know this: human beings are headed into the cosmos ... The vision I outline today is a journey, not a race. And I call on other nations to join this journey, in the spirit of co-operation and friendship ...

Achieving these goals requires a long-term commitment. NASA's current five-year budget is $86 billion. Most of the funding we need for new endeavours will come from re-allocating $11 billion from within that budget. We need some new resources, however. I will call upon Congress to increase NASA's budget by roughly a billion dollars spread over the next five years. This increase, along with the re-focusing of our space agency, is a solid beginning to meet the challenges and the goals that we set today. This is only the beginning. Future funding decisions will be guided by the progress that we make in achieving these goals. We begin this journey knowing that space travel brings great risks. The loss of the Space Shuttle Columbia was less than one year ago ...

Mankind is drawn to the heavens for the same reasons we were once drawn into unknown lands and across the open sea. We choose to explore space because doing so improves our lives and lifts our national spirit.

So let us continue the journey."

The President's new plan left NASA with a dilemma. In the wake of the STS-107 tragedy the remaining three Shuttle orbiters had begun a rolling programme of 2.5-year refits to update their flight systems. Endeavour had begun her refit in July 2003. Discovery would be next, and Atlantis last. However, with the new plan to retire the Shuttle in 2010, there was no longer sufficient time for Atlantis to complete its 2.5-year refit. Therefore, NASA decided to only refit Endeavour and Discovery. Atlantis would now be retired in 2008, and held in the Orbiter Processing Facility at Kennedy Space Centre, where she would be cannibalised and her systems would be used to keep the other two orbiters flying. The decision would leave only two orbiters to complete the construction of ISS.

Within a matter of days the true cost of the President's "Vision for Space Exploration" and the new Crew Exploration Vehicle, which quickly assumed the name "Constellation", became clear. NASA cancelled a Shuttle service mission to the Hubble Space Telescope, although this was later re-instated, and a plan to build a new science centre to support ISS was "postponed", by at least one year.

Meanwhile, it was estimated that meeting all of the CAIB's recommendations and returning the Shuttle to flight would cost a further $280 million through 2004, but that figure was expected to rise as details of the true amount of work required to do so was identified. NASA's figures were

Launch system modifications	$280 million
External Tank modifications	$ 65 million
Additional launch camera coverage	$ 40 million
Thermal Protection System repair kit	$ 57 million
New Engineering Safety Centre	$ 45 million

Consideration was also being given to using an uncrewed Shuttle derivate to carry large amounts of cargo to ISS, alongside the proposed new Orbital Space Plane (OSP), which would have been launched on one of the new Extended Expendable Launch Vehicles (EELVs): Delta-IV or Atlas-V. The Orbital Space Plane would now be replaced by the Crew Exploration Vehicle.

While the press and media sunk their teeth into the contents of the President's speech and produced articles depending on their political points of view, the station's Expedition-8 crew continued with the routine of keeping ISS functioning. Much of the next week was spent loading rubbish into Progress M-48. Kaleri also replaced an electronics box on the Elektron oxygen generator. When the crew reported a noisy air filter on the Elektron system a replacement was added to the Progress M1-11 manifest, along with a replacement vent hose for Destiny's window. Progress M-48 was undocked from Zvezda's wake at 03:36, January 28, and was subsequently commanded to enter the atmosphere, where it burned up.

As the first anniversary of the loss of STS-107 approached, Wayne Hale, Deputy Shuttle Program Manager, sent an e-mail to staff at JSC. It read, in part:

> "Last year we dropped the torch through our complacency, our arrogance, self-assurance, sheer stupidity, and through our continuing attempt to please everyone. Seven of our friends and colleagues paid the ultimate price for our failure. Yet the nation is giving us another chance ... We must not fail ... The penalty is heavy; you can never completely repay it ... Do good work. Pay attention. Question everything. Be thorough. Don't end up with regrets."

PROGRESS M1-11

In the immediate aftermath of the loss of STS-107 in February 2003, NASA had asked the Russians to launch an additional, fourth, Progress in 2003. Despite having requested the extra launch, NASA was unable to offer any payment in support of it, due to the terms of the Iran Non-Proliferation Act, which forbade American companies and government departments from spending their money in countries that supported Iran's attempts to develop nuclear power and, through that programme, nuclear weapons. The launch of the extra Progress had been delayed from November 2003 to January 29, 2004.

On that date Progress M1-11 was launched at 06:58, and followed a standard two-day rendezvous trajectory, before docking to Zvezda's wake at 08:13, January 31. The Expedition-8 crew spent the next week unpacking the new arrival. Following a test-firing of the Progress' thrusters, the crew observed a small strip of material drifting away from the station, but noted that it did not appear to represent any danger to ISS. Two weeks later the object was identified as a bolt and washer used to hold the photovoltaic arrays on Progress M1-11 in their folded position during launch. They served no purpose after the spacecraft's arrays had been deployed. During the week ending January 31, Foale also initiated the ESA PROMISS-3 cell culture growth experiment that had been carried into space on Progress M1-11. It would be run for 30 days and the results would be returned to Earth on Soyuz TMA-3.

February began with a period of silence to mark the first anniversary of the loss of STS-107, and the unloading of Progress M1-11. The crew also began a highly toxic Japanese experiment called the Granada Crystallisation Facility. Kaleri commenced the Russian Mimtec-K protein crystal growth experiment. He also deployed the Russian Brazdoz radiation detectors inside Zvezda. On February 4, Kaleri began repairs of the Elektron oxygen generator and the Vozdukh carbon dioxide remover. A fan in the latter had begun making a loud noise and would be replaced during the following week. The international experiment programme continued throughout the next week alongside routine maintenance and exercise. During the second week of the month the crew completed the ongoing regular collection of air and swab samples from surfaces inside ISS. Progress M1-11 had been emptied by February 11, and the crew began loading the spacecraft with unwanted items. Elektron failed during the day, but Kaleri had it working again within 24 hours. It failed once more on February 16, and it took 3 days of efforts by Kaleri, with Korolev's assistance, to put it back in action. It failed again the following day and the station's atmosphere was topped up using oxygen from Progress M1-11.

On February 16, Korolev released details of the problem with the Soyuz TMA-3 spacecraft. It was

> "... a minute pressure decay in the two helium systems that pressurise the Soyuz propellant tanks and lines ... The pressure decay was first noted on system-2 when the Soyuz arrived at the station in October, and was confirmed on system-1 during a routine thruster test ... Flight directors have concluded that the decay poses no concern. The decay was extremely small and there are no plans to change normal entry and landing procedures."

As they approached 4 months (February 18) in their occupation, Foale and Kaleri commenced early preparations for their only Stage EVA, planned for February 26. They shifted their sleep schedule to accommodate the start of the EVA. Foale also positioned the SSRMS so that its cameras and lights could offer maximum coverage of their external activities. The EVA would take place from Pirs, with both men wearing Orlan pressure suits. It would be the first EVA from ISS where all crew members were outside of the station, and no one was inside monitoring the station's systems and choreographing the EVA against the flight plan, or operating the SSRMS to provide the best possible video coverage. Although NASA was originally unhappy about this arrangement, the Soviets/Russians had used it throughout their Salyut and Mir space station programmes.

The following week was spent preparing for the planned EVA. Both men unpacked their Orlan EVA suits and checked them over, working closely with controllers at Korolev. They also carried out another, successful, exercise that proved they could move between Pirs and Soyuz TMA-3 while wearing Orlan pressure suits, in the event that they had to abandon ISS as a result of an unexpected event occurring during the EVA, which prevented them re-entering the station.

After configuring the station for a period of non-occupation, Foale and Kaleri locked themselves inside Pirs. At 16:17, February 26, they opened the hatch and made their way into open space, leaving ISS unoccupied. They collected their tools, secured their tethers, and made their way to the exterior of Zvezda. There, they replaced one of two cassettes of long-duration material exposure samples that were part of the Japanese Micro-Particle Capture and Space Environment Exposure Devices (MPAC/SEEDs) experiment that had been put in place, to measure micrometeoroid impacts, in October 2001. Next they installed the Russian "Matryoshka" experiment on handrails on Zvezda's exterior. The experiment housed simulated human tissue samples, which would be used to study radiation absorption. As they completed this work Kaleri reported water droplets forming on the inside of his visor and a rise in the temperature inside his suit. It was approximately 18:00. A few minutes later, engineers at Korolev reported problems with the cooling system in Kaleri's suit, leading to a build-up of condensation on the inside of his helmet.

Kaleri returned to Pirs, while Foale replaced one of two cassettes of long-duration exposure material samples on Zvezda's wake airlock housing before joining his colleague in Pirs. The hatch was closed at 20:12, after an EVA lasting 3 hours 55 minutes. After pressurising the airlock Foale climbed out of his suit, so that he could carry out an inspection of Kaleri's suit. During the inspection he discovered a kink in a cooling water tube. When the kink was straightened out the water began flowing freely once more. The two men re-entered ISS, which had performed flawlessly in autonomous mode throughout the EVA. American fears of leaving the station unoccupied had proved groundless. The week following the EVA began with light duties, but the crew also worked on their experiments and routine housekeeping tasks.

Progress M1-11's thrusters were used to raise the station's altitude on March 2. During preparations for the burn a momentary spike was observed in the electrical current reaching CMG-3. After the spike all readings returned to normal and CMG-3 continued to operate as planned. Foale had prepared Destiny's window for the replacement of the jumper hose that had been identified as the source of the pressure leak earlier in the month. The gaps between the various panes of glass in the window were vented over the weekend to disperse condensation that had built up. On Friday March 5, a vacuum was re-established between the panes of glass in the window in advance of replacing the jumper hose. Once installed the new jumper hose was fitted with a new cover, to prevent it being knocked accidentally.

The week ending March 12 began with a three-day weekend off, after which the crew spent three days working closely with Houston to disassemble their exercise treadmill and remove the gyroscope that had malfunctioned in November 2003. They then replaced a bearing within the gyroscope before replacing the gyroscope and reassembling the treadmill. Houston monitored the first few days of treadmill use with the Vibration Isolation System re-activated, to ensure it was functioning correctly. NASA reported:

> "The crew heard noises coming from the treadmill in November, which engineers determined was a failed bearing in the gyroscope that stabilises movement in the roll direction. A repair kit was sent to the station in January [on Progress M1-11] ... After this week's repair work, Foale reported the noises had stopped."

Kaleri continued to work on the Elektron oxygen generator in Zvezda. On March 12, he began a total review of the system, to identify what items needed replacing. Following a final replenishment of oxygen from Progress M1-11, two Russian SFOG oxygen candles were burned each day in Zvezda, commencing on March 13, to supplement the oxygen supply while the Elektron repairs continued. At the end of the week Foale performed further activities while wearing the FOOT experiment to record how he used his limbs in microgravity. He also used a computer training package to familiarise himself with the Advanced Diagnostic Ultrasound in Microgravity (ADUM) experiment. Later in the flight, Foale would use the experiment to make an ultrasound examination of Kaleri. On March 14, CMG-2 in the Z-1 Truss fell off-line for 2 minutes before the emergency system restored it to correct operation. Similar failures and recoveries would occur over the next several days.

The following week Foale and Kaleri spent two days replacing a liquids unit and a water flow system in the Elektron oxygen generator. Following earlier repairs, Russian engineers had decided that air bubbles in the liquids unit had repeatedly caused the Elektron unit to shut down after only a few minutes of operation. The work on Elektron necessitated the re-scheduling of other, lower priority tasks. Following the repair the Elektron unit was activated on Saturday March 20, and performed well. It was left running and the crew stopped burning SFOGs. NASA was at pains to point out that over 100 SFOG candles remained in storage on ISS, along with two tanks of high-pressure oxygen in the Quest airlock, which could supply the crew with breathable oxygen for several months if required. On March 22 and 23, both men participated in noise level measurements. The working week ended on March 26, when Foale carried out a regular inspection of one of the two American EMUs held on the station.

As March turned to April the Expedition-8 crew began their final month onboard ISS. They completed an initial maintenance of the two Russian Orlan suits delivered to the station on Progress M1-11 in January, to replace three older Orlan suits that had been used by earlier Expedition crews. Foale completed a final session of training on the SSRMS on April 2. He used the arm's cameras to complete an external inspection of the station. He also noted that the noise that the crew had reported earlier in the flight was repeated each time he commanded Destiny's external camera to pan up and down. In Russia, controllers were considering having the crew replace the fan in the Soyuz TMA-3 descent module that had failed during the journey up to the station in October. The fan assisted in maintaining the humidity inside the cabin.

American controllers successfully completed tests of the software that would control the Thermal Rotary Radiator Joints on the ITS when its installation continued, following the Shuttle's Return to Flight. The software would be used to automatically position the cooling radiators mounted on the Truss once the station's cooling system was activated, following the flight of STS-116.

Foale spent the last two weeks of his occupation completing the FOOT and PFMI experiments, and Kaleri changed a set of two ventilation and humidity fans in Soyuz TMA-3. They also began preparing the items that they would take back to Earth with them. During the week ending April 16, both men participated in a test of Soyuz TMA-3's thrusters, during which controllers in Korolev noticed the same helium leak that had been monitored during the rendezvous with ISS. Korolev conducted additional tests of the helium system, which was used to pressurise the spacecraft's propellant tanks, to time the leak rate. The crew also completed environmental sampling at various points inside ISS for return to Earth for investigation. Foale also set up the ESA HEAT experiment in the MSG in anticipation of the arrival of Soyuz TMA-4 with Dutch astronaut André Kuipers, who would use the experiment to see if a grooved heat pipe can be used to transfer heat from hot surfaces, such as electronics, to cold surfaces, such as radiators, in microgravity.

SOYUZ TMA-4 DELIVERS THE EXPEDITION-9 CREW

SOYUZ TMA-4	
COMMANDER	Gennady Padalka
FLIGHT ENGINEER	Michael Fincke
ENGINEER	André Kuipers (Holland)

The Soyuz TMA-4/Expedition-9 crew was originally named as Commander Valeri Tokarev and Flight Engineer William McArthur. On January 12, 2004 NASA announced that McArthur had been removed from the crew for "unspecified medical reasons" and replaced by Leroy Chiao. Tokarev and Chiao subsequently proved incompatible as a crew, and in February 2004 they were replaced by the original Soyuz TMA-5 prime crew of Gennady Padalka and Michael Fincke, the original Expedition-10 crew. André Kuipers, a Dutch ESA astronaut, would continue to fly to ISS with the new crew. By the time the crew change was announced McArthur had already recovered his health, but he and Tokarev had lost too much time training, so they were named as the new Soyuz TMA-7/Expedition-12 crew.

Soyuz TMA-4 was launched at 23:19, April 18, 2004. Docking with the nadir port on Zarya occurred at 01:01, April 21, and following pressure checks the hatches were opened at 14:00. The three newcomers entered the station and were given a safety briefing. Padalka and Fincke would spend 6 months on the station as the Expedition-9 crew and make two Stage EVAs from Pirs. Kuipers, the second Dutch national to fly in space, would return to Earth with the Expedition-8 crew, after spending 9 days in flight. While the Expedition-9 crew began their familiarisation and hand-over period, the Expedition-8 crew exercised rigorously prior to their return to Earth. After his recovery, Kuipers would describe his feelings:

> "It was amazing, better than I had ever expected. All the different colours of Earth; such brilliance ... Gliding past huge clouds, and vast expanses of water. You see cities from above and lightning on top of clouds instead of underneath. At times like that it hits you that you really are in space."

Kuipers had begun his Dutch Expedition for Life Siences, Technology and Atmospheric Research (DELTA) experiment programme during the solo portion of Soyuz TMA-4's flight, performing a package of 21 ESA experiments. He commenced his programme on ISS almost as soon as the hatches were open between the two spacecraft. On April 21, two runs of the HEAT experiment in the MGS were terminated automatically when the upper temperature limit was reached, and a third

Figure 44. Expedition-9: ESA astronaut, Dutchman André Kuipers participated in every astronaut's favourite off-duty pastime, watching the Earth turn outside the window, during his short stay on ISS. Arriving with the Expedition-9 crew, he returned to Earth with the Expedition-8 crew.

run of the experiment was cancelled. After troubleshooting of the experimental hardware the following day, a further four runs of shorter duration were completed before the experiment was stored in Zvezda. Kuipers suffered continual problems with some of his experiments' hardware, including a centrifuge. Throughout the remainder of the week the Dutchman became a human subject for a number of microgravity and life science experiments. He also assisted Foale and Kaleri to prepare Soyuz TMA-3 for their return to Earth.

On April 21 the Expedition-9 crew were informed that CMG-2 had gone off-line at 16:20, after the RPCM, a circuit breaker mounted on the S-0 ITS, had malfunctioned and cut power to the CMG. CMG-1 and CMG-2 continued to perform flawlessly and were sufficient to control the station's attitude. Controllers in Houston began planning an unscheduled third Stage EVA, on June 10, to replace CMG-2 with a CMG held in storage on the station.

The official hand-over ceremony between the Expedition-8 and Expedition-9 crews took place in Destiny on April 26. On the same day Foale, Kaleri, and Kuipers spent several hours in Soyuz TMA-3, rehearsing their undocking and re-entry procedures. Two days later propellant was transferred from Progress M1-11 while the crew slept.

After a week-long hand-over, Foale, Kaleri, and Kuipers said farewell to Padalka and Fincke and sealed themselves in Soyuz TMA-3, preparing to return to Earth. Soyuz TMA-3 undocked at 16:52, April 28, retrofire occurred at 18:20, with no difficulties, and the re-entry module landed at 20:12 the same day. The Expedition-8 crew had spent 194 days 18 hours 35 minutes in space. They were just one day behind the Expedition-4 crew. Foale's personal accumulative record stood at 374 days 11 hours 19 minutes, including his Shuttle flights and his stay on Mir. He was the first American astronaut to spend a total of over one year in space. All three men were in good health and excellent spirits following their landing. As usual, the returning Expedition crew members underwent a 45-day rehabilitation programme.

Meanwhile, NASA admitted that they would be unlikely to meet the March 6, 2005 target date for STS-114, the Return to Flight mission. One of the limiting factors remained the development of the OBSS. The OBSS would give access to most of the Shuttle's TPS, but not all of it. It would also be strong enough to support an EVA astronaut if any repairs had to be made to a Shuttle's damaged TPS. It was not even guaranteed that STS-114 would carry the OBSS. If it did not, then the TPS inspection would have to be made by astronauts on an EVA to inspect those areas of the orbiter that could not be seen from the flight deck windows, or with the cameras on the Shuttle's standard RMS.

EXPEDITION-9

Padalka and Fincke began their solo occupation of ISS with three days of light duties to help them get over the hectic hand-over week that had just passed. Padalka described his goals for the occupation as follows:

> "The biggest goal for us as a crew is to keep the Space Station in operational condition and maintain the human presence aboard the Space Station because [the] last malfunction—I mean the situation with the leakage—showed us that if we had not had crew on board, we could have lost the Space Station."

He added:

> "We have the West science program, and currently about 40 experiments on behalf of the American side and the same number on behalf of the Russian side, are scheduled for us, and about 20 experiments for the European Space Agency ... If we conduct all of the science program, if we can keep Space Station in operational condition, if we manage to perform scheduled spacewalks, if we keep our friendship with Mike, if we hand Space Station in operational condition to the next crew, in this case I would say that our mission was successful."

The first week of May was spent carrying out a programme of Russian, American, and European biomedical experiments to determine how their bodies adapted to microgravity. They also completed maintenance on the battery chargers and the batteries in the American EMUs and the station's EVA tools, as well as beginning a procedure to regenerate the canisters that would remove CO_2 in their exhaled breath from the EMUs. The EMUs had not been used for over a year. A second

Figure 45. Expedition-9: Gennady Padalka and E. Michael Fincke pose in the Pirs airlock with two Russian Orlan extravehicular activity suits.

series of battery recharging took place at the beginning of the next week, while Fincke serviced the water-cooled underwear that would be worn under the EMUs. Both men also spent part of the week loading rubbish, including an Orlan EVA suit that was past its maintainable date, into Progress M1-11, which was scheduled for undocking on May 24. During the week American flight controllers up-linked new software to two multiplexer/demultiplexers and two S-0 ITS MDM computers, as part of a major programme to update the software on the station throughout 2004. The crew also installed the EarthKam in one of Zvezda's windows on May 11, and performed biomedical experiments.

The Expedition-9 crew's first month on ISS ended with a week of preparations for the Stage EVA to replace the CMG that had lost electrical power in April. The EVA was originally planned to take place on June 10, with both men wearing American EMUs and egressing from Quest. That changed on May 19, when they began a planned 7-hour check of the two pressure suits. Shortly after beginning the checks, Padalka reported that he had no circulation of cooling water in his EMU. Closer inspection revealed that the water held in the suit was bubbling and frothing. During the same tests Fincke's EMU also suffered difficulties with its water-cooling system, caused by a sticking valve. On May 21 the water tank in Padalka's EMU was drained and refilled. America admitted that only one of the three EMUs on ISS was operational. As a result, the EVA was delayed until June 16, and on that date both men would wear Russian Orlan suits and egress from Pirs. Russia would negotiate "compensation" for the use of their suits and consumables by having American astronauts spend additional hours working on Russian ISS experiments. The new plans complicated the EVA in that the astronauts would now exit Pirs and use one of the two Russian Strela cranes to reach the required area on the S-0 ITS, which was not as readily accessed from Pirs as from Quest. The two ISS Strela cranes had not been human-rated at that time, and the Russians were concerned for their capabilities. Also, the two men would be under Russian control while they exited the station and used the Strela to reach the junction between the Russian and American segments of ISS; once they were on the outside of the American sector and making their way to the S-0 ITS they would pass to American control. They would return to Russian control as they returned to the exterior of the Russian segment for their return to Pirs.

During the week, Progress M1-11's thrusters were fired to raise the station's altitude in preparation for the launch of Progress M-49. The burn also corrected the station's orbital inclination, as NASA explained:

> "Since the last inclination correction, about 3 years ago, [the] orbit inclination has decreased by approximately 0.01 degree, to a current value roughly mid-way between the ground rule range of 51.62 and 51.68 degrees. The lower limit would be approached in approximately 5 years, but doing part of the upward correction right now propellant is more efficiently used. Once the inclination is adjusted to the upper limit, no further inclination adjustments should be required for the remaining life of the ISS."

Progress M1-11 undocked from ISS at 05:19, May 24, manoeuvred clear of the station, and entered a parking orbit. For the next 10 days the spacecraft was monitored to see if future Progress vehicles could be used to support microgravity experiments, although no experiments were conducted on this occasion. Progress M1-11 was commanded to re-enter the atmosphere and burn up on June 3, 2004.

PROGRESS M-49

Progress M-49 lifted off from Baikonur Cosmodrome at 09:34, May 25, 2004, carrying 2,566 kg of food, water, propellant, and equipment for the two men on the station. The spacecraft completed an automatic docking to Zvezda's wake at 09:55, May 27. Following pressure and leak checks the crew spent the next few days unloading the new vehicle, beginning on May 28.

RUSSIA CALLS FOR 12-MONTH EXPEDITION FLIGHTS

In Moscow, Russia was pressing for the Expedition-10 crew to spend 12-months in orbit. While NASA was not prepared to consider such a mission all the time the ISS programme was flying with two-man caretaker crews and relying on the Soyuz TMA for return to Earth, the Russians were adamant that the time had come to advance the Expedition crews' stay time, and increase the amount of science performed by each crew. Yuri Semenov, director of Energia stated:

> "Our position is rigid—the next crew [Expedition-10] must make a long flight. I would urge our American colleagues not to drag their feet on solving this issue. We are ready for long flights. Our equipment is ready and our partners [the Americans] must listen to their Russian colleagues ... Russia is keeping the station running while the USA and Japan are cutting down their budgets. This cannot last for long because Russia has had to freeze the construction of later ISS hardware and stop selling trips to rich tourists in order to mobilise its resources to keep the ISS afloat."

The Russians were careful not to mention that 12-month Expedition crews would leave two seats available for sale to paying passengers on intervening Soyuz TMA "taxi" flights.

NASA replied to Russia's request that "The time is not right." They explained that the Expedition-10 crew would not stay in space for 12 months, but that the Expedition-11 crew might do so in 2005; but only if the Shuttle was flying again by then. In July, ESA managers also began to call for a six-person Expedition crew. They suggested that this could be achieved by using two Soyuz CRVs and having the three extra astronauts finding sleeping accommodation wherever they could. They did not address the question of who would pay for the extra Soyuz spacecraft, but did criticise America for cancelling the American Habitation Module and the X-38 CRV. Meanwhile, NASA had also failed to act on the Young Committee recommendation that two Expedition crews could work side by side on the station for one month, rather than just one week during hand-over periods.

On June 3, Padalka and Fincke began preparation for the EVA that had been rescheduled after the May 19 EMU checks. The EVA was tentatively planned for no earlier than June 15. During the first week of June, their sixth week on ISS, they also performed experiments and routine housekeeping. On June 10, programme managers scheduled the EVA for June 24. The move placed the EVA at a better time in the crew's workday, optimised Russian communications coverage, and offered additional time for preparation. The crew carried out training with the tools that they would use and prepared their Orlan suits; they also performed cardiovascular evaluation sessions on the fixed bicycle in Zvezda. Much of the rest of the first half of the month was spent unloading Progress M-49 and performing their experiment programmes in Destiny and Zvezda.

Fincke's wife, Renita, gave birth to their second child on June 18. The astronaut took the opportunity to remind the world that many men and women in the service of their country were also forced to miss similar important family occasions. Meanwhile, EVA preparations, including fitting American EMU helmet lights on to the helmets of their Russian Orlan suits, continued alongside mass measurement checks and software replacement in three racks of experiments in Destiny.

In his pre-launch interview, Fincke had described how he felt about the two planned Stage EVAs during his occupation of ISS:

> "[O]nly recently did I actually start to think how really exciting that is, to be alone in the cosmos without a spacecraft around me except for this suit that was put together by human hands. It's made out of material and a little bit of metal and a lot of plastic, and yet we'll be able to look out there on our planet below and the stars in the sky and really experience a true spaceflight. [I]t's an honour as a rookie to get a chance to perform two spacewalks, and it's an honour to be able to fly on a Russian Soyuz spacecraft, and to work in a Russian spacesuit. My instructors spent a lot of time with me, and I'm glad I've earned their confidence in the U.S. and Russia to get a chance to do that."

The EVA to repair the RPCM finally began at 17:56, June 24, 2004. Following depressurisation of the Pirs airlock they opened the hatch and made their way outside. Almost immediately, Russian flight controllers noticed that the primary oxygen bottle in Fincke's PLSS was losing pressure much faster than expected. The two men were ordered to return to the airlock and terminate the EVA. The hatch was closed and the EVA ended at 18:10, after just 14 minutes 22 seconds. NASA announced, "The overall pressure in Fincke's suit remained stable at all times and he was not in danger."

Following immediate troubleshooting, the astronauts were instructed to remove their pressure suits, return to the station, and reconfigure it for normal use. Although the Russian controllers were not able to immediately identify the cause of the problem, Fincke thanked both control teams for being alert and noticing the problem so quickly. He informed Korolev that the two of them would sleep late the following morning and then resume their normal work/sleep routine until it was time to prepare for a second attempt at the EVA. Programme managers rescheduled the EVA for no earlier than June 29, as dictated by Russian communication coverage. The problem with Fincke's suit was identified as an injector switch that controlled the flow of oxygen.

Although investigations would continue for the next few days, the crew were informed that they had followed the correct procedures when preparing for the EVA and could expect to wear the same Orlan suits when they completed the rescheduled EVA. On June 29, the EVA was rescheduled again, for the following day. On the same date, Russian engineers confirmed that the selector switch in Fincke's suit had not seated properly when it was set. Energia told the media, "This valve has a particular design feature—whilst it is being closed, one must make sure not only that the signal light goes out, but also that the handle has been locked." New procedures were put in place to confirm the switch's seating when preparing for future EVAs.

The second attempt to repair the RPCM began at 17:19, June 30, 20 minutes ahead of schedule, when the two men left Pirs. Padalka described the scene outside as "Dark, but very beautiful." They moved over to the Strela crane, Padalka turned the hand crank to extend it to its full 15 m length, and Fincke made his way along to the end of it. Padalka manoeuvred Finke to a position where he secured the crane to the handrails on Zarya's wake. Padalka then made his way along the Strela to Fincke's position before they both transferred to the handrails on the exterior of Unity, at 18:09. At that time, control of the EVA passed from Korolev to Houston and the two men stopped speaking Russian and began speaking English. Controllers in Houston guided them to their work position on the S-0 ITS, where, by 18:52, they had completed their tasks to replace the RPCM. Fifteen minutes later, word was passed to the astronauts that power was flowing to CMG-2 once more and that it was spinning at 30 rpm. The two men collected their tools and made their way back to the Strela crane, where control of the EVA was handed back to Korolev, at 21:11, and they began speaking Russian once more. Having traversed the Strela crane, Padalka cranked the telescopic crane back to its stowed position, bringing Fincke back to the exterior of Pirs. They also completed get-ahead tasks in preparation for later EVAs, when they installed two flexible handrails, mounted a contamination monitor to measure station thruster exhaust, and added end caps to two handrails on the exterior of Pirs. Having entered the airlock they closed the hatch at 22:59, after an EVA lasting 5 hours 40 minutes. CMG-2 was powered up to verify its full 6,600 rpm, at 14:30, July 1. After performing tests overnight, the CMG was returned to its role of helping the two working CMGs to control the ISS's attitude, at 07:20, July 2. The fourth CMG remained off-line. The two astronauts spent July 2 tidying up after the EVA, after which they had a three-day weekend off, to celebrate July 4, American Independence Day. Meanwhile, controllers in Korolev pumped air from the tanks in Progress M-49 into the station's atmosphere.

Back at work the crew concentrated their efforts on their experiments and general maintenance of the station. Padalka used the new ultrasound experiment to examine Fincke, and thereby demonstrated a capability to transmit medical data to a flight surgeon on the ground in real time. Meanwhile, Fincke continued to troubleshoot the cooling systems in the two American EMUs. The problems experienced prior to the last EVA were traced to pumps in the cooling system. New pumps would be sent up to the station on the next Progress, due for launch in August. Throughout the third week of July the crew performed more experiments, studying their cardiovascular systems and fluid motion in microgravity. They also spent the week loading rubbish in to Progress M-49 and removing the KURS automatic docking system. Both men donned their Sokol launch and re-entry suits and made their way into Soyuz TMA-4 for fit-checks in their couches, on July 7. A full fire drill at the end of the week was followed by a round of taking air and swab samples around the station. Two false activations of the station's fire alarms led to the crew cleaning the fire alarms. During his weekend off Padalka completed sessions with the Russian Pulse medical experiment and the ESA Eye Tracking Device experiment. The following week saw him beginning a new round of Russian biomedical experiments. Fincke worked on American experiments in Destiny.

On July 16, Russian controllers at Korolev made an unsuccessful attempt to upload software into computers in Zvezda. The software was designed to support the rendezvous and docking of the ESA-developed Automated Transfer Vehicle (ATV) to Zvezda's wake. The ATV would be launched by Ariane-V from the ESA launch site in French Guiana. It would carry 2.5 times the payload that a Progress could carry, including propellants, water, oxygen, and nitrogen. It would also be able to re-boost the station's orbit. At that time the ATV was due to make its maiden flight in 2005 and then operate alongside Progress. During the day, Padalka replaced a pump in Zvezda's cooling system, which had failed two days earlier. The back-up cooling system had continued to work properly throughout the malfunction and its repair.

Padalka and Fincke passed the halfway point of their mission on July 19. Fincke spent the day removing the water pump from one of the failed EMUs. Two spare water pumps would be launched on Progress M-50. The following day, a computer failed on the station's starboard thermal radiator, but had no impact on operations, as the radiator was not active. On July 21, they both participated in celebrations of the 35th anniversary of the Apollo-11 Moon landing. The crew spent the week loading rubbish into Progress M-49. On July 23, they began preparations for their second EVA, before they manoeuvred the SSRMS to a position where it could video the EVA planned for August 3.

The leaders of the international space agencies involved in ISS met at Noordwijk, Holland, on July 23. They reviewed the status of ISS operations, and the final configuration of ISS at the end of the decade. The group reaffirmed their commitment to international co-operation and to the completion of the station's construction and operation. They also committed their agencies to continually review launch schedules and opportunities to accelerate that schedule. American and Russian representatives renewed their commitment to continued occupation of ISS throughout its construction phase. The representatives recognised America's efforts to return the Shuttle to flight status and Russia's commitment to maintaining access to the station and to its re-supply. They also discussed access to the station after Energia's contract to supply Soyuz and Progress spacecraft came to an end. At the beginning of the ISS programme Russia had agreed to supply 11 Soyuz spacecraft free of charge. The last of those spacecraft was due to be launched in October 2005, and recovered in April 2006. The Russians made it clear that they would require payment for their Soyuz spacecraft when a new contract was negotiated. This presented a major problem, in that the Iran Non-proliferation Act banned NASA from paying money directly to Russia. Congress put the Act in place, in fear that President Clinton's relationship with Russia's President Putin was allowing him to overlook Russia's part in the Iranian nuclear programme. Although the Clinton White House fought the new Act, it was voted into law and forced the President to operate sanctions against any country that helped Iran's programme. Following the loss of STS-107 a Democratic Party proposal to change the Act, to allow additional Soyuz spacecraft to be purchased from Russia, had received only 3 signatures in 17 months. Meanwhile, the prospect arose that, after April 2006, America might only be able to fly short-duration Shuttle flights to ISS because they were unable to pay for American astronauts' places on the Soyuz CRV docked to the station for return to Earth in the event of an emergency. In such an event all American astronauts would have to be launched to and returned from ISS by Shuttle. A further problem also arose if the Constellation spacecraft were not ready to start crewed orbital flight before the Shuttle was retired in 2010.

Progress M-49 undocked from ISS at 02:05, July 30. Fincke filmed the spacecraft as it departed. The station's cameras also recorded the Progress re-entry. The Expedition-10 crew spent the remainder of the week preparing for their third EVA. Fincke also spent part of the week completing soldering experiments in microgravity.

Padalka and Fincke wore Orlan suits when they exited Pirs, at 02:58, August 3, and made their way to Zvezda's wake. The EVA was intended to last up to 6 hours. Their first task was to replace the SKK material exposure experiment with a new container full of fresh samples. They also replaced the Kromka experiment, which measured the contamination from the thrusters on Zvezda. Next, the two men made their way onto the wake face of Zvezda, where they installed two antennae and replaced three laser reflectors with more advanced models. The three old reflectors were replaced with a single three-dimentional reflector. All of these items would be used to support the rendezvous and docking of ESA's ATVs. While they worked at the rear of Zvezda the three CMGs that controlled the station's orientation approached saturation level. This condition was anticipated, and the station was placed into free drift. As a result, S-band communication was lost as the antennae drifted. At 05:15, the astronauts were 40 minutes ahead of their flight plan, and were instructed to leave the area. The CMG resumed attitude control at 06:00, and the two men were allowed to return to the area at the rear of Zvezda. Finally, they disconnected a cable on a malfunctioning camera that would be collected on a later EVA, before removing the Platan-M materials exposure experiment. The crew returned to Pirs and closed the hatch at 07:28, after an EVA lasting 4 hours 30 minutes.

The following day NASA pumped additional nitrogen from one of the two high-pressure tanks on the exterior of Quest into the station's atmosphere. The week ended with the crew performing their experiment programmes.

PROGRESS M-50

The Expedition-9 crew was asleep and passing southwest of Baikonur when Progress M-50 lifted off at 01:03, August 11, 2004. The cargo ship, carrying pumps for the two malfunctioning EMUs, clothes for the Expedition-10 crew, and propellant, air, and water; a total of 2,542 kg of cargo. After a standard two-day rendezvous, Progress M50 docked to Zvezda's wake docking port at 01:01, August 14. The crew began unloading the cargo the following day.

Using items delivered on the new Progress, Fincke took 4 hours to replace the water pump in the cooling system of one of the malfunctioning Extravehicular Mobility Unit, before turning his hand to repairing an exercise machine. The following day, the EMU was subjected to several more hours of testing, during which it performed perfectly. Two of three American EMUs on the station were now functioning correctly. The third suit, which sufferred a similar problem in its cooling system was left to later. The crew spent most of the remainder of the week preparing for their fourth EVA. On August 25 the thrusters on Progress M-50 were fired to raise the station's orbit, in preparation for the arrival of Soyuz TMA-5, due for launch in October. Throughout the period the Elektron unit in Zvezda failed, on average, once every three days. NASA referred to the Elektron as "a major source of trouble".

HURRICANE FRANCIS VISITS CAPE CANAVERAL

Fincke had reported photographing Hurricane Francis over the Atlantic Ocean on August 27. The hurricane passed over Kennedy Space Centre on September 7, with winds of 70 mph. Those winds pulled approximately 820 panels off of the side of the VAB along with the insulation beneath them, leaving the building interior open to the weather. The building's roof proved to be weakened when it was inspected after the storm and nets were hung inside the building to catch any falling debris until the roof could be repaired. The two Shuttle ETs and various SRB components inside the VAB were not damaged. The roof was also partially ripped off a building used to prepare heatshield tiles for the Shuttle, but the three remaining Shuttle orbiters were secured within the three Orbiter Processing Facility Buildings and were not damaged. Only one month earlier, Hurricane Charley had caused $700,000 worth of damage to Cape Canaveral, and Hurricane Ivan, one of the most powerful hurricanes on record, also threatened to add to the damage at the site, until it changed course and missed Florida.

The highlight of September 3 came when Padalka and Fincke donned their Orlan pressure suits and depressurised Pirs for their final EVA. Egress occurred at 12:43. Having gathered their tools, the two men made their way to the exterior of Zvezda, where they replaced a flow regulator valve panel and installed three communication antennae at the station's wake. The antennae would be used during rendezvous and docking of the European ATV. Fincke made his way across the exterior of Zvezda to photograph the Japanese MPAC/SEEDS experiment. Upon returning to Pirs they

Figure 46. Expedition-9: Gennady Padalka wears a Russian Orlan suit during the Expedition-9 crew's third two-man extravehicular activity.

Figure 47. Expedition-9: Hurricane Francis was observed from ISS during Expedition-9.

installed covers on the handrails around the airlock hatch to prevent EVA astronauts' tethers becoming ensnared during future EVAs. Pirs' hatch was closed at 18:04, after and EVA lasting 5 hours 21 minutes.

The Elektron unit failed during the night of September 6–7. The new failure involved the unit's hydrogen gas analyser and had nothing to do with the ongoing problem of air in the water loop. On September 8, Padalka replaced one of the Elektron liquid units with one he had repaired using spare parts. The crew then flushed the Elektron through with water, cleaned a mounting plate and, after it had shut itself down several more times after only short periods of operation, they removed the gas analyser. The unit was then turned on and run for a few days. It was turned off once more before the crew went to sleep on September 17. On the ground engineers began studying the data relayed from the partially repaired unit. Korolev announced that the latest failure may have been caused by crystalline deposits of potassium hydroxide in the oxygen supply line of the liquid unit. While the Elektron was turned off, the station's atmosphere was repressurised using oxygen carried into orbit on Progress M-50 and nitrogen from the tanks on Quest.

Even as repairs to the Elektron unit continued, on September 17 Fincke depressurised the area between the window panes in Destiny and replaced the flex hose, which had malfunctioned and allowed air to enter the space between the panes of glass. He also installed a cover that he had made previously at the workbench in Destiny. They tested the communications systems in the Soyuz TMA-4 spacecraft and Fincke videoed those areas of the station's exterior that were visible from windows in the various modules and transmitted the images to Earth.

During the week ending September 24, Padalka and Fincke performed the regular 6-month preventative maintenance of the station's treadmill. They also continued to troubleshoot the Elektron oxygen generator, working on the assumption that the hydrogen line was being prevented from pressurising correctly by contamination in the line. During the week, the two crewmen cleaned out the line in question. Meanwhile, the station's atmosphere was repressurised twice, using oxygen from the tanks in Progress M-50. The crew also began storing some items for their return to Earth during the second week of October. On the ground, NASA had begun talking publicly about evacuating ISS, if the onboard stock of breathing oxygen fell below 45 days.

The Elektron repairs continued into October. Under instruction from Korolev, the crew disconnected the unit's hydrogen vent pipe from its overboard vent valve, and Padalka jury-rigged a hose to redirect the vented hydrogen through Zvezda's micro-purification unit. The unit then operated correctly during several days of testing. Fincke also fitted a mass spectrometer unit to the Major Constituents Analyser in Destiny. Progress M-50 had delivered the mass spectrometer, and its installation allowed the Analyser to be operated continually, rather than only periodically, as it had been up until that time. During the week, Fincke also carried out a series of soldering experiments. Engineers in Houston carried out a remote test of the Thermal Radiator Rotary Joint, which would allow the radiator to rotate to the best position for loosing heat when more SAWs were added to the station, after the Shuttle resumed flying, probably in 2005. Alongside the numerous equipment repairs, regular maintenance, and daily exercise, the crew still found time to perform a number of experiments on themselves.

As their flight approached its end, during the second week of October both men donned their Sokol launch and re-entry suits and entered Soyuz TMA-4 for routine checks. Fincke replaced the gas trap and pump inlet filter in the still malfunctioning EMU. He also replaced the cycle ergometer control panel with one that had been brought up on Progress M-50. Both men collected samples of potable water for in-flight analysis.

SOYUZ TMA-5 DELIVERS THE EXPEDITION-10 CREW

SOYUZ TMA-5	
COMMANDER	Salizhan Sharipov
FLIGHT ENGINEER	Leroy Chiao
ENGINEER	Yuri Shargin

In March 2004 the Russians had suggested to NASA that the Expedition-10 crew should double the standard 6-month Expedition crew duration to one year. Such a flight would build on Russian medical experiments from their Salyut and Mir stations. It would also clear a second couch in Soyuz "taxi" spacecraft to be sold to visiting astronauts. NASA refused the proposal, after heated debate. While some NASA employees argued that the Administration was not ready to support a 12-month flight, others pointed to Russia's experience on Mir, when several cosmonauts had approached 12-months in space and a few had surpassed it. Despite the discussions, Expedition-10 would fly a standard 6-month caretaker mission. Chiao described the role of a two-person caretaker crew in the following terms:

> "[I]t is a very demanding timeline for a crew of two, and the past two-person crews have shown that they can accomplish those timelines and remain healthy and well-rested and things like that. [O]ur flight will be the same . . . and we have a very full schedule; we'll be doing a lot of work. But at the same time, we'll be having scheduled time off, where we can kind of re-energize and recharge, and so I really don't see a problem with that. Now, of course, the thing that suffers sometimes when we are scheduled in like this, is we don't get as much science done as we'd like. The purpose of the International Space Station is to do all kinds of cutting-edge science that can't be done on the ground, but our goal right now is to kind of keep that laboratory going . . . until we can get the Shuttle flying again and . . . we get the laboratory finished . . . Neither Salizhan nor I have flown a long-duration flight; however, between us, we have four Shuttle flights, and so we have a wealth of experience being in space and operating in space. We work very well together; our personalities complement each other, and we both have the same views on how things ought to be done. And so I think that everything will be just fine."

The third couch on Soyuz TMA-5 became available when prospective spaceflight participant Sergei Polonsky was grounded for unspecified medical reasons. Unable to sell it at short notice to another spaceflight participant, to ESA, or to the French National Space Agency, the Russian Federal Space Agency allocated it to Yuri Shargin, a member of the Russian Rocket Forces, who would complete the usual short visit to ISS, returning to Earth with the Expedition-9 crew.

On September 15 the Russians announced that the launch of Soyuz TMA-5, which had been scheduled to lift off on October 9, had been delayed for 5 to 10 days. The cause of the delay was the premature firing of an explosive bolt on the spacecraft. The few details announced at the time suggested that the bolt was one of those used to separate the orbital compartment prior to re-entry. The Russians subsequently announced that it was actually one of a ring of bolts used to separate the docking system from the front of that compartment in the event of the docking system failing to release as the spacecraft tried to undock from ISS. In that event the spacecraft's docking system could be explosively severed and left attached to the station's docking system while the Soyuz returned to Earth. On September 22 the launch was reset for October 11. Six days later the launch was delayed a second time "for a few days". RSC Energia officials did not release details of what had caused the delay, but it has been suggested that it was a leaking membrane in a hydrogen peroxide tank. The launch was rescheduled to October 13.

Soyuz TMA-5 lifted off at 23:06, October 13, 2004. A communications problem involving a Russian Molniya satellite delayed each of the first two orbital correction burns by one orbit. Two days later, as the Soyuz approached within 100 metres of the station, an alarm sounded suggesting that the Kurs automatic rendezvous and docking system had malfunctioned. A Russian investigation would show that a forward firing thruster on the Soyuz was producing less thrust than expected and Soyuz TMA-5 therefore approached ISS too fast, causing the docking attempt to be aborted. Sharipov assumed manual control, backed his spacecraft out to 200 metres and then executed a perfect manual docking at 00:16, October 16, thereby earning himself a bonus. Following leak checks the Expedition-10 crew entered ISS at 03:13, and was greeted by their predecessors. Shargin transferred his couch liner to Soyuz TMA-4.

For the next week Sharipov and Chiao spent 2 or 3 hours a day working closely with Padalka and Fincke to ensure a smooth hand-over, the remainder of the time they spent setting up their own experiment programme. During the hand-over period Sharipov assisted Padalka in making final repairs to the Elektron unit before it was powered on during the hours the crew was awake, and powered off while they were asleep. Meanwhile, Chiao and Fincke worked to replace the rotor pump in the EMU that Fincke had begun repairs on a few days earlier. The new crew also took the opportunity to gain hands-on experience with the SSRMS and several of the ISS systems. Shargin participated in a number of medical experiments, with assistance from Sharipov and Padalka. He also took a number of photographs of the Earth's surface. Although Shargin was the first member of the Russian Rocket Force to fly in space, none of his experiments was identified as military in nature. He completed his experiment programme, primarily in the Russian sector of the station.

Figure 48. Expedition-10: Leroy Chiao wears his Sokol launch and re-entry suit while posing alongside a Russian Orlan extravehicular activity suit.

Figure 49. Expedition-10: having arrived with the Expedition-10 crew, Yuri Shargin worked on his experiment programme before returning to Earth with the Expedition-9 crew. The individual behind him is not identified.

Fincke had described his feelings on returning to Earth before his flight began:

> "For the return, I'll be the Flight Engineer, sitting in the left seat, pushing all the buttons, and Gennady will have his command panel in front of him. [W]e'll work together as a team to bring ... the ship safely home ... We do all of our systems checks, and if everything looks good and we have concurrence from the ground, we undock from the Space Station—some springs push us off, and we're on our way home. It's going to be a very bitter-sweet moment. I'll be so excited to go home and see my family ... so, that'll be the sweet part. The bitter part is leaving our home for six months. And, it'll all happen in just a few hours. We make sure we have a successful undocking from the Space Station, we wait one orbit to upload our commands from ... Moscow, who gives us all of our vectors so that we can come in and land right on target. [O]nce we get the 'go', we start our de-orbit burn; it slows us down by several hundred meters per second. Then ... we ... enjoy the ride. I think it's going to be very ... exciting just watching to see if anything goes wrong and to be there ready to solve any malfunction. And once we start to get close to the atmosphere, we'll separate all three modules ... Then the real fun begins. The other Americans that have flown in the Soyuz ... have called it an incredible ride because we tumble end-over-end until we stabilize in the atmosphere. And I'll be sitting right next to a window, and I'm just looking forward to seeing what that would look like. Once we stabilize in the atmosphere, the automatic re-entry system engages and brings us close to our point, and then we have a set of primary parachutes that will open up and slow us down. And then right before we land ... a series of retro-rockets ignite and soften our blow as we return home to the planet. We open up the door and, hopefully, the helicopters will be outside waiting for us."

Following an official hand-over on October 22, Padalka, Fincke, and Shargin locked themselves in Soyuz TMA-4, and undocked from Zvezda's wake at 21:08, the following day. After retrofire, and a routine separation, Soyuz TMA-4 re-entered the atmosphere and landed at 20:36, the same day. Locally it was dawn, and recovery crews had seen the plasma sheath surrounding the re-entry module as it passed through the atmosphere. The landing took place in semi-darkness. The Expedition-9 crew had spent 187 days 21 hours 17 minutes in space. Shargin's first spaceflight had lasted 9 days 21 hours 21 minutes.

As October drew to a close NASA announced that the Shuttle's Return to Flight launch would slip until May or June 2005.

EXPEDITION-10

Following the departure of Soyuz TMA-4 Sharipov and Chiao began the standard 3 days of light duties to allow them to get over the rushed workload of the past week. Their occupation would receive two Progress cargo vehicles and they planned to make two Stage EVAs before Soyuz TMA-6 delivered the Expedition-11 crew, in

April 2005. The new crew activated EarthKam during the first days of their occupation. In 8 days over 800 images were exposed. At the outset of their mission, word was received from the ground that the Elektron unit was performing well in tests and was cleared for permanent use. The joint repairs carried out by Padalka and Sharipov had finally returned the unit to operational use.

On November 4, Chiao used the Advanced Ultrasound in Microgravity experiment (ADUM) to make ultrasound scans of Sharipov and the positions were reversed the following day. The experiment relayed the scan directly to doctors on the ground, who could use it to make a real-time diagnosis. On November 5, both men took part in emergency medical drills and collected air and swab samples in Zarya. Chiao used the BCAT experiment. He also practised using the SSRMS on November 8. During the exercise the arm's video cameras were used to image a possible indentation that had been observed on the exterior of Destiny by the last Shuttle crew to visit ISS (STS-113). The images proved that the dent was not caused by a micrometeorite, or a debris strike. The flat area on one of Destiny's protective panels appeared similar to flat areas observed on the protective exterior of Unity. The flattening of the panels was thought to be caused by flexure with changes in temperature.

Throughout the week, the two men carried out work with the Binary Colloids Alloy Test (BCAT) experiment. Chiao also worked on the faulty pump in the second of the American EMUs that had suffered a cooling failure during the Expedition-9 occupation. The work had to stop when a metal shim could not be located. Following a search, plans were put in place to launch a replacement on Progress M-51. The American EMUs were not due to be used until the Shuttle resumed flight, in mid-2005.

Chiao explained how he and Sharipov had trained to receive STS-114 during their occupation of ISS:

> "That's something we're hoping for—we'd love to have STS-114 come up and visit us during our flight. [W]e'd welcome them and well, we're keeping our fingers crossed. It'll be a really neat event having the Space Shuttle return to flight and come up during our increment and do some construction work while we're there, and our two crews will work together, and we sure hope that'll happen . . .we have been doing the training for these events. Salizhan and I have both received MPLM training and we also received a lot of photography lessons on how to take pictures of the Shuttle tiles and leading edges to inspect the heat shielding. [S]o we're just keeping our fingers crossed it'll work out for 114 to come up."

On November 8, NASA announced that they had initiated a study to consider how best to continue development of ISS after the Shuttle resumed flying. At the same time, they confirmed that the Shuttle's return to flight had been delayed from March to May 2005. Bill Readdy, Associate Administrator for Space Operations, told the media, "After four hurricanes in a row, we could not make the March launch." The new schedule called for three Shuttle launches in 2005 and five each year thereafter, until ISS was complete and the Shuttle was retired, in 2010. Readdy stated that, if the first two daytime Shuttle launches were successful, then the third

Figure 50. Expedition-10: Salizhan Sharipov floats inside Zvezda wearing his Sokol launch and re-entry suit.

flight might carry the first three-person Expedition crew to ISS since the loss of STS-107. To cover the continued delay in the Shuttle's Return to Flight, Russia agreed to continue supplying Soyuz TMA spacecraft for crew transport through 2006. This was a negotiated agreement that allowed Russia to recover the experiment hours it had given up to NASA in the early years of the programme when Zvezda was under-funded and launched late. Twenty-eight Shuttle flights were considered necessary to complete ISS. Studies considered transferring as many as 17 ISS payloads to expendable launch vehicles, leaving only 11 Shuttle flights. Among ideas being considered was a plan to launch the European Columbus and Japanese Kibo modules on Russian Proton launch vehicles. Initial development of the CEV design was underway in a multi-contractor competition.

On November 11, a circuit breaker tripped out on ISS and stopped the supply of electrical power to a number of crew equipment items. After checking the equipment involved the crew were able to power the circuit breaker back on. The following day, Chiao moved the SSRMS to a position that allowed its cameras to view the transfer of Soyuz TMA-5 from Pirs' nadir to Zarya's nadir, a manoeuvre planned for November 29. The change of location would allow Pirs to be used for two Stage EVAs, in January and March 2005.

The following week, controllers in Korolev commanded Progress M-50's rocket motors to fire, to raise the station's orbit. Although the burn lasted the correct amount of time, it left the station in a slightly lower orbit than had been planned.

Controllers at Korolev later blamed the shortfall on "human error". Rather than fire a correction burn, it was decided to delay the launch of Progress M-51 by 24 hours to compensate for the station's lower orbit. Throughout the week, the crew completed a varied science programme. Working with the American ADUM and Serial Network Flow Monitor (SNFM), which used computer software to track the communications and data flow between payloads in Destiny. Sharipov collected samples for the PLANT experiment and worked with the Russian experiments Hematokrit, which counted red blood cells, and Sprut, a study of human body fluids. Sharipov also checked out a new Russian Orlan suit before discarding an old Orlan suit that had exceeded its on-orbit life, on Progress M-50. Both men also participated in routine housekeeping tasks.

A NEW RUSSIAN LAUNCH SCHEDULE

On November 20, managers at Roscosmos decided to develop the FGB-2 module as a Multi-purpose Laboratory Module (MLM), rejecting RSC Energia's proposal to develop the FGB-2 as the Enterprise Module. RSA announced that MLM would be launched by a Proton launch vehicle in 2007 and would dock to Zarya's nadir, the European Robotic Arm would then be attached to its side. The MLM would be launched with some scientific equipment pre-installed, but the remainder would be launched separately over the following 2 years. It would also house support equipment, hygiene facilities, a sauna, and an additional sleeping room, as well as a storage area to hold spare parts and other cargo on-orbit. An aft docking port would support Soyuz and Progress vehicles, while a lateral port would house a scientific airlock, to be delivered by Shuttle.

The next major Russian launch would be the Scientific Energy Module (Russian initials: MEM), originally called the Scientific Energy Platform (Russian initials: MEP). This would be mounted on Zvezda's Zenith after delivery by Shuttle in 2010. The new MEM would be downsized from the original MEP and would consist of a pressurised section containing gyrodines, a boom, and eight SAWs. The pressurised section would provide a new docking location for Pirs, which would be moved from its present location on Zarya's nadir port by means of the SSRMS.

Plans were less substantial and less certain for a Russian dedicated science module using the FGB design, due for launch in 2011. It would be docked to Zvezda's nadir and two smaller Russian scientific modules built around the Pirs design.

Meanwhile, Energia unveiled plans for the Soyuz replacement, a 13-tonne spacecraft to be called "Kliper". The lifting body design of the re-entry module would be enhanced by a Soyuz-style orbital compartment and would carry six people. Uncrewed flight-tests on a Zenit launch vehicle were planned for 2010, with the first crewed flight to ISS in 2012. Again, no government funding existed for Kliper and attempts to convince ESA to help fund its development were unsuccessful.

While the plans sounded optimistic, there was no budget to build any of this equipment. Russian participation in the ISS programme would remain restricted to those modules already in orbit and any Soyuz and Progress vehicles that were purchased, either by the Russian government or the other ISS partners in support of the Expedition crews throughout the life of the station, which was originally intended to end in 2016.

NASA'S BUDGET APPROVED

As November drew to a close, Congress approved NASA's budget request for FY2005. The final figure allocated was only $44 million short of that requested by President Bush when he announced the new Vision for Space Exploration. $4.3 billion was allocated to the Shuttle, but NASA was mandated to report back to Congress on a regular basis regarding the cost of returning the Shuttle to flight. The budget also included money to begin development of the CEV, which would replace the Shuttle.

SOYUZ TMA-5 RELOCATED

On November 24, Sharipov test-fired the thrusters on Soyuz TMA-5 in anticipation of its relocation to Zarya. During the test, controllers in Korolev noticed the same reduction in pressure in the fuel system that had been observed during the initial approach to the station. As the thruster problem had no effect on the future use of Soyuz TMA-5 it was decided to carry on with the relocation manoeuvre and a second test-firing of the thrusters during that manoeuvre was cancelled. Two days later, the crew reviewed the plans for the relocation manoeuvre and closed the hatches between Progress M-50 and Zarya. November 25 was the American Thanksgiving Day holiday and the Expidition-10 crew had a day off.

Having configured ISS for automatic operation, both astronauts sealed themselves in the descent module of Soyuz TMA-5 on November 29. Sharipov undocked the Soyuz from Pirs at 04:29, and backed away to a distance of 30 metres. He then flew the spacecraft 14 metres along the length of the station before rotating it through 135° and held it in position for 8 minutes of station keeping. Docking with Zarya's nadir port took place at 05:53. Following the standard leak and pressure checks the crew opened the hatches between the two vehicles, returned to ISS, and reconfigured it for occupied operation. The crew had November 30 and December 1 off. On November 30 the missing metal shim from the EMU pump drifted past Chiao, who was working inside Pirs. He recovered it, but pressure suit engineers decided not to use it, preferring to use a new shim, to be delivered on the next Progress. During the month the crew had carried out three audits of the food onboard, which showed that the previous Expedition crew had eaten into the current crew's rations. New supplies would be launched on Progress M-51, but if the launch failed, or the spacecraft failed to dock, the food currently on ISS would run out in mid-January, requiring the crew to evacuate the station. As the food ran out the crew reduced the amount they were eating, until Sharipov told an interviewer, "We are short of food. We are eating less than half of our normal rations ..."

Returning to work on December 2, they swapped out the filter cartridges in the Elektron oxygen generator. The crew spent the next 3 weeks filling Progress M-50 with rubbish to be taken away when it undocked, on December 22. A laptop computer failed in Zvezda, and was replaced by one from Sharipov's sleep station. Three new laptops would be in Progress M-51, to bring the Station Support Computer network back up to full capability.

The second week in December was spent preparing Destiny for additional experiment racks. Chiao worked with the ARIS experiment. He also photographed the BCAT experiment and worked with the in-space soldering experiment. Meanwhile, Sharipov donned the Chibis suit, which simulated forces on the musculo-skeletal system using suction as part of a study of the cardiovascular system. The crew also completed a range of housekeeping duties.

Sharipov and Chiao carried out a routine practice with the SSRMS and left it with both ends holding grapple fixtures on the exterior of the station. It was a precautionary measure, in case the food shortage, coupled with a failure of the Progress M-51 flight, led to the evacuation of ISS, during the week ending December 17. They installed cables and a switching unit for the docking system that would guide the European ATV to docking with Zvezda's wake. The week was a busy one, with experiment periods using the ADUM. The crew also carried out an inventory of all items loaded into Progress M-50. At one point, the Elektron oxygen generator was deliberately turned off, to allow the last of the oxygen in Progress M-51's tanks to be used. They also carried out maintenance on the ventilation system in Zvezda and replaced the batteries in the station's heart defibrillator. A major task during the week was a top-to-bottom inventory of every item held in the Quest airlock. This was carried out in advance of three EVAs planned from Quest during the STS-114 Return to Flight mission. With their loading tasks complete, the crew closed the hatches between Progress M-50 and Zvezda. The Progress undocked at 14:37, December 22, and backed away from Zvezda's wake. A separation manoeuvre placed it in its own orbit before it was de-orbited to burn up in Earth's atmosphere.

On December 13, Sean O'Keefe announced his retirement as NASA Administrator. He would remain in place until a replacement was named. The new Administrator would face the following tasks relating to human spaceflight:

- Return the Shuttle to flight.
- Complete ISS before the Shuttle was retired in 2010.
- Re-structure NASA to bring the Shuttle programme to an end.
- Downsize NASA as the Shuttle programme wound down.
- Develop Project Constellation.
- Define Project Constellation hardware.
- Restructure the NASA field stations for Project Constellation.

PROGRESS M-51

Progress M-51 was originally scheduled for launch on December 22, 2004, but was delayed by 24 hours as a result of the off-nominal ISS orbital boost carried out on

November 19. The Progress was launched at 17:19, December 23, and followed a standard rendezvous pattern to dock to Zvezda's wake at 18:58, December 25. Docking was delayed for approximately 30 minutes so that it would occur over Russian ground stations. Throughout the approach Sharipov stood ready to take over manual control using the TORU system, but his skills were not needed. After the docking, Korolev's controller told the crew, "Hip, hip, hooray; Congratulations." Chiao replied, "We're looking forward to our big Christmas present, the arrival of Progress. Merry Christmas, to all of the people keeping us safe up here."

The crew took the remainder of the day off to celebrate the Christmas holiday and then had an extended sleep period that night. They opened the hatches into the new Progress the following day. In a Christmas message, Chiao remarked, "Salizhan and I are privileged to be the only humans off our beautiful planet. Although we miss our families and friends, our role on board this international complex is a message for all of us with a fundamental curiosity to explore."

On December 26 an earthquake in the seabed below the Indian Ocean sent a tsunami crashing into the coastlines of several Asian countries, causing thousands of deaths and millions of dollars worth of damage. Around the world a major relief effort swung into action. The Expedition-10 crew took photographs of the disaster

Figure 51. Expedition-10: Salizhan Sharipov holds a docking probe removed from a Progress cargo vehicle. It would be stored in Zarya until the Shuttle resumed flying, at which time it would be returned to Earth for refurbishment and re-use. The doorway on the left-hand edge of the frame gives access to one of Zvezda's two sleeping quarters. Note the picture of Russian spaceflight pioneers Konstantin Tsiokolvsky and Yuri Gagarin on the wall behind him.

area to show coastal change. Those pictures were sent down to Houston for analysis. Sharipov and Chiao then spent the remainder of the year unloading the new Progress and completing routine maintenance. Houston played a recording of the song *Auld Lang Syne* over the radio link to mark the moment of New Year. The crew also received greetings from controllers in Korolev.

In a press conference from space the two cosmonauts talked about the weight they had lost as a result of having to ration their food intake until Progress M-51 had arrived at the station. Chiao remarked:

> "Both of us ended up losing a few pounds, but I guess that's nothing we can really complain about . . . A lot of people would be happy to lose about 5 or 10 pounds. We looked at it as kind of a challenge, kind of a camping adventure, roughing it, I guess."

He added:

> "All throughout this whole thing, we kept real good spirits. Salizhan and I have been keeping each other up, joking around, and it's been very pleasant even with some of the shortages."

Chiao blamed the food shortage on Foale and Padalka, who had sought permission to eat some of the next crew's food when their own return to Earth was delayed:

> "The last crew had gotten into our food and had failed to actively report to the ground what they had taken out of our allotment."

During January 1, 2005, air bubbles in the plumbing caused the Elektron oxygen generator to shut down again. Sharipov carried out a number of repairs over the following days, but the unit performed erratically. The crew purged gases from the system's lines and replaced the electrolyte in the unit, and finally had it running again by January 6. Later that day it was switched from the primary pump to a back-up and continued to function intermittently. On January 4, while the Elektron was not operating, oxygen was added to the station's atmosphere from tanks in Progress M-51.

The crew observed the Russian Orthodox Christmas on January 7, but both men performed experiments. The first week of the year saw the crew follow a light maintenance schedule, including the NEUROCOG experiment, which studied how the body's sense of balance adapted to microgravity. The astronauts sent a message of condolence to the victims of the tsunami. Throughout the following week, the crew assisted Houston in updating 1.5 million lines of computer code. They also began preparations for the Stage EVA planned for January 26. They had a day off on January 15, although controllers in Korolev fired the rockets in Progress M-51 to raise the station's altitude and place it in the correct position to receive Progress M-52, due to dock on March 2. These burns are not really designed to change the altitude, but by setting the altitude to achieve a specific orbital period the phasing could be arranged to suit the rendezvous requirements of the ship about to launch.

The third week of the month was spent preparing for the EVA, charging pressure suit batteries, preparing tools, checking out their Orlan suits, and spending time on the station's stationary bicycles to complete cardiovascular evaluation exercises. Meanwhile, controllers in Houston completed vibration and electrical current tests on CMG-2, which was run at each of its 15 available speeds over a period of 4 hours. Solar storms on January 18 and 20 led to instructions that both men should shelter in Zvezdawhenever the station passed above the most heavily affected regions of Earth's atmosphere. Chiao and Sharipov marked their 100th day in space on January 21. The previous night the Elektron unit had shut down and only began working again after air bubbles in the water loop had been purged.

On January 26, both men donned their Orlan EVA suits and sealed themselves in Pirs. As on previous EVAs undertaken by caretaker crews, ISS' systems were either powered off or placed into automated mode, and the hatches between the Russian and American sectors of the station were closed. After venting the atmosphere in Pirs they opened the outer hatch at 02:43 and exited. Their first tasks were to set up the tethers and tools for the tasks ahead. Next they installed a Universal Work Platform (Russian initials: URM-D) and a base plate (FP-20) at the forward end of the large conical section of Zvezda, before mounting the German Robotics Component Verification on ISS (ROKVISS) experiment on the platform along with its antenna and cabling. When the 0.5 m long, two-jointed manipulator arm, designed to test small robotic joints in the vacuum of space, was first powered on it did not receive power. Chiao had to return to the location and re-seat the two power plugs. The system would be operated from the ground in Germany, or from a workstation inside Zvezda. The two astronauts then moved the tray holding the sole remaining Japanese MPAC/SEEDs experiment mounted on the exterior of Zvezda during the Expedition-3 occupation, in October 2001. They removed it from its present mounting bracket and moved it to an adjacent bracket, in order to facilitate the installation of the ROKVISS antenna. Moving on again, they inspected vents on the exterior of Zvezda used by a number of environmental control systems, including the Elektron unit. Sharipov reported seeing both brown and white residues near the Elektron and Vozdukh carbon dioxide scrubber vents, and an oily substance on the thermal insulation surrounding the vents. Returning to Pirs, they installed the Biorisk-MSN experiment near the airlock's hatch. Biorisk contained bracket-mounted canisters holding micro-organisms that would be exposed to the space environment before being returned to Earth for analysis. Entering the airlock, they sealed the hatch at 08:11, after an EVA lasting 5 hours 28 minutes. Returning to the station, they removed their suits, opened the hatches to the American sector, and reactivated the systems. They began a sleep period early in the afternoon, had the following day off, and spent the following week reconfiguring ISS for crewed operations. The last day of January was spent dressed in Sokol pressure suits, rehearsing re-entry procedures in Soyuz TMA-5.

As February began the two men began preparing for the Space Shuttle's return to flight. On February 1, Chiao reported that he had finished packing all American articles due to be returned to Earth in Discovery's MPLM on the STS-114 flight. Work was continuing to pack Russian equipment. Chiao also completed an inventory of the food remaining onboard. The crew also worked on their national experiments. These included three runs of the Russian Plasma Crystal experiment, EarthKam, and the Space Experiment Module Satchel (SEMS). The last contained six school experiments in separate vials and had been delivered on the last Progress. Chiao also updated the software in the malfunctioning Space Integrated GPS/Inertial Navigation System, which supplied Global Positioning Satellite information to the station's guidance and navigation system.

On February 8 the two Asian crewmembers celebrated Chinese New Year. On the same day there was a temporary loss of attitude control, when a wrong command sent from the ground caused one of the CMGs to become saturated. The Russian thruster system on Zvezda assumed the responsibility for attitude control, until the CMG could be reset. The crew spent several hours auditing the EMUs and their supporting equipment held in the Quest airlock. Chiao also replaced the metal oxide (METOX) canisters, which scrubbed the used oxygen in the suits. Both men completed their ongoing experiment programme throughout the week.

The Elektron oxygen generation system was powered off, after failing twice overnight, to allow the station to be repressurised, three times, using the oxygen remaining in the tanks on Progress M-51. The motors on Progress M-51 were used to adjust the station's orbit on February 15. Remaining propellant was then pumped from the Progress to tanks in Zvezda later in the week. Sharipov removed the KURS docking system from the Progress for return to Earth on Discovery when Shuttle flights resumed. The crew also spent several days carrying out the semi-annual inspection and service of their treadmill, which involved partially disassembling the machine. In Houston, flight controllers powered up the Mobile Base System on the truss, and found they could not receive video from one of the cart's cameras.

Most of the following week was spent packing rubbish into Progress M-51 prior to its departure. The spacecraft's hatch was closed on February 25. On the same day American controllers in Houston completed 2 days of tests on the SSRMS, using the new software that had been loaded the previous month. This was the first time the SSRMS had been able to be operated from Earth. Throughout the test Chiao stood at the arm's operating station in Destiny, ready to assume control if required. The test was successful and Chiao's intervention was not required. This upgrade in capability would enable the ground to manipulate the arm to observe future two-person EVAs.

Progress M-51 was undocked at 11:06 on February 27. The cargo vessel was manoeuvred to a safe distance where Russian flight controllers used it in a 10-day long series of engineering tests.

PROGRESS M-52

Progress M-52 lifted off from Baikonur at 14:09, February 28, 2005, and was soon in orbit with its antennae and photovoltaic arrays deployed. As the launch occurred, ISS was over the Atlantic Ocean, west of Cape Town, South Africa. Among its 2.4

tonnes of cargo, Progress carried 160 days of food for the crews of ISS. Following a standard 2-day approach the spacecraft docked to Zvezda's nadir, at 15:10, March 2. This was the first Progress docking to occur outside of direct-line-of-sight communications with Russian ground stations. All telemetry and video links were routed to Korolev through American communications systems.

Meanwhile, NASA had voiced a concern over a projected shortfall of ISS logistics flights in the long term. In 2005 the schedule stood as follows:

- Three Shuttle flights in 2005 and five Shuttle flights per year from 2008 through 2010.
- Two Soyuz flights per year to 2008 and four Soyuz flights per year from 2009 through 2015.
- Four Progress launches per year to 2010 and five per year from 2011 through 2015.
- One European ATV launch per year from 2006 through 2013.
- One Japanese HTV launch per year from 2009.

NASA stated that this "does not meet the projected re-supply and return needs."

MICHAEL GRIFFIN TAKES OVER AT NASA

As March began, Michael Griffin was named as NASA's new Administrator. Although he had worked for NASA in the past, he was not doing so when he accepted the new position. In the past Griffin had been vocal in his criticism of ISS and his belief that the Shuttle should be grounded before 2010. He made no secret of his dislike of O'Keefe's plans for an Atlas-V or Delta-IV launched Crew Exploration Vehicle, to be developed in three spirals (phases). In 1989, Griffin had been NASA's Associate Administrator for Exploration, when President George Bush Senior attempted to send NASA back to the Moon. Now he would oversee the beginning of NASA's attempts to make President Bush Junior's Vision for Space Exploration a reality.

In orbit, the Expedition-10 crew spent the next few days unloading the supplies that Progress M-51 carried. Chiao also moved the SSRMS to the positions required for its cameras to view the exterior sites where he and Sharipov would perform work during their second EVA, planned for March 28. He left the SSRMS parked in the best position for the camera to view their work. Although a successful test had been made, the SSRMS was not yet certified to be controlled from the ground. The week was marred by the intermittent performance of the Elektron after it was powered on, on March 2. Sharipov performed several hours of maintenance work on the device, but failed to completely correct the problems. Further attempts at repair, between March 16 and 18, also failed and the unit was powered off until after the crew's final EVA.

The following week Chiao installed a new heat exchanger in Quest, thereby returning the airlock to full operation. The heat exchanger had been delivered on

Progress M-52. Meanwhile, Sharipov worked in Pirs, preparing it for their forthcoming EVA, planned for March 28. American controllers tested the SSRMS for a second time on March 23. On the same day, Russian controllers fired the station's thrusters to adjust its orbit in advance of the launch of Soyuz TMA-6. The RPC replaced by the Expedition-9 crew in 2004 failed on March 16, causing CMG-2 to stop working once again. The ISS returned to the minimum attitude control capability of just two working CMGs in the Z-1 Truss. If a third CMG failed, attitude control would be passed to the thrusters on Zvezda. Cables would be re-routed to bypass the faulty RPC during an EVA by the STS-114 crew, after the Shuttle returned to flight later in the year. Two Shuttle astronauts would also replace the CMG that had failed in June 2002. On March 25, the engines on Progress M-52 were used to raise the station's orbit. The following day a cooling loop panel failed in Pirs. The crew had to replace the panel before their planned EVA could proceed.

Having configured ISS for autonomous flight and sealed all of the internal hatches, Chiao and Sharipov exited the Pirs airlock dressed in Orlan suits at 01:25, March 28. They collected their tools and Sharipov activated a Russian nano-satellite for later deployment. Their first task was to install three WAL S-band low-gain antennae on the conical section at Zvezda's ram. The antennae were part of the Proximity Communication Equipment (PCE) to be used by ESA's ATV. Approximately 2 hours into the EVA, Sharipov stood on a ladder mounted on the exterior of Zvezda and launched the nano-satellite by hand. He released the satellite, which was designed to test new attitude sensors and small satellite control systems, towards the station's wake.

Russian controllers in Korolev inhibited the station's thrusters before the two men made their way towards Zvezda's wake. There, they installed a GPS receiver, which would provide the ATV with its position relative to ISS during rendezvous. They then installed cables for the GPS receiver and photographed the position of another antenna for Russian engineers. They also secured cables along the exterior of the station as they made their way back to Pirs. During this work the station drifted out of alignment and Russian controllers re-activated the thrusters, to resume the correct attitude, as soon as the two men were clear of the area. Having stored their tools, Chiao and Sharipov re-entered Pirs and closed the hatch at 05:55, bringing the EVA to an end 1 hour earlier than planned, after 4 hours 30 minutes. Shortly after the EVA ended a series of spikes in vibration were detected in CMG-3, one of the station's two functioning CMGs. Engineers in Houston began troubleshooting the vibration immediately and ISS was re-positioned so as to minimise demands on the two CMGs.

The crew spent much of the next week preparing the station for re-occupation, stowing tools and cleaning and venting unused oxygen in their Orlan EVA suits into the station's atmosphere before storing the suits. They also continued stowing equipment that would be returned to Earth in Discovery, during the STS-114 Return to Flight mission, in July. They also tested the cameras that they would use to photograph the approaching Shuttle's heatshield. The cameras had been delivered on Progress M-52. In the week ending April 8, the Expedition-10 crew began packing for the end of the flight. The station was repressurised using oxygen from the tanks in

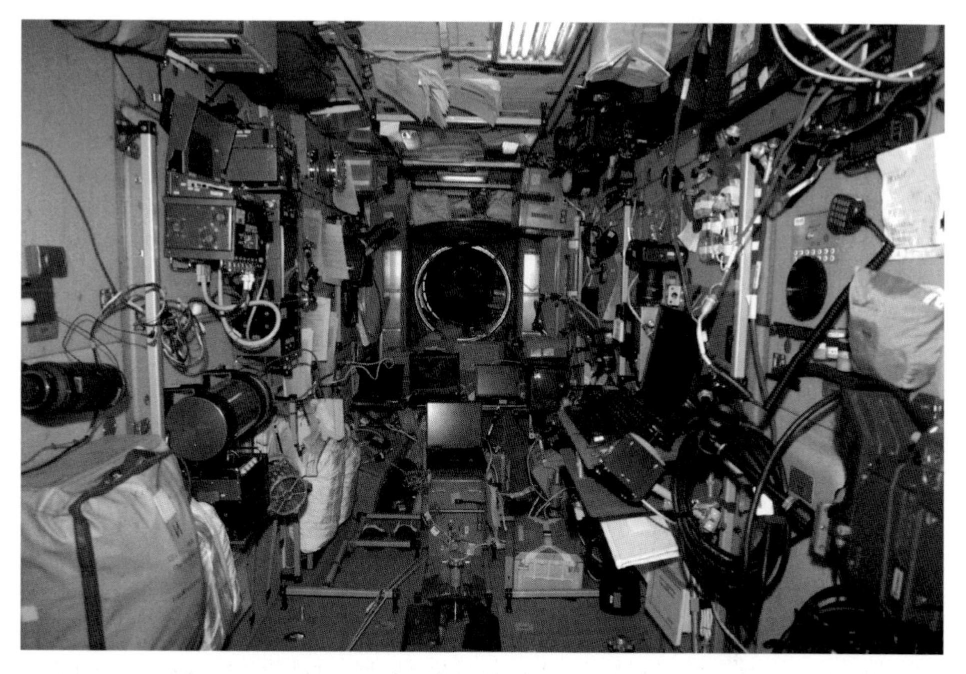

Figure 52. Expedition-10: Zvezda began to fill with equipment and rubbish during the 3-year period when the Shuttle was grounded.

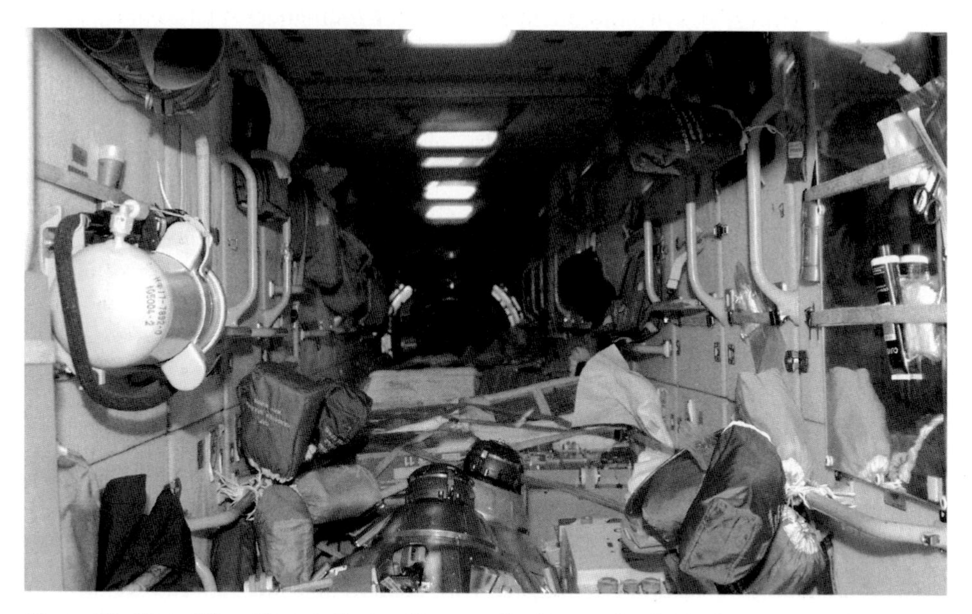

Figure 53. Expedition-10: note the two Progress docking probes in the foreground of the Zarya image. Only the Shuttle could return bulky items such as these to Earth. Zvezda was equally cramped by this time.

Progress M-52, while American engineers continued to work on the CMG vibration spikes. Sharipov continued to work with the Elektron system, but the final repair still eluded him. The unit was powered on, on April 13, and was deliberately powered off on April 16, in advance of the Soyuz TMA-6 docking.

SOYUZ TMA-6 DELIVERS THE EXPEDITION-11 CREW

SOYUZ TMA-6	
COMMANDER	Sergei Krikalev
FLIGHT ENGINEER	John Phillips
ENGINEER	Roberto Vittori (ESA)

Soyuz TMA-6 was launched from Baikonur Cosmodrome at 20:46, April 14, 2005. At the time, ISS was over the southern Atlantic Ocean. Following a standard rendezvous the Soyuz docked to Pirs at 22:20, April 16. After routine pressure checks the hatches between the two vehicles were opened at 00:45, and the Expedition-11 crew of Krikalev and Phillips entered ISS, where they were greeted by the Expedition-10 crew and given a safety brief. Krikalev and Phillips would complete 6 months on the station, with Krikalev passing 800 days of cumulative time spent in space during the flight and making a Russian record-breaking sixth spaceflight. This would be his third visit to ISS, and his first as Commander. Asked how being Commander differed from being Flight Engineer, he has remarked:

> "I think that's a very subtle issue, because when you fly a crew of two or three, the difference between the Commander and every other crewmember is very subtle ... In this case you work as much, and maybe even harder, than your partners because you know more, you have more experience. I think for every Commander safety of the mission, mission success, is a primary goal. Mission success is again a very complicated issue. To say after a flight that a mission was successful, you have to know that all experiments were completed, all the work that was scheduled was done, but most importantly, to know that the crew returned safely to the ground. When you become Commander, you have a responsibility not only for mission success but for your crew, basically for the life of your crew."

Phillips had also visited ISS on an earlier Shuttle flight. For him the aims of Expedition-11 were fairly basic, but also vital to the programme:

> "For me the first goal is to keep the Station in good shape. That is, basically routine maintenance and contingencies—if something comes up that we have to fix, we need to be able to keep it in good shape. We want to leave it in at least as good shape as we found it. Second is to carry out a program of scientific research. Even with only two people on board, where maintenance is a large piece of our

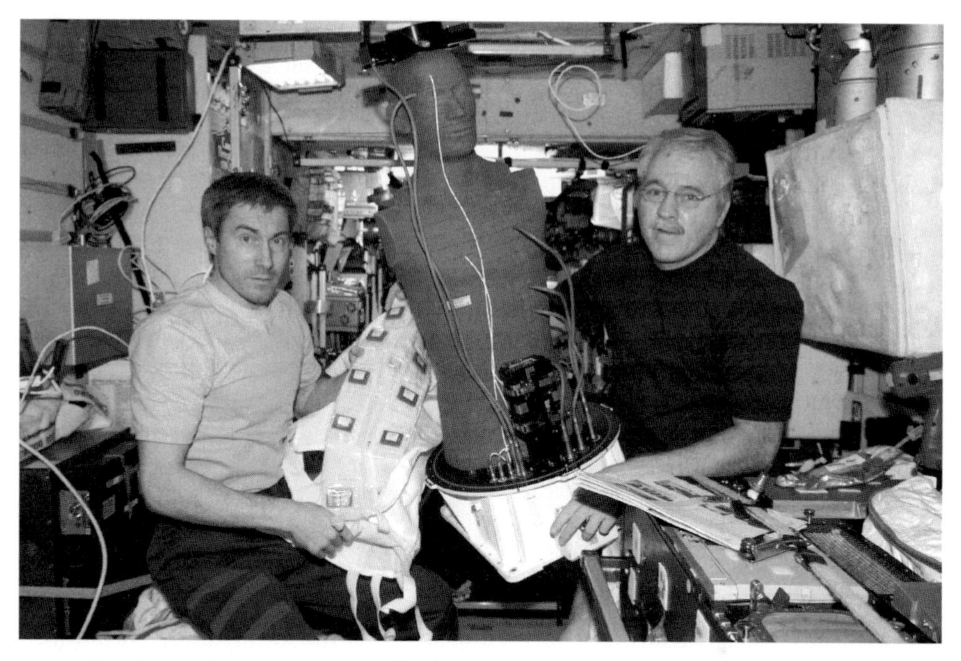

Figure 54. Expedition-11 (L to R) Sergei Krikalev and John Phillips work with a Russian radiation exposure experiment in Zvezda.

Figure 55. Expedition-11: ESA astronaut, Italian Roberto Vittori uses a communication system on ISS. He arrived with the Expedition-11 crew and returned to Earth with the Expedition-10 crew.

working day, we still have time to do scientific research. But third, and maybe the most dramatic part of all this, is that we're going to have the privilege and the challenge of being there when the American Space Shuttles return to flight, hopefully in May of this year."

For the next week the two Expedition crews worked together preparing for the Expedition-10 crew's departure. Safety and equipment briefings were dispersed between sessions unpacking the new Soyuz and preparing the old one for return to Earth. The two crews also worked together in a further attempt to repair the Elektron oxygen generator and the cooling system in Quest. Krikalev and Phillips also had a training period on the SSRMS and received additional briefings on the station's experiments. Vittori spent much of his time completing the "Endine" suite of 23 ESA, Italian Ministry of Defence, and Italian Chamber of Commerce experiments, before returning to Earth in Soyuz TMA-5, with the Expedition-10 crew. In the meantime, he would become the first ESA astronaut to make a second visit to ISS.

On April 20, Chiao and Phillips worked together to flush the cooling system and replace an umbilical in the Quest, in advance of its return to use during the Expedition-11 occupation. The following day they worked to re-size the EMUs stored in the airlock. Meanwhile, they both worked with Vittori to prepare Soyuz TMA-5 for their return to Earth. On April 22 the two crews joined together for the official change of command ceremony.

During the following week Krikalev worked on the condensate removal system of the Elektron oxygen generation system, which remained off-line after a further 12 hours of work. Krikalev also completed the transfer of water to the station from the docked Progress vehicle. Phillips spent his time installing the Expedition-11 software in the station's computers. Both men also worked to prepare for the arrival of STS-114.

Chiao, Sharipov, and Vittori separated from ISS in Soyuz TMA-5 at 14:41, April 24. Sharipov completed the undocking manually in order to reduce the drain on the back-up battery, which had been showing reduced current throughout the Expedition-10 occupation. Soyuz TMA-5 landed at 18:08, the same day, after a flight lasting 192 days 19 hours 2 minutes. Vittori had been in flight for 9 days 21 hours 21 minutes.

EXPEDITION-11

During their first week alone on ISS, Krikalev worked on the condensate removal system of the Elektron oxygen generation system, which continued to be off-line. He also completed the transfer of water to the station from the docked Progress M-52. On April 29, the Expedition-11 crew were informed that the launch of STS-114, then planned for May 22, had been delayed until "no earlier than July 13". This was due to a requirement to spend additional time studying the potential for ice damage on the underside of the orbiter caused by ice falling from the Shuttle's External Tank. Onboard ISS, both men continued to prepare for the arrival of STS-114. Included in the preparations was work with a digital camera that would be used to photograph the underside of the approaching orbiter, Discovery. They also cleared cargo away from the hatch in Unity that would be used to access the MPLM that Discovery would bring up to the station. Meanwhile, the Soyuz TMA-5 crew had been returned to the Gagarin Cosmonaut Training Centre outside Moscow, where they were reunited with their families.

In the following week Krikalev cleared a blockage in the Russian de-humidifier system and transferred waste water from the station's storage tank to Progress M-52. He also replaced a liquid-processing component of the Elektron oxygen-generating system, but it failed almost immediately. The crew continued to burn two SFOGs per day, as had occurred during the hand-over period when additional oxygen was required. Phillips packed items to be returned to Earth on STS-114, and repaired a treadmill that had stopped working. Both men participated in experiments and practised with the SSRMS, putting it through a series of manoeuvres that would allow operators on the ground to operate the arm remotely in the future. During an exercise period at the end of the week the treadmill stopped working when a circuit breaker tripped.

The crew observed Victory Day, the Russian 60th anniversary of the end of the Great Patriotic War (World War II) on May 8. In the following week, a monthly inspection of the treadmill revealed a broken restraint cable and the crew were instructed to use alternative exercise equipment until they could repair the cable, on May 16. Oxygen from Progress M-52 was used to repressurise the station, and on May 11 the thrusters on Progress M-52 were used to raise the station's orbit. The remainder of the week was spent preparing for the arrival of STS-114, updating software on the station's computers and performing experiments.

At this time, future access to ISS was the subject of an announcement by a spokesman from Roscosmos Space Agency. Russia had carried the financial responsibility of keeping ISS occupied and supplied since the loss of STS-107 and were seeking recompense. With the Russian contract to provide Soyuz and Progress spacecraft to the station outside of the Iran Non-proliferation Act due to end in April 2006, the Russians repeated that after that date Soyuz spacecraft would only carry Russian cosmonauts and paying passengers into space. They also demanded that Russian cosmonauts no longer be made to perform American experiments in lieu of the American money used to complete the building and launch of Zvezda. The relationship between the two major partners was beginning to change.

On May 20, *Florida Today* carried an article entitled, "NASA may have to abandon station." The article quoted US Representative Sherwood Boehlert as saying, "If we don't have agreement with the Russians, then we won't be able to have people in space for long periods of time." The article went on to explain the details of how and why the Iran Non-proliferation Act had been set up in 2000, but 11 Soyuz spacecraft for ISS were exempted because they were already contracted for. Luckily, those spacecraft had been available to keep ISS occupied following the loss of STS-107 in February 2003, but the last of those 11 Soyuz was due to be launched in

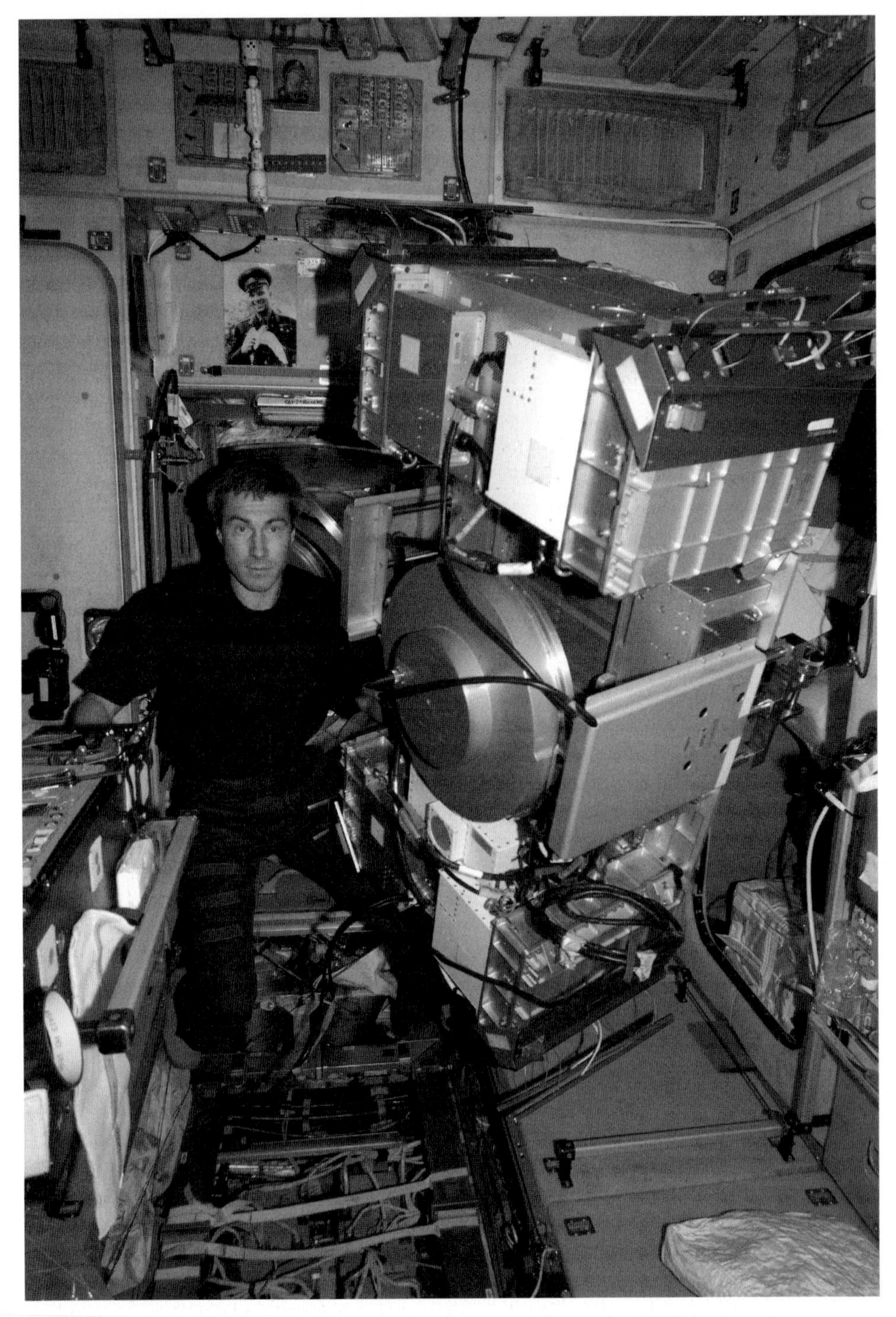

Figure 56. Expedition-11: Sergei Krikalev works on the TVIS in Zvezda.

September 2005 and return to Earth in April 2006. After that date Russia was under no obligation to launch American astronauts to ISS, or recover them from the station, including acting as CRV for American astronauts. The article suggested that, if no new contract with the Russians was signed, then American astronauts would only be able to occupy ISS when the Shuttle was present and even then the Russians would be under no obligation to recover American astronauts if the Shuttle, or ISS, malfunctioned while they were onboard. If the Russians chose to apply the letter of their contract then American astronauts would be restricted to occupation periods of just 2 or 3 weeks and would have no access to ISS between the Shuttle programme ending in 2010 and the first flight of the new Crew Exploration Vehicle.

On May 17 the Expedition-11 crew removed the contents of the Quest airlock and PMA-2 before depressurising both modules in a rehearsal of procedures that would conserve the station's nitrogen supply during the visit of STS-114. Over the following two days the station was repressurised using the last of the oxygen held in Progress M-52. They also conducted test ignitions on two of the station's SFOG oxygen-generating candles. Throughout the week both men performed a variety of experiments and Phillips wore the FOOT instrumented leggings, recording data for that experiment.

During the third week of May, both crew members performed a number of medical and microgravity experiments. They also practised further with the cameras that they would use to photograph STS-114 during its approach to the station. They also continued to burn SFOG candles to replenish the oxygen inside the station. Krikalev worked on the Elektron oxygen generator, providing additional data for the engineers at Korolev. His work identified that the electrolyser showed no voltage and was presumed to have failed. Krikalev also worked to bypass a blockage in the condenser system of the Russian modules.

On June 3 the final series of tests were completed to give the Robotics Officer in Houston control of the SSRMS. During the tests the arm was commanded to move out, latch onto a fixture on the exterior of ISS, then unlatch, and move back to its parked position. Phillips monitored the test from the SSRMS station in Destiny. Following instructions from Korolev, Krikalev continued to work on the Elektron, to add to the data available to Russian engineers. Meanwhile, the Expedition-11 crew continued to burn two SFOG candles a day. Phillips also continued to work on a number of microgravity experiments.

Preparations for the undocking of Progress M-52 began during the second week of June. The crew spent the week packing the Progress with items of rubbish. Progress transferred 217 kg of propellants to Zvezda's tanks on June 6. Four days later, the crew were able to photograph Tropical Storm Arlene. Krikalev swapped the liquid unit in the Elektron, in preparation for the arrival of filters and gas lines due to be delivered on Progress M-54. Phillips performed medical experiments to measure muscle tone, for comparison with data recorded before he was launched into space. Progress M-52 undocked at 16:16, June 15 and was de-orbited, to burn up in the atmosphere, later the same day. In Moscow, Expedition-9 Commander Gennady Padalka told journalists, "The station is very overloaded, with cosmonauts using all nooks for keeping cargoes which await return trips in US Shuttles."

Figure 57. Expedition-11: John Phillips works on the Elektron oxygen generator in Zvezda. Note the window beneath his right foot.

PROGRESS M-53

Progress M-53 was launched from Baikonur at 19:09, June 16, 2005. Ten minutes later the Expedition-11 crew were informed of its successful launch. Following a routine 2-day rendezvous the KURS automatic system failed when a Russian ground station malfunctioned and prevented data being up-linked to the Progress. Krikalev had to use the TORU flight station in Zvezda to manually dock the new Progress to Zarya's wake, at 20:42, June 18. The crew equalised the pressure between the two vehicles and opened the connecting hatches just hours after the docking, but did not begin unloading until the following day.

In the week following the new arrival, Krikalev and Phillips spent much of their time unloading the new supplies and preparing for the launch of STS-114 in July. Krikalev installed the newly delivered electrolyte and filters in the Elektron and powered the oxygen generator on. It failed almost immediately. After a second activation the Elektron operated for almost 30 minutes before shutting down. PMA-3 was pressurised and opened for the first time in four years before being utilised as storage space. The crew also cycled the docking system on Destiny's nadir, in preparation for the docking of the MPLM that STS-114 would carry to ISS. The system had not been cycled in 2.5 years.

On June 30, NASA Shuttle programme managers set an official launch date of July 13, for the STS-114 Return to Flight mission. Discovery's lift-off was set for 15:51, with docking to ISS at 12:27, July 15. For the first time in the Shuttle programme, Atlantis would be ready to launch a rescue mission if required during the duration of the STS-114 flight. This was one result of the "if it is damaged there's nothing we can do to help the crew" attitude shown by some programme managers during STS-107, which had been heavily criticised by the Columbia Accident Investigation Board.

In preparation for that launch, Phillips installed a camera used to align the MPLM correctly for docking with Destiny. The ISS crew also trained with the SSRMS, which would be used for the first time to lift an MPLM out of the payload bay. During the session they commanded the SSRMS to "walk" from Destiny's operating base and install itself on the MBS mounted on the ITS on June 29, and back to Destiny's operating base on June 30. The manoeuvre would be used to allow the SSRMS' camera to observe heatshield tiles on Discovery one day after docking. The MPLM transfer manoeuvres would all be carried out with the SSRMS on Destiny's operating base. Krikalev also tested radio guidance equipment intended for use with the European ATV. On June 30 the engines on Progress M-53 were used to adjust the station's orbit in preparation for Discovery's arrival. Oxygen in the Progress tanks was used to repressurise ISS, allowing the crew to stop burning SFOG candles. Both men continued to operate experiments from their respective countries.

The following week both men collected items to be returned to Earth in Discovery's MPLM once the new stores had been removed from it. Phillips conducted the routine charging and discharging cycle on the batteries in the American EMUs held in Quest, in preparation for the three EVAs planned during the visit of STS-114. On June 5, Progress M-53 was employed in another orbit-raising manoeuvre, while

Krikalev continued to oversee the use of oxygen and the transfer of water from the tanks in Progress M-53. Both men also continued their experiment programmes, including their second of three sessions with the Renal Stone Experiment. Throughout the week they photographed Tropical Storm Denis. In the second week of July they continued to prepare items for return to Earth with Discovery, while continuing their experiment and personal exercise programmes.

RETURN TO FLIGHT

From the moment that STS-107, Columbia, was lost, on February 1, 2003, NASA had been planning towards the Shuttle's Return to Flight. NASA Administrator Sean O'Keefe had immediately established the CAIB to identify the cause of Columbia's loss and identify the actions NASA would be required to take before Discovery, Atlantis, and Endeavour could return to space. On July 28, 2003, O'Keefe also established the Return to Flight Task Group (RTFTG), to oversee NASA's preparations for the Shuttle's Return to Flight and the correct implementation of the CAIB's recommendations. The CAIB published its report five weeks later, on August 26, 2003. It listed the 15 recommendations for the Shuttle's safe return to flight contained earlier in this manuscript (see pp. 136–137).

Meanwhile, the loss of the second Space Shuttle orbiter with a full crew of seven astronauts had caused a re-think of America's human spaceflight policy. On January 14, 2004, President George W. Bush announced the Vision for Space Exploration. It called for the Shuttle's Return to Flight and the completion of ISS to "Core Complete", with all American and International Partners' modules in place. Additional Russian modules would be added as and when the Russian economy allowed. "Core Complete" was to be achieved in 2010, at which time the Shuttle fleet would be retired. NASA would develop by 2014 a new Crew Exploration Vehicle (CEV), designed to act as a CTV and CRV for the station before returning humans to the Moon by 2020. In order to complete ISS by 2010, NASA would have to return the Shuttle to flight in 2005 and fly five successful missions per year for the next five years thereafter. It was a bold, if not impossible launch schedule. Despite all of their good words about the new priority to be given to crew safety to the media, NASA would have no choice but to be budget and schedule-driven.

Throughout 2004, NASA set a succession of target dates for the Return to Flight launch of STS-114, but they passed, one after the other, with no launch. Again and again the RTFTG reported to NASA Headquarters that the Administration was not sufficiently advanced on the CAIB's 15 recommendations to attempt the launch. In January 2005, the RTFTG reported that NASA had still failed to meet 8 of the 15 recommendations. The following month Sean O'Keefe resigned as NASA Administrator. His position was filled by Michael Griffin, a former senior NASA engineer, on April 14, 2005.

By that time the STS-114 launch was set for May 15, and the vehicle was sitting on LC-39, Pad-B. Griffin delayed the launch and STS-114 was returned to the VAB, to have its ET and SRBs exchanged with those that had been delivered for the next flight by Atlantis. The new ET contained anti-ice formation heaters that were part of the attempt by Lockheed-Martin to prevent foam and ice shedding from the ET during launch. STS-114 had originally been stacked to an ET that did not have these heaters.

The RTFTG final report Executive Summary was published on June 28; it stated that NASA had still failed to meet three of the CAIB's recommendations for a safe return to flight. They had:

1. Failed to prove they had significantly reduced, or stopped foam and ice shedding from the ET during launch. NASA Administrator Michael Griffin had stated that the Administration might have to accept that it was impossible to stop all foam shedding from the External Tank.
2. Failed to harden the orbiter against strikes by foam and ice shed from the ET.
3. Failed to demonstrate heatshield tile repair methods and prove that repaired tiles could survive re-entry heating. (Ironically, STS-114 was due to demonstrate tile repair methods on deliberately damaged tiles carried into orbit for that specific purpose in the payload bay.)

STS-114 was moved back to the launchpad and prepared for launch on July 13. On that date the countdown was stopped at $T-2$ hours when one of four fuel sensors malfunctioned. The launch was cancelled and an investigation began. Rather than replace the sensor, NASA informed the media that they fully understood the problem and would continue with the new launch if the exact problem was repeated during that attempt. Launch was set for July 26.

On July 18, Krikalev and Phillips tested the motion control system on Soyuz TMA-6. With Discovery's launch delayed, NASA managers brought the transfer of Soyuz TMA-6 from Pirs to Zarya forward. The transfer was required so Pirs could be used as an airlock for an EVA by the Expedition-11 crew, originally planned for August. The transfer took place on July 19, with undocking from Pirs occurring at 06:38. Krikalev backed the Soyuz away 30 m, flew laterally along the length of the station, and, after 14 minutes of station keeping, docking to Zarya took place at 07:08. Fifty-two minutes later the crew entered ISS and began reconfiguring it for normal use. July 23, 2005, was the Expedition-11 crew's 100th day in space. Having completed the packing of items for return to Earth on STS-114, they began to pack items for return to Earth on STS-121, due to fly in September 2005.

In advance of his present launch, Krikalev had said:

> "We have a pretty well-developed process and pretty good calculations of how much water the human body consumes, how much food you need to continue operation on board the Station, and of course this kind of supply is first priority. That's why, when Shuttle was not able to fly, all this kind of load was taken by Progress. But as a result we were not able to deliver as much equipment for scientific experiments. So increasing the variety of means to deliver cargo on orbit increases not only amount of food (we don't need food more than we can eat—we can increase our margins in case of emergency again, but we don't need much

> more food than was delivered before), but we would be able to deliver more equipment for experiments. Returning [the] Shuttle back to flight would mean more scientific capabilities because we would be able to have three crewmembers after that. We would be able to conduct more experiments."

Phillips was also looking forward to the Shuttle fleet returning to flight:

> "[S]ince the Columbia accident, the Russian Space Agency has been literally carrying the load and bringing us all the supplies we need, mostly on the Progress vehicle, [and] smaller amounts on the Soyuz vehicles. One impact of that is that we've only had a crew of two instead of a crew of three, which, of course, reduces the amount of science we can do. Another impact is that we've frankly been operating on pretty thin margins of certain consumables—food, water and oxygen. Once the Space Shuttles start flying, they carry a huge amount of mass to orbit, so they can bring our reserves of food, oxygen, and water back up to where they should be. The Shuttle makes water with its electrical power generation system [fuel cells]. We should get well with water, food and oxygen, as well as spare parts that we haven't been able to bring back.
>
> Another thing that people don't often think of is the Shuttle also carries a tremendous amount of down-mass. We've been accumulating a lot of equipment, some of which is equipment that needs to be returned to Earth and some of which is just plain trash, and there are limited amounts we can get rid of on the Progress vehicles. We should be able to load some of that stuff on the Multi-Purpose Logistics Module that'll be in the payload bay of the Shuttle, and help clean out the Station a little bit."

STS-114: A TEMPORARY RETURN TO FLIGHT

STS-114	
COMMANDER	Eileen Collins
PILOT	James Kelly
MISSION SPECIALISTS	Wendy Lawrence, Andrew Thomas, Charles Camarda, Stephen Robinson, Soichi Noguchi (Japan).

The STS-114 Return to Flight crew was made up of two pre-STS-107 crews. Lawrence, Thomas, and Camarda were assigned to the original STS-114 crew, which should have carried the original three-person Expedition-7 crew to ISS and returned the Expedition-6 crew to Earth. Collins, Kelly, Robinson, and Noguchi were the original STS-116 crew, which should have carried the original three-person Expedition-8 crew up to ISS and returned the Expedition-7 crew to Earth. The new STS-114 would retain much of the original flight's logistics mission, carrying

Figure 58. STS-114 crew (L to R): James Kelly, Andrew Thomas, Wendy Lawrence, Charles Camarda, Eileen Collins, Soichi Noguchi.

an MPLM to the station and a replacement CMG. For the first time the MPLM would be lifted out of the payload bay and returned to it using the SSRMS.

After a 2.5-year delay, STS-114 was finally launched at 10:39, July 26, 2005, and entered orbit a few minutes later. Throughout the launch over 100 high-resolution cameras filmed the ascent from all angles, in order to identify any debris that might shed from the vehicle. Noguchi used a hand-held video camera to film the ET as it was jettisoned. Before beginning their sleep period at 17:00, the crew un-berthed the RMS and used its cameras to view the clearances between Discovery's Ku-band antenna and the new Orbiter Boom Sensor System (OBSS) moored along the starboard sill of the payload bay.

Meanwhile, processing of the launch imagery showed a small piece of heatshield tile departing from close to Discovery's nosewheel doors. A tile in that area had been damaged and repaired during vehicle preparation, and this repair may have been shaken loose during launch. At SRB separation, a large piece of debris was seen departing from the ET and moving away without striking the orbiter. Both events were videoed by new cameras mounted on the ET. Subsequent review of the images would show that the debris seen departing from the ET was the foam protuberance air load ramp: in short, it had the potential to fatally damage Discovery, just as the foam bipod ramp had doomed Columbia during the launch of STS-107. Only the airflow over the vehicle at SRB separation had prevented the large block of foam striking Discovery. Despite all of the work re-designing the area surrounding the bipod after STS-107, cameras revealed that a strip of insulation some 15 cm to 17 cm long had peeled away from the bipod itself. Also, two small dents were observed where the bipod ramp had been fitted up until STS-107. The dents should have been filled prior to the ET's delivery to KSC, but apparently had not been, nor was the fact picked up during post-delivery inspections in the VAB, or during vehicle stacking and preparation for launch. On the subject of foam loss, NASA Administrator Michael Griffin said:

> "Our guys are going to take a professional look at every frame of footage we have from every camera we have. These are test flights. The primary object under flight test is the external tank and all of the design changes we have made so we would not have a repeat of Columbia."

Discovery's crew were told of the two major debris-shedding events and that flight controllers would continue to review the various images at their disposal. Those images would ultimately reveal some 25 impacts on the orbiter during launch. Six areas would receive further inspection, including two regions where the felt-like material that was used to fill the gaps between the Shuttle's tiles was seen to be protruding. It was feared that re-entering with the filler strips protruding might cause boundary layer turbulence and higher localised heating levels during re-entry if they were not removed.

Despite everything, shortly after achieving orbit, Collins paid respect to the seven astronauts lost on STS-107 saying, "We miss them, and we are continuing their mission. God bless them tonight, and God bless their families." The crew began their 8-hour eat–sleep period just after 16:00.

STS-114's first full day in space began at 00:39. The crew downloaded their images of the ET separation. Kelly and Thomas activated the RMS, using it to pick up the OBSS, which was then employed to carry out a thorough laser-video scan of Discovery's exterior, which was downloaded to Houston in its turn. Following the survey, the OBSS was re-berthed and the RMS cameras were used to video the Thermal Protection System on the crew compartment. With the external surveys completed the crew began preparations for docking with ISS by extending the docking ring and inspecting equipment that they would use during the closing manoeuvres. They also completed tests on the two EMUs that would be used during the mission's three EVAs.

Flight Day 3 began at 23:39, July 27. During the final rendezvous with ISS Collins slowed Discovery's approach and performed a nose-over-tail pitch manoeuvre (official name: r-bar pitch manoeuvre) at a distance of 200 metres below the station. This allowed the Expedition-11 crew to obtain high-resolution digital photographs of Discovery's underside. Docking occurred at 07:18, July 28, by which time NASA had already announced that the remainder of the Shuttle fleet had been grounded indefinitely, as a result of the debris shed from STS-114's ET during launch. Shuttle programme manager William Parsons announced, "Until we're ready, we won't go fly again. I don't know when that might be." He continued, "You have to

Figure 59. STS-114: as the first Shuttle flight after the loss of STS-107 three years earlier, STS-114 performs the first r-bar pitch manoeuvre to allow the Expedition-11 crew to photograph the orbiter's Thermal Protection System. The Multi-Purpose Logistics Module Leonardo is in the payload bay.

Figure 60. STS-114: the underside of the orbiter was photographed during the new r-bar pitch manoeuvre, to allow engineers in Houston to study the Thermal Protection System for damage incurred during launch.

admit when you're wrong. We were wrong ... We need to do some work ... foam should not have come off. It came off. We've got to go do something about that."

Following the usual greeting and safety brief, the two crews got to work carrying out additional inspections of Discovery. With the station structure obstructing direct use of the RMS, Kelly, Lawrence, and Phillips used the SSRMS to lift the OBSS from its storage location and hand it to Discovery's RMS, which was operated by Thomas. The OBSS was then used to continue observations of the orbiter. Robinson and Noguchi spent time preparing for their three EVAs. After a joint meal both crews began their sleep period at 03:40, July 29. While they were asleep mission control cycled Unity's nadir CBM in anticipation of the following day's activities.

The Shuttle crew were woken up at 23:39, with the Expedition crew following at 00:09, July 30. Following breakfast, Lawrence and Kelly used the SSRMS to grapple the MPLM Raffaello, lift it out of Discovery's payload bay, at 02:00, and dock it to Unity's nadir. Electrical power from the station was applied to the MPLM at 08:50, and the hatches were opened to allow unloading to commence just after 10:00.

In the meantime, Kelly and Phillips walked the SSRMS off Destiny and on to the MBS at 05:39. They then used the arm's cameras to provide situational awareness views of the survey that Camarda and Kelly would perform using the OBSS. Beginning at 07:00, the latter pair used the OBSS mounted on Discovery's RMS to view the six areas of special interest on Discovery's heat protection system highlighted by Houston following review of the images already downloaded. Programme managers told the media that they were "feeling good about Discovery coming home." Noguchi and Robinson continued to prepare their equipment for their first EVA, planned for July 30.

Discovery's crew sealed the hatches between the two vehicles when they returned to the Shuttle for their sleep period. Before they went to bed they lowered the internal pressure to allow Noguchi and Robinson to acclimatise to the lower pressure at which their EMUs would operate. The air removed from Discovery was used to replenish the station's atmosphere. Discovery's crew were awoken at 23:43, and after their breakfast they prepared for the first EVA. Noguchi and Robinson began their pre-breathing of pure oxygen at 00:39. At the same time Krikalev and Phillips walked the SSRMS off the MBS and back on to the exterior of Destiny, from where Lawrence and Kelly would operate it in support of the EVA.

The EVA began at 05:46, July 30, when Noguchi and Robinson exited through Discovery's airlock. Noguchi remarked, "What a view." Robinson countered with, "There are just no words to describe how cool this is."

Once they were outside, Quest's outer door was also opened to provide an emergency return to the station. Internally the hatches between Discovery and ISS had been closed while the two astronauts made their egress. They were now opened, to allow Collins and Camarda to transfer items between the two vehicles. During the EVA, Camarda also assisted Kelly in operating Discovery's RMS, to use the OBSS to view seven areas of interest along the leading edge of Discovery's port wing.

Having collected their tools, Noguchi and Robinson began a demonstration of how damaged heatshield tiles might be repaired on a future Shuttle flight. Working side by side, and using deliberately damaged tiles and RCC panels for the demonstration, one astronaut repaired damaged tiles using the Emittance Wash Applicator (EWA) while the other attempted to repair RCC panel samples using the Non-Oxide Adhesive Experiment (NOAX). With that important task complete they moved to the exterior of Quest, where they installed the External Stowage Platform-2 (ESP-2) Attachment Device (ESPAD) and associated cabling.

Noguchi's next task was to replace a GPS antenna mounted on the ITS, which he completed without difficulty. Meanwhile, Robinson collected the tools they would require on their second EVA, and also re-routed electrical power plugs to direct power to CMG-2, which had been off-line since a circuit breaker had tripped in March 2005. Power flowed to CMG-2 at 10:20, and controllers in Houston began the gyro's spin-up to 6,600 revolutions per minute before bringing it back on-line as part of the ISS attitude control system. With time to spare at the end of the EVA, the two men recovered two long-term exposure experiments and photographed some disturbed insulation on the exterior of Discovery's crew compartment. Quest's hatch was closed, as were the internal hatches between Discovery and ISS, while Noguchi and Robinson returned to Discovery's airlock. The EVA ended after 6 hours 50 minutes, at 12:36.

The day ended with Houston declaring Discovery's tiles and thermal blankets fit to withstand re-entry. Their review of the RCC areas along the wings' leading edges was still continuing. During the day Collins voiced her concerns over the foam shedding from the ET during launch saying, "Personally, I did not expect any large piece of foam to fall off the External Tank. We thought we had that problem licked." During the day, the Shuttle's flight had also been extended by 24 hours. The extension would allow for the transfer of more water and additional supplies, including two of the orbiter's laptop computers, from Discovery to ISS, to support the new delay before the next Shuttle's arrival at the station.

July 31 was a day of relatively light duties, including the transfer of equipment between the MPLM and the station, as well as interviews with journalists. Noguchi and Robinson reviewed plans for their second EVA, during which they would attempt to replace CMG-1, which had failed in June 2002, and had been off-line since that time. In preparation for the EVA, Lawrence and Kelly walked the SSRMS to the correct position on the exterior of Destiny.

Following their sleep period the crew were woken up at 23:09, July 31, and began preparation for the EVA. Noguchi and Robinson opened Discovery's airlock hatch at 04:42, August 1, and set about preparing their tools. Meanwhile, controllers in Houston had turned off the electrical power to CMG-1. Both men made their way hand over hand to the Z1 Truss, on Unity's zenith, where Noguchi mounted the work platform on the end of the SSRMS, which was operated by Lawrence and Kelly. During the ride out to his work place Noguchi remarked, "Oh, the view is priceless. I can see the moon." The two men removed CMG-1, and Noguchi held it in his arms while he was manoeuvred down to Discovery's payload bay. Robinson also made his way back to the payload bay. There, Noguchi temporarily stowed the old CMG while its replacement was removed from a crate and then replaced by the old unit. Noguchi held the replacement CMG-1 while he was manoeuvred back up to the Z1-Truss, where he waited for Robinson to arrive. The two men then worked to install the new

Figure 61. STS-114: Robinson rides the SSRMS during the flight's second extravehicular activity.

CMG. With the installation complete, the new CMG-1 was spun up and, after several hours of monitoring, brought on-line as part of the station's attitude control system. Discovery's hatch was closed at 11:56, after an EVA lasting 7 hours 14 minutes. The remainder of the day was spent transferring equipment and rubbish between the two spacecraft.

On the ground, Flight Director Paul Hill described the new attitude for keeping the crew informed on the state of their spacecraft:

> "Our intent is never to hide the state of the vehicle from the crew. But has our threshold changed for TPS damage assessment? You bet it has. There are some things that we are significantly smarter on today than we were two and a half years ago, and I don't know how we could be in any different place today, since we all know that TPS damage cost the lives of the last crew."

On August 2, Houston gave permission for Robinson to venture beneath Discovery and attempt to remove, or cut away the two protruding gap fillers during the third EVA, planned for the following day. The crew spent the day preparing for this additional task, which included Lawrence and Kelly practising the intended SSRMS manoeuvres using software carried in one of the station's laptops. To make time for Robinson to work with the gap fillers, Lawrence and Kelly used the SSRMS to unstow the EPS-2 from its position in Discovery's payload bay. This had originally been included in the timeline for the third EVA.

Following another sleep period, the Shuttle crew were awoken at 23:09, August 2. Noguchi and Robinson exited Discovery's airlock at 04:48, August 3, to begin their third EVA. Their first task was to make their way to the ESPAD that they had installed on the exterior of Quest during their first EVA. Lawrence and Kelly then manoeuvred ESP-2 into position on the ESPAD using the SSRMS. With the ESP secured in place the SSRMS was walked off Destiny and on to the MBS, mounted on the ITS.

Noguchi installed the MISSE-5 exposure experiment and removed the Rotary Joint Motor Controller from the ITS and placed it in storage. In the meantime, Kelly and Camarda used Discovery's RMS, with the OBSS still attached, to view the tile and RCC repair experiments that Noguchi and Robinson had completed during their first EVA. With his tasks completed Noguchi moved over to offer whatever support he could to Robinson in his final task. Robinson mounted the SSRMS and was manoeuvred beneath Discovery's nose, where he was easily able to pull the two protruding "gap fillers" out from between TPS tiles. Robinson told Houston, "It looks like the big patient is cured!" With that task complete, Noguchi and Robinson returned to Discovery's airlock and sealed the hatch at 10:49, after 6 hours 1 minute of exposure to space.

Following their hectic schedule, August 4 was planned as a relatively easy day. Lawrence and Kelly walked the SSRMS off the MBS and back to the exterior of

Figure 62. STS-114: in an unanticipated activity Stephen Robinson extracted two "gap fillers" from the underside of the Shuttle's nose.

Destiny. They then attached the free end to Raffaello, in advance of its undocking. Noguchi spoke to the Japanese Prime Minister by video link and the American astronauts received a call from President Bush, who told them:

> "I just wanted to tell you all how proud the American people are of our astronauts. I want to thank you for being risk-takers for the sake of exploration. And I wish you Godspeed in your mission. I know you've got very important work to do ahead of you. We look forward to seeing the successful completion of this mission. And obviously, as you prepare to come back, a lot of Americans will be praying for a safe return."

A day of more equipment stowage was followed by a joint meal and a commemoration of the STS-107 crew. During the day the Shuttle's crew paid their respects to the American astronauts and Russian cosmonauts who have been lost in the exploration of space. Lawrence remarked:

> "Even if the future is equally unimaginable to us, we can be sure that future generations will look upon our endeavours in space as we look upon those early expeditions across the seas. To those generations, the need to explore space will be as self-evident as the need previous generations felt to explore the Earth and the seas."

Krikalev and Phillips ended their day by preparing Unity's CBM for Raffaello's undocking, before both crews went to sleep in their respective spacecraft.

Discovery's crew was woken up at 22:15, August 4. Kelly and Lawrence used the SSRMS to undock Raffaello from Unity's nadir, at 07:34, August 5. Raffaello now contained 3,175 kg of items dating back to Expedition-6, the last Expedition crew to be supported by a Shuttle flight. With the MPLM secured in Discovery's payload bay at 10:03, Camarda and Thomas joined Kelly and Lawrence to locate the OBSS along the starboard payload bay door sill. The remainder of the day was spent stowing equipment on Discovery's mid-deck. Both crews went to bed at 14:09.

Awake again at 22:09, the two crews shared a farewell ceremony for the leaving visitors at 00:36, August 6. Discovery's crew returned to their spacecraft and the hatches were sealed at 01:14. Kelly was at the controls when Discovery undocked, at 02:24, and moved away to a distance of 122 m. He began a fly-around manoeuvre at 03:54, and finally manoeuvred clear of ISS at 05:09. The crew were given the remainder of the day as free time, going to sleep at 00:39, August 7. The ISS crew went to bed at 14:09.

Discovery's crew were woken up at 20:39, and spent much of their work day stowing equipment for re-entry. The one remaining question was the area of TPS below Collins' window which had been under review since it was first seen in video footage. Julie Payette called from Houston to say, "We have good news. The MMT just got to the conclusion that the blanket underneath ... the window is safe for return. There is no issue." During the day, Collins, Kelly, and Robinson tested Discovery's aerodynamic surfaces and fired the orbiter's thrusters. The day ended at 12:39, August 7. Meanwhile, the Expedition-11 crew spent a quiet day and adjusted their schedule back to their normal routine. They were woken up at 02:00, August 7.

Discovery's crew began August 7 by waking up at 20:39. They commenced their final preparations for the de-orbit burn and re-entry. The crew had two opportunities to land in the pre-dawn darkness at KSC, at 04:47 or 06:22, August 8, but Discovery was waved off for 24 hours, due to unpredictable weather. Following the wave-off, Discovery's engines were fired at 08:19 to optimise the landing opportunities on August 9. The crew's day ended at 00:39, August 9.

Up at 08:39, the crew repeated their final preparations for re-entry, going through the checklist once again. All three landing sites, KSC, Florida, Edwards Air Force Base, California, and White Sands, New Mexico, were activated for this second attempt to land. Meanwhile, the world's media were baying like dogs, waiting for Discovery to burn up, so they could repeat their tired calls for the space programme to be cancelled. Persistent thunderstorms over Florida led to the two KSC landing opportunities being cancelled. Discovery would now land in California. Travelling with its three SSMEs forward, Discovery's thrusters performed retrofire at 07:06. Turning, the orbiter assumed the usual nose forward and high position for re-entry. Collins brought the flight to a successful close, gliding Discovery to a perfect landing at Edwards Air Force Base in the pre-dawn darkness, touching down at 08:11.

Krikalev and Phillips sent their congratulations from ISS, but the successful landing meant that the Shuttle fleet was now grounded once more.

In her post-landing speech Collins remembered the crew of STS-107 once more:

> "Today was a very happy day for us, but we have mixed feelings. We have very bittersweet feelings as we remember the Columbia crew ... I thought about them the whole mission—what their experiences were. The Columbia crew believed in what they did, they believed in the space mission. I know if they were listening to me right now, they'd want us to continue this mission."

NASA Administrator Michael Griffin reminded the media at a press conference a few days later, when the crew returned to JSC in Houston:

> "For two-and-a-half-years we have been through the very worst that manned spaceflight can bring us. Over the last two weeks, we have seen the very best."

He continued:

> "Essentially, this was a test flight. It has provided data that we can use going forward. The bad news is there were three or four things we didn't get. The good news is we hugely reduced any damage to the orbiter through the engineering measures we took to improve the tank. We specifically said the return to flight test sequence was two test flights. We plan for the worst and we hope for the best and that's how we conduct business."

Collins was more personal when she spoke to the engineers who had returned the Shuttle to flight,

> "Getting the Shuttle flying again was difficult work, but it was a labour of love. Words cannot describe how much my thanks go out to you for putting your heart and soul into what you believe."

EXPEDITION-11 CONTINUES

On August 9, Krikalev and Phillips began preparation of the Pirs airlock and the tools that they would use in their next Stage EVA, planned for after the STS-114 flight, the work continued throughout the week. In Zvezda, the Russian Vozdukh carbon dioxide removal system stopped working on August 11. While Russian engineers began troubleshooting, controllers at Houston activated the American Carbon Dioxide Removal Assembly to scrub the station's atmosphere. At 01:44, August 12, Krikalev surpassed the 747 days 14 hours 14 minutes 11 seconds human spaceflight endurance record held by Sergei Avdeyev. He discussed the moment in his pre-launch interview:

> "I probably never paid enough attention to this record-setting subject because [the] job itself is very interesting for me. Being there and being able to look back to Earth, to do something challenging; that was really important. How many days was not as important."

The 62nd EVA in the ISS programme began at 03:02, August 18, 2005, when Krikalev and Phillips exited Pirs wearing Orlan suits. After preparing their tools, their first task was to recover three canisters from the Biorisk experiment that had been installed on the exterior of Pirs, in January 2005, during the Expedition-10 flight. Next they moved to the exterior of Zvezda and prepared the MPAC/SEEDs exposure samples for removal. Leaving the samples in position for the time being, they moved to Zvezda's wake, to install a back-up television camera to assist in the docking of ESA's ATV during its first flight, planned for 2006. They also photographed the condition of the Kroma experiment, which measured the residue from Zvezda's thrusters as well as replacing the sample containers in the Russian material exposure experiment. When they returned to Pirs with their tools and the MPAC and SEED experiments, they were running 45 minutes behind schedule. As the next task, the relocation of the grapple fixture that had originally held a Strela crane from the exterior of Zarya to PMA-3 on the exterior of Unity, was estimated to take 2 hours, Russian controllers decided to delay the task to a later EVA. Krikalev and Phillips closed the hatch on Pirs at 20:00, ending the EVA after 4 hours 58 minutes.

Following the loss of the Russian Vozdukh system, the American Carbon Dioxide Removal Assembly also failed, on August 18, due to a stuck valve.

Following the landing of STS-114, Administrator Griffin let it be known that he still hoped to solve the foam-shedding problem on the ET and launch STS-121 on schedule, in September 2005. That changed on August 18, when the launch was moved back to March 2006. At the same time the decision was made to swap orbiters and free Atlantis for STS-115. STS-121 was scheduled as the second test-flight, following the flight of STS-114. It would carry an MPLM full of equipment and stores to ISS before STS-115 launched the next element of the ITS and resumed the construction of ISS. Meanwhile, those ETs that had been previously delivered to KSC were returned to the Michoud Assembly Facility, Louisiana for testing and any modifications identified as a result of the post-STS-114 investigation.

With their EVA behind them, Krikalev and Phillips completed unpacking the supplies delivered by STS-114. They also commenced filling Progress M-53 with rubbish, in preparation for its undocking, on September 7. On August 23, Krikalev replaced a faulty valve in the Vozdukh carbon dioxide removal system, returning the unit to full operation. The following day, Phillips replaced a laptop computer used as part of the station's inventory system. On the same day the crew exercised a depressurisation emergency on the station.

The following week the two men prepared the station's laptop computers for a software update that would be up-linked later in the month. They also rehearsed an emergency evacuation on Soyuz TMA-6 and completed new medical experiments, delivered by STS-114. Progress M-53 was fully packed by the end of the first week in September, by which time the crew had also upgraded their treadmill, during two days of work using parts delivered by the Shuttle.

PROGRESS M-54

Krikalev and Phillips watched Progress M-53 undock at 06:26, September 7. The fully laden Progress was commanded to re-enter Earth's atmosphere, where it was heated to destruction. Progress M-54 was launched at 09:08, September 8, 2005. The heavily loaded spacecraft docked to Zvezda's wake at 10:42, September 10. Among its 2,414 kg of cargo, it carried a new liquids unit for the Russian Elektron oxygen generator, oxygen, water, propellants, clothing, food, and experiments for the Expedition-12 crew, William McArthur and Valeri Tokarev, scheduled to launch in Soyuz TMA-7 on October 1. The following week began with a day of unloading Progress, followed by a day of entering everything on the station's computerised inventory using the barcodes on each item. The third day was spent dismantling the ESA Martoshka experiment retrieved from the station's exterior during the EVA. On September 22, the crew replaced the liquids unit in the Elektron, thereby returning the unit to full use.

Meanwhile, Hurricane Katrina had flooded New Orleans, including the plant where the Shuttle's ETs were made. As the month ended, Hurricane Rita threatened JSC, in Houston, which was evacuated and Korolev assumed primary control of the station. JSC resumed normal operations on September 26. The following day, Krikalev celebrated his birthday. He spoke to his family in Korolev and opened private packages that had been delivered on Progress M-54. The Expedition-11 crew spent the week ending September 30 unloading Progress M-54, and the following week preparing for the arrival of Soyuz TMA-7 with the Expedition-12 crew, as well as preparing for their own departure from ISS. Throughout everything they continued their daily exercise regime, regular maintenance of the ISS systems and the station's experiment programme.

SOYUZ TMA-7 DELIVERS THE EXPEDITION-12 CREW

SOYUZ TMA-7	
COMMANDER	William McArthur
FLIGHT ENGINEER	Valeri Tokarev
SPACEFLIGHT PARTICIPANT	Gregory Olsen

Soyuz TMA-7 was launched from Baikonur at 23:55, September 30, 2005, while ISS was over the Pacific Ocean, off the coast of Chile. McArthur and Tokarev were the Expedition-12 crew, and planned to spend 6 months on the station, while American Gregory Olsen was a paying passenger making a 10-day flight under a commercial contract with the Russians. A medical condition had grounded the American businessman from a Soyuz flight in 2004, but he returned to the programme the following year. Olsen insisted that he would be doing good work on the station and to that end he would perform a number of experiments for Russia and ESA.

The spacecraft docked to Pirs nadir at 01:27, on October 3. Following pressure and leak checks the hatches between the two vehicles were opened at 04:36. As with all new crews aboard the station, the newcomers were treated to the traditional Russian greeting of bread and salt upon their entry into the station. The usual safety briefing and emergency evacuation exercise followed before the new crew were allowed to settle in and Olsen's couch liner was transferred to Soyuz TMA-6. American ISS commander Bill McArthur had arrived on ISS by Soyuz, but at that time, there was a faint possibility that he might have to stay on the station until a Shuttle could return him to Earth. During his occupation Russia's original contract to supply Soyuz spacecraft for ISS would come to an end. If America could not find a away around the Iran Non-proliferation Act then the Russians would be under no legal obligation to return McArthur to Earth, even in the event of an emergency evacuation of the station! NASA made it clear that they expected to find a solution to the problem and negotiations were underway to overcome the problem.

A week of joint hand-over activities followed, along with the performance of a number of short-term experiments carried to the station in the new Soyuz. Meanwhile, Olsen completed his 8-day experiment programme. Krikalev and Olsen even filmed a television commercial for a Japanese company while on the station. On October 4, Phillips and McArthur reviewed the software for the SSRMS, before they performed several manoeuvres with the SSRMS itself the following day. The crew also took a message from Mikhail Fradkov, the Russian Prime Minister.

Figure 63. Expedition-12: Sergei Krikalev is on the right, with Expedition-12 crew member William McArthur in the centre and American spaceflight participant Gregory Olsen on the left. Olsen flew to the station in a Soyuz with the Expedition-12 crew and returned to Earth with the Expedition-11 crew.

Following the official hand-over and final farewells Krikalev, Phillips, and Olsen sealed themselves inside Soyuz TMA-6. Prior to undocking the crew discussed a pressure leak between the re-entry module and the orbital module with engineers in Korolev. Krikalev finally undocked the Soyuz under manual control, at 17:49, and backed it away from the station. Throughout the separation manoeuvre and de-orbit burn the pressure leak continued to cause concern until the orbital module was jettisoned, at which time it ceased. The re-entry module landed at 21:09, on target 85 km northeast of Arkalik. The Expedition-11 flight had lasted 179 days 23 minutes and Krikalev's personal endurance record now stood at 803 days 9 hours 39 minutes. Phillips felt light-headed after being helped out of the spacecraft by the recovery forces, but could not remember afterwards if he actually blacked out. The long-duration crew were subjected to the usual 45-day medical rehabilitation. Olsen had been in flight for 9 days 21 hours 15 minutes. At a post-flight press conference he repeated his dislike of the term "space tourist", used by much of the media to describe commercial passengers on Soyuz taxi flights. He explained, "I dedicated 2 years of my life to this. It's not a hop-on-and-go kind of thing."

Meanwhile, McArthur had described his hopes for the Expedition-12 increment before launching to the station, "Above all that we launch and land safely; that we conduct this mission in a safe manner. Having said that, I think for it to be considered a success my criteria is that we will complete meaningful science during our stay, and that we will leave the Station more capable than we found it."

During October, the Russian government approved funding for the national space programme through 2015. It was supposed to include the joint development, with ESA, of the Kliper spacecraft and the development by Krunichev of the new Multi-purpose Science Module to be launched to the Russian sector of ISS.

EXPEDITION-12

After the standard weekend of light work McArthur and Tokarev began working full time on their experiment programme, as well as commencing their housekeeping, maintenance, and daily exercise regimes. They reviewed procedures for an emergency escape from the station, changed a battery in Zvezda, and rearranged the items stowed inside Unity. McArthur began work with the Pulmonary Function Facility in Destiny. Both men also began the first of a series of Renal Stone Experiment food logs and gave urine samples for the same experiment. By the second week of October the new crew were beginning preparations for their first Stage EVA, which would be made from Quest and would be the first to use American EMUs since 2003. The 5.5-hour EVA was planned for November 7. The Elektron oxygen generator in Zvezda shut down unexpectedly on October 13. The problem was a result of a partially filled source water tank being connected to the system, rather than a full tank.

Progress M-54's engines were to be used to boost the station's orbit on October 18. The engines began thrusting at the correct time but the procedure was aborted when a Russian navigation computer lost telemetry and shut them down. Trouble-shooting began at Korolev. On October 17 and 21 the station's atmosphere was repressurised using oxygen from Progress M-55. Meanwhile, planning was underway at Korolev for another attempt to repair the Elektron unit. Tokarev purged the air bubbles from the Elektron oxygen generator's systems during a 5-hour work session on October 22, thereby restoring the unit to use. During the same week, McArthur checked out the second Pulmonary Function Facility, developed by ESA for use inside their Columbus laboratory module and carried to ISS on STS-114, the unit had been installed in HRF-2 in Destiny.

On October 25, the two men carried out routine tests of the two EMUs that they would wear during their first EVA. The following day they reviewed the procedures for donning and operating the EMUs. On October 27, they donned the suits and rehearsed their EVA activities inside the station. Meanwhile, on October 26, Russian controllers had performed a test-firing of Progress M-55's engines, using a different manifold to that used during the aborted re-boost firing. The engines operated normally and there was no loss of telemetry. Tokarev celebrated his birthday on October 29. The following day, both men worked to strip down and sample the airflow in the Trace Contaminant Control System. Engineers had noticed a reduction in the airflow and the astronauts' work led to the conclusion that replacement parts might be required. Following re-assembly, the unit continued to work at a reduced airflow rate. During the week they also replaced a faulty pump in a thermal control loop in Zvezda, and replaced smoke detectors, also in Zvezda.

McArthur and Tokarev marked the fifth anniversary of permanent human presence on ISS on November 2, 2005. They sent messages to everyone who had flown to the station and to the engineers and scientists from 16 nations who supported its activities.

At 10:32, November 7, McArthur and Tokarev began their first EVA as they placed their EMUs on to battery power and began depressurising Quest. During the preparations they had to repressurise the airlock and re-enter the inner chamber of the two-chamber module and reset a misaligned valve. They then had to seal themselves back in the outer chamber and depressurise it for a second time. Exiting the airlock, they collected their tools and retrieved a stanchion for a television camera from a toolbox mounted on the exterior of Quest, before making their way to the outer limit of the Port-1 ITS, where they installed a television camera on a stanchion and installed this on the outer limit of the Port-1 ITS. When power was applied to the camera the first pictures were received just before 13:00. The new camera would be used during future assembly tasks, when additional SAWs would be added to the port side of the ITS. The camera should have been installed as part of the final STS-114 EVA, but the installation was delayed to allow for the removal of the two gap fillers from the underside of the orbiter. Their next job was a "get-ahead" task. They removed a failed Rotary Joint Motor Controller (RJMC), a box of electronics. It had not yet been used, and was to be returned to Earth on the next Shuttle for evaluation of why it had failed.

Both men then used their hands to make their way to the top of the P-6 Truss, the "highest" point on the station. There, McArthur removed the now defunct Floating Potential Probe and pushed it away from the station. It would burn up when it re-entered Earth's atmosphere, in approximately 100 days. It had been installed by the STS-97 crew in December 2000, to help define the electrical environment around the station's SAWs. Images taken on STS-114 had shown it to be breaking up, so the decision was taken to remove it. With both of their primary tasks completed, the crew received permission to progress on to a second "get ahead" task. They removed a failed circuit breaker controlling redundant heating on the Mobile Transporter, and installed a new one. The two astronauts then returned to Quest after an EVA lasting 5 hours 22 minutes. In the days following the EVA both men spent time servicing the suits they had worn.

On November 10, Progress M-54's thrusters were fired to boost the station's orbit. The 33-minute, two-stage re-boost was the longest yet carried out using the engines of a Progress spacecraft, and was designed to place the station in the correct orbit for the arrival of Progress M-55, in December. During the week the station toilet control panel malfunctioned and Tokarev replaced it. The following week, McArthur spent several hours photographing the Binary Colloidal Alloy Test experiment that had been undisturbed in microgravity for over a year.

After configuring the station for automatic function the crew sealed themselves inside Soyuz TMA-7 on November 18. At 03:46, Tokarev undocked the Soyuz from Pirs and manoeuvred along the station to dock at Zarya's nadir, at 04:05. McArthur and Tokarev returned to ISS just after 10:00. The newly installed Port-1 ITS television camera transmitted images of the manoeuvre, which cleared Pirs' nadir for the crew's second Stage EVA, during which they would wear Orlan suits. The EVA was originally planned for December 7, but was under review as the crew moved their spacecraft. Mission managers were considering delaying the EVA to early 2006, in order to give the crew more time to unload Progress M-54 and prepare it for undocking.

McArthur powered up the SSRMS on November 21, and put it through a series of engineering tests. He left it in a suitable position for its cameras to monitor the crew's second EVA, which had been rescheduled to February 2, 2006 by that time. A possible third Stage EVA was cancelled, because the "get ahead" tasks had been achieved during their first EVA, in November. November 24 was a day off for the crew to celebrate the American Thanksgiving Holiday.

During the week ending December 2, McArthur worked with the HRF-2 experiment rack in Destiny. He set up a refrigerated centrifuge and worked with the BCAT-3 and InSPACE Magnetic Materials experiments. He also replaced fuses in a Trace Contaminant Monitor in Destiny. At the same time Tokarev used oxygen contained in Progress M-54 to repressurise the station. Propellant was also transferred from Progress to Zarya. Tokarev also installed a muffled adjustable fan in the crew quarters to reduce noise in that region. Both men spent time collecting rubbish for disposal in Progress M-54. Oxygen from the spacecraft was pumped into the station's atmosphere and the 221 kg of propellant that it carried was transferred to Zvezda's tanks. McArthur replaced an air circulation fan in one of Destiny's experiment racks and updated the software used by all five experiment racks in the laboratory module. Tokarov repaired air ducts in the American sector, thereby improving airflow in the modules. He also installed muffled fans in the sleeping quarters, thereby reducing the noise that the fans in that important area produced. As part of the preparation for Progress M-54's undocking they removed the spacecraft's Kurs automatic docking system for return to Earth. Ultimately, plans to undock Progress M-54 on December 20 were cancelled in favour of keeping the craft docked to the station for several more months, thereby allowing the crew to continue to use its oxygen supply and to load it with additional rubbish. The second week of December was taken up with biomedical experiments and maintenance work. On December 16, one of two cables carrying power, command data, and video to and from the Mobile Transporter was severed, causing loss of data. Telemetry suggested that the cable had been deliberately cut by the disconnect actuator system, designed to cut the cable if it became snagged or tangled. This was a malfunction of the cutting system. The cable being severed resulted in one of two redundant electrical power circuit breakers being tripped. The second cable, on the other side of the ITS, remained undamaged.

As the year drew to a close, ESA announced that technical difficulties had led to the first Ariane-V/ATV flight being delayed by almost a whole year, to 2007. ESA had also transferred the launch of the European Robotic Arm from the American Shuttle to a Russian Proton launch vehicle. A third ESA announcement gave details of how the organisation had refused the requested $51 million to undertake a joint

Figure 64. Expedition-12 (L to R): Valeri Tokarev and William McArthur pose at the Zvezda mess table with Christmas tree, stockings, and a Russian doll.

Preparatory Design Study with Roscosmos of Russia's proposed Kliper spacecraft. The vote went against the proposal because ESA would have no control over the programme and would receive only minor industrial contracts. NASA also had an announcement: with the redirection of the American human spaceflight programme towards the new Project Constellation, the Administration had already begun cancelling some experiment projects designed to be flown to ISS. On the positive side, Congress had approved the purchase of additional access to Russian Soyuz spacecraft, despite the wording of the Iran Non-proliferation Act. The new spacecraft would provide access to ISS and CRV responsibilities for American astronauts through 2012. It was accepted that some of the "taxi" flights would carry spaceflight participants in the third couch. At the same time, Russia agreed to double the production rate of Soyuz spacecraft from 2009, thereby allowing the ISS Expedition crew to be made up to six people, supported by two Soyuz CRVs docked to the station at all times. Seven Soyuz spacecraft would be flown in the period 2008–2011. Eight Progress spacecraft would fly in the same period.

PROGRESS M-55

Progress M-55 was launched from Baikonur, at 13:38, December 20, 2005 with 2,490 kg of cargo. The new spacecraft carried food, including 14 kg of fresh fruit and vegetables, water, oxygen, propellant, spare parts, and experiments for ISS. Progress docked to Pirs' nadir automatically at 14:46, December 23. The crew began unloading the dry goods over the holiday week.

In the week running up to Christmas, McArthur and Tokarev performed experiments and recorded educational films. McArthur checked the hatch seals in the American modules of the station. The Christmas break represented the halfway point in their 6-month mission. On Christmas Day they were able to speak to their families, ate a traditional Russian meal, and opened the gifts that had been delivered on Progress M-55. They had a day off on December 26, and again on January 2, in keeping with standard American government employment rules that gave employees a day off in lieu when a national holiday falls on a Sunday. Meanwhile, Zvezda's Elektron oxygen generator had been performing flawlessly since its last repair. It was deliberately shut down between December 28 and February 9, to allow the crew to burn Russian SFOG oxygen candles in order to re-certify that method of producing oxygen. On December 31, the last of the oxygen carried in Progress M-54 was used to re-supply ISS.

McArthur and Tokarev spent the first week of 2006 performing experiments, including the ground-commanded BCAT-3, placing the "phantom Torso" with its 370 radiation detectors in Pirs, and locating the EarthKam in one of the station's windows in advance of the new school term. They also installed batteries in the American EMUs in Quest. January 9 was the final day of the Russian holiday and the crew had the day off. During the week, they installed the Recharge Oxygen Orifice Bypass Assembly (ROOBA), a method of allowing EVA astronauts to pre-breathe oxygen from the Shuttle's supply rather than the station's tanks in Quest. Two days later the Elektron oxygen generator was powered on. On January 12, McArthur manoeuvred the SSRMS to provide views of the Interface Umbilical Assembly (IUA) on the S-0 ITS, which held the cable cutter for the MT's uncut power cable. The following day he manoeuvred the SSRMS to view the CBM at Unity's nadir and ensure that it was clear of debris. The SSRMS was left in a position where its cameras could view the crew's up-coming EVA.

On the ground, the Houston marathon was taking place on January 15. In orbit, McArthur ran a half-marathon on the treadmill while ISS circled Earth. Over January 17–18, they rehearsed the procedures to be followed in the event of a rapid pressure leak requiring evacuation of the module in question. The following day, programme managers delayed the next EVA from February 2 to February 3, in order to ease the astronauts' preparation schedule. The remainder of the week was taken up with both men performing experiments for their national programmes.

In the week commencing January 23, both men began preparing for their EVA. On January 31 they prepared an old Orlan suit, mounting a radio and slow-scan television transmitter on the helmet. The system transmitted messages in six languages that could be received by amateur radio operators. Now called "RadioScaf", the suit had last been worn by Michael Foale in February 2004. It was filled with rubbish and would be jettisoned from ISS during the EVA. After preparing ISS for automated flight regime and shutting down the Elektron generator, the oxygen delivered in Progress M-55 was used to pressurise the ISS.

Figure 65. Expedition-12: A "past its sell-by date" Orlan pressure suit fitted with a radio transmitter was jettisoned to become a satellite in its own right. The experiment was named "RadioScaf", but in their wisdom the media called it "Suitsat".

McArthur and Tokarev left Pirs wearing Orlan suits at 17:44, February 3. Having prepared their tools, they removed RadioScaf from the airlock and mounted it on a ladder on the exterior of the module, before releasing the suit into orbit with the words, "Goodbye, Mr. Smith," from Tokarov. They photographed the suit as it drifted away. It transmitted its greeting messages for two orbits before the transmitter stopped working.

Moving away from Pirs, the two men made their way to the exterior of Zarya, where they removed a grapple fixture adapter for the Russian Strela crane and moved it to PMA-3, mounted on Unity. The adapter was removed to prepare Zarya for the temporary stowage of debris shields, prior to their deployment on a later Shuttle flight. Command of the EVA passed from Korolev to Houston as the astronauts passed from the exterior of Zarya on to the exterior of Unity. Next, they moved to the S-0 ITS, where they attempted to drive home a safety bolt in the cutting device in the IUA that McArthur had filmed on January 13. Despite several attempts with a high-tech tool, the safety bolt could not be installed to prevent the blade from falling and cutting the cable. Instead, as a temporary measure, McArthur removed the cable from the cutting mechanism and tied it to a handrail with a piece of wire. The cut cable on the other side of the ITS would be repaired during an EVA by the crew of STS-121. After transferring control of the EVA back to Korolev, the final task for this EVA was to photograph the exterior of Zvezda before returning to Pirs, where Tokarev recovered the Biorisk-2 experiment. The airlock hatch was closed after an EVA lasting 5 hours 43 minutes.

Two days later NASA Administrator Michael Griffin made a speech from NASA Headquarters, Washington, in which he stated:

> "... The greatest management challenge the agency faces over the next five years is the transition from retiring the Shuttle to bringing the Crew Exploration Vehicle on-line ... We are delving more deeply into the strategic implications of using Shuttle-derived launch systems for the Crew Launch Vehicle and Heavy-Lift Launch Vehicle ... Thus, we are applying some funds from the exploration budget profile between now and 2010 to the Shuttle's budget line to ensure the Shuttle and Station programmes have the resources necessary to carry out the first steps of the Vision for Space Exploration. NASA has asked industry for proposals to bring the CEV on-line as close to 2010 as possible, and not later than 2012 ..."

Griffin had initiated studies inside NASA to replace the Delta-IV and Atlas-V launch vehicles proposed for the CEV with a launch vehicle derived from proven Shuttle technology. Those studies would lead to the new Ares class of launch vehicles.

In the days following the EVA the crew took part in the standard debriefings with experts on the ground. McArthur also conducted a video tour of the station. The crew also continued their experiments; McArthur participated in the FOOT experiment, and Tokarev performed two ESA experiments. Both men performed Russian biomedical experiments and monitored the numerous experiments that ran automatically. They also gave the treadmill its 6-monthly overhaul. The motors in Progress M-55 were used to boost the station's orbit on February 11. This was the first time that a Progress docked to Pirs had been used for an orbital re-boost. Over February 16–17, McArthur worked to replace the spectrometer inside the Mass Constituent Analyser (MCA) in Destiny. This measured the composition of the station's internal atmosphere. An attempt to power the device up on the second day failed and McArthur was requested to perform troubleshooting.

As a new week began plans for the crew to "camp out" in Quest for their February 23–24 sleep period were delayed until March. The experiment called for the two men to spend a single sleep period in the airlock with the hatches sealed and the pressure reduced. Although they would not be wearing their EVA suits, the camp-out was seen as a way of reducing the pre-breathing of pure oxygen required before an extravehicular activity, by having future EVA astronauts spend their sleep period prior to an EVA camping out in Quest, and remaining in the airlock when they wake up, in order to don their EMUs and prepare for the EVA. The camp-out procedure would be used for the first time by EVA astronauts on STS-115. The delay was called as a direct result of McArthur's failure to repair the MCA and, pending its repair, the camp-out was rescheduled for March 23.

McArthur and Tokarev began preparations for the next Shuttle flight, STS-121, now planned for no earlier than July 2006. McArthur made space in Destiny's experiment racks for the equipment that Discovery would deliver to ISS. Both men worked to load Progress M-54 with additional rubbish, in preparation for its undocking planned for March 3. Meanwhile, on February 21, Progress M-55 performed a second re-boost manoeuvre. Progress M-54 was finally undocked from

Figure 66. Expedition-12: William McArthur works within the rear of an experiment rack inside Destiny.

Zvezda's wake at 06:06, March 3, 2006. Three hours later controllers in Korolev commanded it to re-enter the atmosphere, where it burned up over the Pacific Ocean. With the Progress gone, McArthur returned to his work on the MCA.

Meanwhile, on March 2, the programme managers from all of the ISS nations agreed a revised launch schedule to bring the station to "Core Complete", plus the International Partner modules before the Shuttle was grounded in 2010. The new schedule required 16 Shuttle flights, but almost as many Shuttle flights were wiped off the launch manifest. A number of Utility flights were removed, as were Russian plans to launch a smaller than originally planned MEM. The Americans would now supply additional electrical power to the Russian sector from the SAWs on the ITS, through 2015. The European and Japanese experiment modules were advanced, now launching in late 2007 and early 2008. Russian, European, and Japanese robotic cargo vehicles would now deliver many of the items originally scheduled to be launched on the cancelled Shuttle flights. NASA administrator Michael Griffin told reporters at KSC:

> "The main thing you are seeing here today is the decision to put together an assembly sequence that allows us to have very high confidence we will finish the space station by the time the Shuttle must be retired ... It's the same station. The end product is very much as we envisioned it."

McArthur completed proficiency training on the SSRMS on March 8. Over the next two days Controllers in Houston moved the SSRMS to view the IUA that had malfunctioned in December, cutting the back-up cable to the MT. They also viewed a valve on Destiny, looking for contamination. The valve was used to vent carbon dioxide overboard and appeared to be clean. This was the first time the SSRMS had been operated from the ground in an operational, rather than an experimental capacity.

The decision to retain Progress M-54 meant that Progress M-55 had had to dock to Pirs. The crew now had to move Soyuz TMA-7 back to Zvezda's wake, thus allowing the Expedition-13 crew to dock Soyuz TMA-8 at Zarya's nadir, where it would remain for the duration of their occupation. In the final hours of March 19, the Expedition-12 crew configured ISS for un-crewed flight and sealed themselves in Soyuz TMA-7. At 02:49, March 20, Tokarev undocked the Soyuz from Zarya's nadir and manoeuvred it to dock with Zvezda's wake, at 03:11. Korolev congratulated them for being the first crew to dock at all three Soyuz docking ports on ISS. The transfer was followed by a day of light duties, before the crew began reconfiguring the station for full occupation.

In March, a problem arose when blisters were found on EVA handrails during production on the ground. The blisters led to a questioning of the strength of the handrails on ISS and an instruction to only attach EVA tethers to the base of the station's handrails, rather than the rail itself. Meanwhile, the crew spent time locating lithium hydroxide canisters to fit the Orlan EVA suits on ISS. New canisters would be launched on Progress M-56. They spent the remainder of their time on ISS preparing for the end of the Expedition-12 occupation, and their return to Earth. They also performed their last experiments. On March 29, they observed and photographed the total solar eclipse.

SOYUZ TMA-8 DELIVERS THE EXPEDITION-13 CREW

SOYUZ TMA-8	
COMMANDER	Pavel Vinogradov
FLIGHT ENGINEER	Jeff Williams
ENGINEER	Marcos Pontes (Brazil)

The Expedition-13 crew was launched in Soyuz TMA-8, at 21:30, March 29, 2006. Korolev lost all telemetry links with the spacecraft for 10–15 minutes, shortly after it achieved orbit. The problem was caused by a communication outage with a Russian Molniya satellite. Onboard Soyuz TMA-8 all systems were performing as planned. The spacecraft docked to Zarya's nadir at 23:19, April 1. Following systems checks the hatches between the two vehicles were opened at 00:59 the following day. Following the usual safety briefings Pontes transferred his couch liner from Soyuz TMA-8 to Soyuz TMA-7. He had trained for his flight as a temporary NASA astronaut, in return for an experiment rack to be mounted on the exterior of ISS. Budgetary difficulties prevented the Brazilians from producing the rack, but the flight of the Brazilian astronaut went ahead. In the next few days the two Expedition crews worked together to complete their hand-over while Pontes performed his own "Centenário" experiment programme. In all, Pontes completed 31 sessions with the 8 experiments, all in the Russian sector of the station.

Vinogradov was typically sincere in describing his duties as Commander of Expedition-13:

> "[A]s the crew Commander, first and foremost I'm responsible for the safety of the crew, and my main task, the most important task is that we ... return safe and sound back to Earth and have completed the flight program ... The second thing is the state of the station. We understand that two or three of us are entrusted with the vehicle that's valued at maybe hundreds of millions in dollars. It's not even the question of its specific monetary value but tens of thousands of people worked on it and provided their labour and knowledge for its creation and we're entrusted to control this complex setup. So that's a responsibility that is quite significant for us, that we would not break anything or make it perform worse. So, that is quite a significant responsibility and that's the second function of the crew Commander. The third important issue is our relationship as a crew, as a team. A small crew of two people creates a situation where even the smallest detail gains significant importance. Of course, spaceflight is different depending on its duration—during a short flight you can sort of do it as one feat, but a long spaceflight, you have to make sure that you pace yourself, that you distribute your strength evenly

throughout the flight, and build the proper relationship with your crewmates. Even the smallest thing becomes quite important."

He added:

"We will certainly be expecting the Shuttle and I'm hoping that it's not going to be the only Shuttle that will visit us. Our task is to prepare the station to the maximum for the arrival of the Shuttle, and to be as effective as possible in terms of using the Shuttle flight ... Those are our main tasks."

On April 3, McAthur and Williams "camped out" in the Quest airlock during their sleep period. Quest was isolated from ISS at 19:45, and the pressure was lowered as planned. The two men began their sleep period, but 4 hours later they were woken by an error tone issued by software monitoring the atmospheric composition on ISS. Controllers in Houston decided to end the experiment at 01:43, at which time the pressure was raised and the internal hatches opened.

The following day McArthur spoke to journalists during his preparations for return to Earth. He told them:

"By golly, it's time to go home and spend some time with the family ... It's an absolute thrill and joy to live and work in space. But we miss the richness, the texture, the three-dimensional nature of living on our home planet. The coffee [on

Figure 67. Expedition-13: Pavel Vinogradov works on the lighting inside the Pirs airlock.

> ISS] tastes good, but it's all in bags, and I'm really looking forward to smelling my cup of coffee. The next thing is food that crunches, like a good chef salad, and the sensation you get when you bite down of crunching into nice fresh lettuce or a raw carrot."

Earlier in the week Tokarev had commented, "We are ready to go home ... We accomplished all of our tasks. We are happy, and we feel good."

Williams completed a training session with the SSRMS on April 5. He also received a briefing from McAthur on payload operations in Destiny. Meanwhile, Tokarev stowed equipment in Soyuz TMA-7 and reviewed undocking and re-entry procedures.

After a week of joint operations, Tokarev, McArthur, and Pontes sealed themselves in Soyuz TMA-7 for their return to Earth. Undocking occurred at 17:28, April 8. TMA-7 landed in Kazakhstan at 20:48, after 189 days 18 hours 51 minutes in space. Pontes had been in flight for 9 days 21 hours 17 minutes. Following their recovery McArthur described re-entry for reporters, saying:

> "It was wild ride, we loved it."

Talking about the end of his flight he added:

> "I feel an overwhelming sense of satisfaction and maybe even closure ... I have a little muscle and joint soreness, but I feel strong."

On his return to Houston, he told the crowd:

> "There are a lot of people here I've known for a very, very long time. That is the thing I missed, family and close friends, and I look forward to spending time with all of them."

EXPEDITION-13

With the departure of Soyuz TMA-7, the Expedition-13 crew began their 6-month occupation of ISS. If all went well the third member of this crew would be launched on STS-121, on April 1, 2006. German astronaut Thomas Reiter would become the first Expedition crew member that was neither American nor Russian. His arrival on ISS would raise the Expedition crew to three for the first time since May 2003, when the crew was reduced to conserve supplies in the wake of the loss of STS-107, Columbia. Reiter, an ESA astronaut who had spent a tour on Mir, was a commercial passenger flying under contract to the Russians.

Figure 68. Expedition-13: Brazilian spaceflight participant Marcos Pontes works in Destiny. He rode to the station in Soyuz TMA-8 with the Expedition-13 crew and returned to Earth with the Expedition-12 crew.

Vinogradov and Williams had a light workload during their first weekend alone on the station before picking up the pace on the following Monday. On April 11, they had spent 3.5 hours performing maintenance on the station's toilet. The following day, April 12, Russian President Vladimir Putin spoke to the crew from the Kremlin in Moscow, to mark the 45th anniversary of the first human spaceflight, made by Yuri Gagarin, in 1961. Vinogradov jokingly invited his President to visit the station. The same date was the 25th anniversary of the first Shuttle flight in 1981. The crew continued the task of loading rubbish in to Progress M-55, and completed the routine emergency evacuation drill early in their occupation.

Their third week on the station was filled with Williams performing American experiments and both men performing the first of three sessions on the Renal Stone Experiment. They also spent time preparing for the arrival of Progress M-56, including Vinogradov practising with the TORU system. On July 18, Korolev transferred propellant from Progress M-55 to Zvezda's tanks. The following day a planned re-boost of the station's orbit was cancelled when telemetry showed that one of the sunshades on Zvezda's thrusters had not opened to its full extent. The malfunction was detected by the station's software which then inhibited the ignition of the thrusters. The firing had been designed to test-fire two thrusters that had not been used since Zvezda docked to ISS, in July 2000. Engineers at Korolev began reviewing the problem and discovered that the opening of the sunshade had been impeded by an ATV antenna installed in September 2004.

PROGRESS M-56

Progress M-56 was launched from Baikonur at 12:03, April 24, 2006. The spacecraft docked automatically to Zvezda's wake at 13:41, April 26, bringing with it 2,597 kg of propellants, water, air, and dry cargo, including items from the crew's families. The crew began unloading the new arrival the following day. During the week Vinogradov completed routine maintenance on the Elektron oxygen generator, which was deliberately powered off for the majority of the week. Williams dismantled his sleeping quarters in order to reach and replace a Remote Power Control Module in Destiny. The crew's first month in orbit drew to a close with a series of routine tasks. Williams checked out the refrigerated centrifuge, sampled the potable water, and replaced the cooling water in the American EMUs. Vinogradov inspected the pressure hull in Zvezda and performed maintenance on the ventilation system. Both men installed new software in the station's laptop computers and both also spent time packing unwanted items into Progress M-55. The week ended with a small reduction in nitrogen pressure in the Elektron unit. It was powered off and would remain off until after the Stage EVA planned for June 1. Additional oxygen from tanks in Progress M-55 was used to supplement the station's atmosphere in the meantime. On May 4, Russian engineers fired Progress M-56's thrusters to raise the station's orbit. Four days later the crew had a day off, to celebrate Russia's Victory Day. Williams trained with the SSRMS on May 11, using its cameras to return views of the station's exterior. He left the SSRMS where its cameras could cover their EVA. Controllers in Korolev transferred propellant from Progress M-56 to Zarya on May 17.

The week had begun with a malfunction of the Vozdukh carbon dioxide removal system in the Russian sector. The carbon dioxide removal system in Destiny was activated while Russian engineers worked on the problem. Following its resolution, both systems were run in tandem until a new gas analyser was installed in the Vozdukh the following week. The remainder of the week saw Williams experimenting with a small satellite called Synchronised Position Hold, Engage, Reorient Experiment Satellite (SPHERES) inside Destiny. The satellite, the first of three, was programmed to perform a series of manoeuvres in anticipation of "constellation" flying at a later date, after two more satellites had been delivered on STS-121. The manoeuvres, using small carbon dioxide thrusters, consisted of 15 ten-minute trials during which the satellite went through a series of pre-programmed flight manoeuvres, including object avoidance and station keeping. The technology might be developed in future as automated assistants for space crews. NASA reported:

> "Each satellite is about 8 inches in diameter, weighs about 7 pounds, and has its own internal avionics, software and communications systems. They are powered by 2 AA batteries and will use carbon dioxide gas thrusters to manoeuvre through the Destiny lab. As the satellites fly through the station they will communicate with each other and the ISS laptop through a wireless link."

Elsewhere, Vinogradov reconfigured ventilation lines associated with the Elektron oxygen generator. The maintenance had been planned before the unit was powered off due to the nitrogen pressure drop. The crew spent time packing items for return to Earth on STS-121 when Discovery visited the station.

The last full week in May was spent preparing for the EVA planned for July 1. The crew gathered together the equipment they would use, charged the batteries in their Orlan suits, and checked out Pirs. During the week Vinogradov replaced a gas analyser in the Vozdukh system, returning it to full working order. They were also the first people to spot a new eruption of a volcano on the Aleutian Islands in Alaska. Williams began using the InSPACE experiment, last used by the Expedition-7 crew, studying the behaviour of fluids that change their properties when subjected to a magnetic field.

At 18:48, June 1, Vinogradov and Williams opened the hatch on Pirs to begin their first EVA, the 65th devoted to the construction and maintenance of ISS, and the last planned to be completed by a two-person Expedition crew. After gathering their tools they used a Strela crane to place themselves at Zvezda's ram, where Zvezda docked with Zarya's wake. In that location, Vinogradov installed a new nozzle on a valve used to vent hydrogen overboard from the Elektron. The original valve had become clogged, causing Elektron to use the same vent line used by a contamination-monitoring device. Two weeks earlier Vinogradov had installed an internal line inside Zvezda as a precursor to fitting the new valve on the exterior. Elektron would be powered on, on July 7.

With that task complete, they moved to Zvezda's wake, where they photographed one of the antennae designed for use by the European ATV. The antenna's cable was suspected of having been the cause of the thruster sunshade not fully opening during their third week of occupation, thus causing the aborting of a re-boost manoeuvre. Vinogradov then recovered the Kroma device, designed to collect thruster residue, from Zvezda's exterior. Meanwhile, Williams collected the third Biorisk package from the exterior of Pirs, along with a contamination-monitoring unit, also from the exterior of Pirs. With those three units safely stored inside Pirs, the EVA was extended, before handing over control from Korolev to Houston. Using the Strela crane once more, they manoeuvred themselves to the area where the Russian and American sectors of the station met before moving past Quest and on to the ITS. Control of the EVA passed from Korolev to Houston. In that location, Williams installed a foot restraint and removed a video camera from the MBS replacing it with a new one. The original camera had failed in February 2005. Korolev resumed control of the EVA once more as the two men used the Strela to return to Pirs and re-enter the airlock. The hatch was closed at 01:19, July 2, after an EVA lasting 6 hours 31 minutes.

On the same day Programme Manager Wayne Hale reviewed progress on the Shuttle's foam-shedding problem for the media saying:

> "We are on a road for continuous improvement ... We are trying to eliminate the biggest hazards and work our way on down ... We have found no showstoppers. We believe we have made significant improvements ... There will continue to be foam coming off the external tank. What we have done in a very systematic manner is eliminate the largest hazards."

Figure 69. Expedition-13: Jeffrey Williams rides the exercise cycle [veloergometer] in Destiny.

The following week was spent cleaning, servicing, and storing the suits they had worn during their EVA, before concentrating on preparations to jettison Progress M-55, receive Progress M-57, and STS-121, Discovery's Return to Flight mission, which was due to deliver an MPLM full of logistics to ISS. They also had to work on the Elektron unit, which only restarted after several attempts and then failed 7 hours later. Russian engineers decided that the unit had a malfunctioning power unit, which would require replacement with a spare already held on the station. The malfunction had no immediate impact on operations. The crew also ran their regular experiment programme. On June 9, Progress M-56's thrusters were fired in a re-boost manoeuvre, placing the complex in the correct altitude to receive Progress M-57.

The crew spent much of the second week of June preparing for Discovery's arrival, practising the photography session during which they would take images of Discovery as she performed the new r-bar pitch manoeuvre prior to docking with the station. June 12 was a day off, celebrating Russia's Independence Day. Two days later the last of the propellant was transferred between Progress M-55 and Zarya. Both men also spent time packing rubbish into Progress M-55 and sealed its hatches on June 16. It was undocked at 10:06, and commanded into the atmosphere where it burned up on June 19. Vinogradov spent time practising with the TORU manual docking equipment in case he might have to assume control of Progress M-57 during its final approach. The end of the week involved performing experiments. Vinogradov also replaced internal panels and smoke detectors inside Zvezda, while Williams trained with the SSRMS, practising the manoeuvres he would use to remove the MPLM Leonardo from Discovery's payload bay and dock it to Unity.

PROGRESS M-57

Progress M-57 was launched at 11:08, June 24, 2006 and entered orbit a few minutes later. Following a standard 2-day rendezvous, the new Progress docked to Pirs' nadir at 12:25, June 26. Unloading of the 2,578 kg of cargo, including 1,161 kg of dry goods, began the following day. With Progress in place, the last week of June was spent preparing for Discovery's arrival. The crew flushed the pipes in Quest, in preparation for the Shuttle crew's planned EVAs. Progress M-57 would be unloaded after Discovery had returned to Earth; its pressurised compartment would then be used as a temporary storage area, to hold many of the items delivered in Discovery's MPLM.

5

Recovery and restructuring

STS-121 RETURNS THE SHUTTLE TO FLIGHT

STS-121	
COMMANDER	Steven Lindsey
PILOT	Mark Kelly
MISSION SPECIALISTS	Michael Fossum, Lisa Nowak, Stephanie Wilson, Piers Sellers
EXPEDITION-13 & 14 (up)	Thomas Reiter (ESA)

STS-121 was the first Shuttle flight since the fleet had been re-grounded in the wake of STS-114. The flight would carry 4,000 kg of cargo to the station in the MPLM Leonardo, to re-stock the supplies that had been used during the intervening period. It would also deliver Thomas Reiter, the third member of the Expedition-13 crew, to the station. Reiter was the first ESA astronaut to serve on an ISS Expedition crew. The agreement to fly Reiter was signed in May 2003. Director of Human Spaceflight at ESA Daniel Sacotte explained:

> "It covers the ESA astronaut's flight in a crew position originally planned for a Russian cosmonaut. He will perform all the tasks originally allocated to the second Russian cosmonaut on board the ISS and, in addition, an ESA experimental programme."

Reiter added:

> "... ESA is making important contributions to the ISS and its scientific capabilities. We are assuming significant operational responsibilities in this programme and I am confident that this mission will give Europe valuable

operational experience and scientific results which will further prepare us for the exciting and challenging times ahead."

STS-121 would be the first launch to be controlled from the new Firing Room 4 in the Launch Control Centre, at KSC. The new Firing Room would become the principal launch control room for the remainder of the Shuttle programme, while the original Firing Rooms were converted for use in Project Constellation.

The first attempt to launch STS-121 was made on July 1, 2006. On that occasion the countdown reached the planned hold at $T-9$ minutes, when it was held due to anvil clouds, potential thunderstorms, in the area. The launch was subsequently scrubbed and recycled to the following day. On July 2 the countdown was scrubbed for a second time, due to anvil clouds in the area and recycled for a further 48 hours. The new launch date was set for July 4.

On July 4 a crack in the insulating foam on the Shuttle's ET had to be filled and an investigation was required when a triangular piece of foam fell away from the ET while Discovery stood on the launchpad. It was decided that the launch could proceed as planned. In February, Lindsey had warned the media, "We will lose foam on this flight, just like every other. The key is to make sure that any foam we do lose is of a small enough size so it can't hurt us if it hits the vehicle."

Discovery finally lifted off at 14:38, July 4, 2006 and was in orbit a few minutes later. Film from the numerous cameras on the vehicle showed that the ET had continued to shed foam and the crew were even informed that MCC-Houston thought that one piece might have struck the underside of the orbiter. The crew began their sleep period at 20:38.

Eight hours later, they were awake and ready for their first full day in space. Beginning work shortly after 07:00, Nowak and Wilson used the RMS to lift the

Figure 70. STS-121 crew (L to R): Thomas Reiter, Michael E. Fossum, Piers J. Sellers, Steven W. Lindsey, Mark E. Kelly, Stephanie D. Wilson and Lisa M. Nowak.

OBSS from the opposite edge of the payload bay and manoeuvred it so that the Laser Dynamic Range Imager, the Laser Camera System, and the Intensified Television Camera mounted on the end could image the leading edge of both wings and the orbiter's nosecap. Meanwhile, Sellers set up Discovery's computers and Reiter prepared the mid-deck for the transfer of equipment and stores to ISS. Sellers and Fossum were assisted by Kelly as they checked the EMUs they would wear on two, or possibly three EVAs.

On ISS, Vinogradov and Williams prepared cameras, with the 400 mm and 800 mm lenses, that they would use to photograph Discovery during its approach to the station. They also pressurised PMA-2 in preparation for Discovery's docking.

As Discovery approached ISS, Lindsey commenced station-keeping at a distance of 200 metres, he then performed a nose-over-tail pitch manoeuvre so that Vinogradov and Williams could photograph the underside of the orbiter. The photographs were down-linked to Houston for scrutiny. The rendezvous then continued and Lindsey docked Discovery to PMA-2 at 10:52, July 6. Following pressure checks the hatches between the two vehicles were opened and Discovery's crew entered the station at 12:30. Vinogradov and Williams greeted them and then issued the standard safety briefing. Williams told Houston, "It's a full house. The climate has changed significantly." The first action after that was to transfer Reiter's couch lining and Sokol pressure suit from Discovery to Soyuz TMA-8, thereby signalling his transfer from the STS-121 crew to the Expedition-13 crew. For the first time since the Expedition-6 crew, who had been in orbit when STS-107 was lost in February 2003, the Expedition crew on ISS now consisted of three people. Vinogradov had the following to say about the arrival of the European astronaut:

> "I think that it is a very important milestone . . . at this stage we can support a crew of three or more. But from the human standpoint, it's important because we do have to notice that over the last two years the ISS program is kind of slowing down and the interest is not what it used to be on the part of the Russian government and Congressmen in the United States. There are certain notes of dissatisfaction on the part of the people who are working on the science and experiments on board the station because unfortunately the rate of station assembly and deployment is quite different from what they expected. And so the arrival of the third person, Thomas Reiter in our case, greatly improves the capability of the crew in terms of performing science program and experiments. The other thing is that Thomas Reiter is a representative of Europe. Europeans are important and we are working with them very closely. The European Space Agency contributes considerable effort from the standpoint of research."

Williams was equally keen to see Reiter join the crew:

> "We're really looking forward to the Shuttle arriving, and Thomas joining us as a crew of three. It's obviously very significant. Since Expedition 7 we've been flying and sustaining the Space Station with a crew of two. Those who track that sort of thing say that it takes more than two people just to run the station, so it leaves no excess crew time for the other things. Getting back to a crew of three will help us

accomplish more. It's also significant in that we will be continuing with the assembly of the Space Station, to get it up to its full capability with the resuming of regular Space Shuttle flights, which is important, of course, to meet the vision of space exploration."

The EMUs that Fossum and Nowak would use during their planned EVAs were also transferred, from Discovery to Quest. In preparation for the first EVA, Williams and Wilson used the SSRMS to lift the OBSS from its storage position and hand it over to Discovery's own RMS. With Discovery docked to ISS there was insufficient clearance for Discovery's RMS to retrieve the OBSS. During the EVA, the 33-metre long combination would undergo tests as a work platform giving access to areas of the Shuttle that were previously inaccessible.

The following day, Nowak, Wilson, Fossum, and Sellers used the SSRMS to lift Leonardo out of Discovery's payload bay and dock it to Unity. Docking occurred at 08:15, July 7, following an initial concern that straps on Unity's CBM might prevent a perfect air-tight seal. Following pressure and leak tests Lindsey, Wilson, and Reiter opened the hatches between the two modules and began several days of equipment transfer. Vinogradov was in no doubt of the importance of the arrival of Leonardo at the station:

> "That period of joint flight with the Shuttle is quite busy in terms of the crew of the station and the Shuttle crew working together. It's quite intense work. First you have to move a very considerable amount of cargo, you have to get it out of the MPLM and stow it on the station. It's quite important—extremely important, I would say—because that provides the supplies for our continued flight."

Williams has said:

> "It's also important to get a lot of the equipment that is no longer required on station off and packed into that empty MPLM so it can come home. The station's getting pretty crowded here in recent months and years ... It's going to be important for both crews to be very disciplined in the transfer of equipment, both to the station and returning to the Shuttle. To do that, we have a flight plan onboard. Part of that flight plan is a transfer plan. It's a detailed choreography of all of the transfers, everything that goes across the hatches between the Shuttle and the Station, developed by the folks on the ground and trained by both crews."

Nowak, Kelly, and Wilson also used the RMS/OBSS combination to carry out further inspections of Discovery's exterior, finding six areas requiring further investigation, although none of them were areas of major concern. Areas of particular attention were parts of the nosecap that had been missed on July 5, and a piece of fabric near the orbiter's nose. Fossum and Sellers spent the day preparing for their first EVA. During the day, mission managers made the decision to extend Discovery's flight by one day, including a third EVA. Engineers in Houston also reviewed the initial photographs and laser scans of Discovery's exterior.

EVA-1 began at 09:17, July 8, when Sellers and Fossum left Quest. Their first task was to repair the MT mounted on the S-0 ITS. The emergency cable cutter had malfunctioned and cut one of two cables that moved the MT along the truss. During a Stage EVA the Expedition-12 crew had removed the second cable from the cutter after failing to install a bolt to prevent the blade from falling and cutting it. After collecting their tools together, Sellers and Fossum made their way to the S-0 ITS, where they installed a device to block the cable cutter blade on the MT, thereby denying the ability to sever the cable in an emergency. After installing the block, they reinstalled the cable in the cutter, thereby repairing the MT and making it available to move the SSRMS along the truss during EVA-2. The second portion of EVA-1 included simulating the use of the RMS/OBSS as a workstation. Nowak and Wilson operated the RMS from Discovery's aft flight deck, while Kelly served as intra-vehicular officer, the EVA astronauts' guide, offering whatever assistance he could from inside Discovery. With Sellers standing on the foot restraint mounted on the end of the OBSS the combination was put through a series of pre-planned manoeuvres while sensors recorded the forces involved. Sellers remarked, "Just a general comment. It gets easier as you go along doing all the tasks on the end of a skinny little pole. A little practice makes perfect." In Houston the Flight Director commented at his end-of-shift press conference, "The arm damped a lot quicker than we thought, based on our analysis ... That gives us very good confidence we could use this as a platform for repairs." In space, Fossum joined Sellers at the end of the OBSS, which was then manoeuvred to three different simulated work positions. The last of these lifted Fossum to a position where he could push up with his hands against the P-1 ITS. The EVA ended at 16:48, after 7 hours 31 minutes.

While the EVA was taking place, Vinogradov and Reiter began unloading Leonardo and transferring stores to the station, including a new sample freezer and a new oxygen generator. When installed in Destiny, at a later date, the generator would upgrade the station's oxygen capacity to the point that it could support up to six people on long-duration Expedition crews.

As the day's activities came to an end mission managers cleared Discovery's heatshield for re-entry. The following day, July 9, was spent unloading stores from Leonardo and preparing items on Discovery for return to Earth. Sellers and Fossum spent the day cleaning their EMUs and preparing their tools and Quest for their second EVA, on July 10.

With the crew awake at 02:08, preparations for the second EVA began immediately after breakfast. Sellers and Fossum left Quest at 08:14, and climbed down into Discovery's payload bay. There, they lifted the pump module from its stowage location so that Nowak and Wilson could grapple it with the SSRMS and lift it into position. Meanwhile, Sellers and Fossum remained in the payload bay preparing for the primary task of the EVA, replacing the Mobile Transporter's Trailing Umbilical System (TUS), the power and data cable that had been cut during the Expedition-12 occupation. Both men made their way to the S-0 ITS, where Fossum disconnected electrical cables, and Sellers then replaced the Interface Umbilical Assembly (IUA) with a new one, without a cutting blade. By that time the SSRMS had manoeuvred the pump module to External Stowage Platform 2. Sellers and Fossum made their way to that location and secured the module to the platform, thus allowing the SSRMS to release it. The pump module was a spare, now available if it should be needed in the future.

Sellers and Fossum returned to their work on the TUS. Now working from the end of the SSRMS, Fossum removed the TUS reel assembly and carried it down to the payload bay. While he was doing that, Sellers worked in the payload bay to unpack and prepare the new reel assembly. While Fossum returned to the work site Sellers stowed the old reel assembly in Discovery's payload bay. Back on the S-0 Truss, Fossum was joined by Sellers and they worked together to install the new reel assembly and routed it through the IUA. The work ensured that the MT would have the required redundancy to enable it to support future assembly flights. Having stowed their equipment, both men returned to Quest and the EVA ended at 15:01, after 6 hours 47 minutes. During the EVA Fossum had to twice stop working and secure a loose connection on Seller's SAFER. While the work was taking place outside, Vinogradov, Williams, and Reiter continued to unload Leonardo, and Lindsey transferred two bags of water from Discovery to ISS.

The following day, July 11, was spent transferring equipment and rubbish from ISS to Leonardo, for return to Earth. Wilson served as loadmaster, ensuring everything was secured in the correct place, thereby retaining Discovery's correct centre of balance for re-entry and landing. Sellers and Fossum cleaned their EMUs and prepared them for their third EVA.

That EVA began at 07:20, July 12. Sellers and Fossum left Quest, made their way into the payload bay, collected their tools, and installed a foot restraint on the SSRMS. After Sellers had mounted the SSRMS, Nowak and Wilson manoeuvred him to a point close to Discovery's starboard wing's leading edge, where he recorded several seconds of infrared imagery. The imagery, which recorded temperature differences, would help to identify any internal damage to the area. The manoeuvre simulated lifting an astronaut to a position where he could try to repair damage to the wing's leading edge, as suffered by STS-107. Returning to the payload bay, Sellers joined Fossum at a workstation where both men trialled a variety of methods for repairing damage to 12 samples of Reinforced Carbon–Carbon, similar to the panels on the wing's leading edge. Using a space-cleared caulking gun and a series of spatulas, they pumped a carbon–silicon polymer called NOAX into the simulated damaged tiles in an attempt to repair them. Over almost two hours they repaired three gauged tiles and two cracked tiles. They also imaged four of the repair samples with the same infrared equipment that Sellers had used of the starboard wing. An additional get-ahead task was added to the EVA: Sellers used a pistol grip tool to remove the fixed grapple bar used to move the pump module during the second EVA, moved over to the S-1 Truss, and installed it on an ammonia tank that was due to be moved during the Expedition-15 occupation. The EVA ended at 14:31, after 7 hours 11 minutes. As usual, the remainder of the crew spent the day loading Leonardo.

Following their hectic first eight days, Discovery's crew were given July 13 off, with the exception of a few interviews. July 14 was also spent in interviews and completing the final loading of Leonardo. The MPLM was de-activated and was undocked from Destiny at 09:32, and lowered into Discovery's payload bay, where it

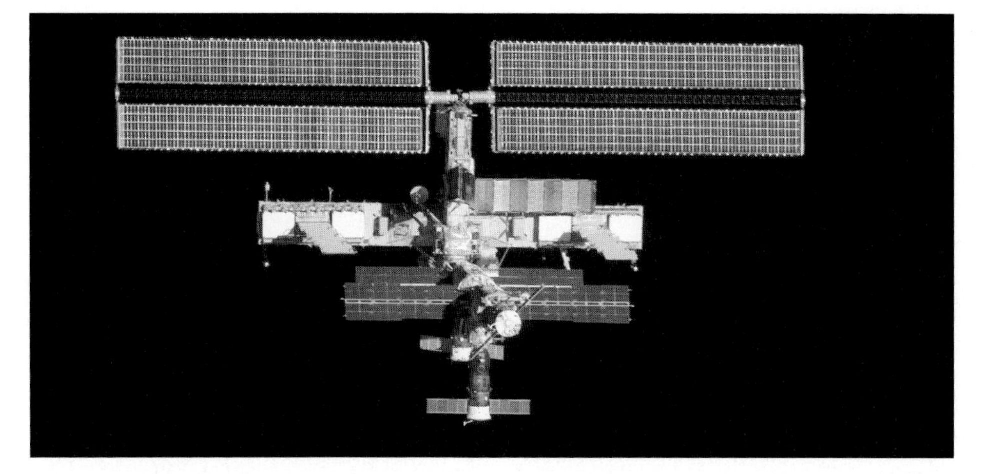

Figure 71. STS-121: A view from the station's wake, as STS-121 completes its fly-around, shows the station as it was before construction was resumed.

was secured at 11:00. Lindsey and Reiter then used the RMS, with the OBSS attached, to view Discovery's heatshield and search for any damage that had occurred during orbital operations. None was found.

July 15 was the final day of joint operations as Discovery's crew prepared for their departure. Undocking took place at 06:08 and was followed by the crew imaging the leading edge of the starboard wing and the nosecone. Discovery performed station-keeping while mission managers reviewed the new images and cleared the heatshield for re-entry. The Shuttle fell behind and above ISS with the minimum of manoeuvring. Discovery's crew spent the following day preparing for re-entry, while, on ISS, Williams depressurised PMA-2 and Vinogradov and Reiter continued the station's maintenance and experiment programmes.

Discovery's payload bay doors were closed at 05:27, July 16, and retrofire occurred at 08:07. Thereafter, Lindsey turned his vehicle so the nose was forward and high, until gravity pulled it out of orbit. The orbiter glided to a perfect landing at KSC, at 09:15, after a flight lasting 12 days 18 hours 38 minutes. It had six astronauts onboard, one less than at launch; the seventh, Reiter, was still in orbit, as part of the now three-person Expedition-13 crew.

At the landing site Michael Griffin told the press, "Obviously this is as good a mission as we've ever flown ... But we're not going to get overconfident." The Shuttle was finally back in business and the construction of ISS was set to continue. Meanwhile, NASA was already busy defining the vehicle that would replace the Shuttle when it retired in 2010. When the crew returned to Houston, Lindsey told the waiting crowd, "In terms of human spaceflight ... we're back." The crowd applauded loudly. Lindsey was more sombre, "I think it's more like the beginning of the next phase ... I don't think we ever want to put Columbia behind us."

As Discovery landed, the published flight programme for the completion of ISS looked like this:

Launch date	*Flight*	*Vehicle*	*Flight details*
August 28, 2006	12A	Atlantis STS-115	• Second port truss segment (ITS P-3/P-4) • Second set of solar arrays and batteries
December 14, 2006	12A.1	Discovery STS-116	• Third port truss segment (ITS P-5) • SpaceHab Single Cargo Module • Integrated Cargo Carrier (ICC)
February 22, 2007	13A	Atlantis STS-117	• Second starboard truss segment (ITS S-3/S-4) with Photovoltaic Radiator (PVR) • Third set of solar arrays and batteries
June 11, 2007	13A.1	Endeavour STS-118	• SpaceHab Single Cargo Module • Third starboard truss segment (ITS S-5) • External Stowage Platform 3 (ESP3)
Under review	ATV1	Ariane 5	• European Automated Transfer Vehicle
August 9, 2007	10A	Atlantis STS-120	• Node 2 • Sidewall, Power and Data Grapple Fixture (PDGF)
September 27, 2007	1E	Discovery STS-122	• Columbus European Laboratory Module • Multi-Purpose Experiment Support Structure, Non-Deployable (MPESS-ND)
November 29, 2007	1J/A	Endeavour STS-123	• Kibo Japanese Experiment Logistics Module, Pressurized Section (ELM-PS) • Spacelab Pallet, Deployable 1 (SLP-D1) with Canadian Special Purpose Dexterous Manipulator, Dextère
February 7, 2008	1J	Atlantis STS-124	• Kibo Japanese Experiment Module Pressurised Module (JEM-PM) • Japanese Remote Manipulator System (JEM-RMS)
June 19, 2008	15A	Endeavour STS-119	• Fourth starboard truss segment (ITS S6) • Fourth set of solar arrays and batteries
August 21, 2008	ULF2	Atlantis (last flight) STS-126	• Multi-Purpose Logistics Module (MPLM)
Under review	3R	Russian Proton	• Multipurpose Laboratory Module with European Robotic Arm (ERA)

Launch date	*Flight*	*Vehicle*	*Flight details*
October 30, 2008	2J/A	Discovery STS-127	• Kibo Japanese Experiment Module, Exposed Facility (JEM-EF) • Kibo Japanese Experiment Logistics Module, Exposed Section (ELM-ES) • Spacelab Pallet, Deployable 2 (SLP-D2)
January 22, 2009	17A	Endeavour STS-128	• Multi-Purpose Logistics Module (MPLM) • Lightweight Multi-Purpose Experiment Support Structure Carrier (LMC) • Three crew quarters, galley, second treadmill (TVIS2), Crew Health Care System 2 (CHeCS2)
Establish six-person crew capability			
Under review	HTV-1	H-IIA	• Japanese H-II Transfer Vehicle
April 30, 2009	ULF3	Discovery (last scheduled flight) STS-129	• EXPRESS Logistics Carrier 1 (ELC1) • EXPRESS Logistics Carrier 2 (ELC2)
July 16, 2009	19A	Endeavour STS-130	• Multi-Purpose Logistics Module (MPLM) • Lightweight Multi-Purpose Experiment Support Structure Carrier (LMC)
October 22, 2009	*ULF4	Discovery STS-131 (if needed)	• EXPRESS Logistics Carrier 3 (ELC3) • EXPRESS Logistics Carrier 4 (ELC4)
January 21, 2010	20A	Endeavour STS-132	• Node 3 with Cupola
July 15, 2010	*ULF5	Endeavour STS-132 (if needed) (final Shuttle flight)	• EXPRESS Logistics Carrier 5 (ELC5) • EXPRESS Logistics Carrier 1 (ELC1)
ISS assembly complete			
Under review	9R	Russian Proton	Research Module

* Two Shuttle-equivalent flights for contingency.

Notes: Soyuz flights for crew transport schedule at approximately 6-month intervals beginning in September 2006. Additional Progress flights for logistics and re-supply are not listed.

At the Farnborough International Air Show, in England, held during July 2006, ESA Director Jean-Jacques Dordain stated at a press conference, "I have a wish that Europe participate in one of the two next-generation transportation systems [the American Orion, or Russia's Kliper]. If we don't, I fear we will always be a second-class partner." America had already made it clear that NASA intended to develop Orion as an all-American spacecraft, so ESA had agreed to undertake a 2-year feasibility study on crewed spacecraft architecture, for a spacecraft to be launched by a Soyuz launch vehicle from Baikonur, or from the new pad for such vehicles at ESA's launch site in Kourou. Meanwhile, Energia had to admit that Kliper would cost more than the entire Russian space programme budget for the period 2006–2015. Therefore, the new spacecraft would not be built. Rather, the Russians would update Soyuz yet again, making it capable of Earth orbital and lunar orbital flight. At the same meeting NASA Administrator Michael Griffin explained to the audience, "Our plan is to have one more daylight [Shuttle] launch before resuming night operations ... We do need to resume night operations to complete the Space Station, we've always known that."

In America, NASA had named the two new Shuttle-derived launch vehicles that would be used to support Project Constellation. The Crew Launch Vehicle would be called Ares-1 and the Heavy Lift Launcher would be Ares-5. The numerical designations were salutes to the Apollo Saturn-class launch vehicles. Meanwhile, the CEV had been named "Orion". Lockheed-Martin was named as prime contractor for the development of the Orion spacecraft in August 2006. It would be a ballistic capsule, superficially similar to the Apollo Command Service Module, and would carry a crew of up to six people. It would be launched by an Ares-1 launch vehicle consisting of a first stage derived from a Shuttle SRB and a new liquid propellant second stage. The new vehicle would use the old Apollo/Shuttle facilities at LC-39, KSC.

NASA had also launched a quest for commercial cargo access to ISS, the Commercial Orbital Transportation Service (COTS). Two industry partnerships, led by SpaceX and Rocketplane Kistler, would use NASA funding, along with private funding to develop an automated vehicle to deliver cargo to ISS and to carry away rubbish. The new vehicle would be heated to destruction during re-entry. The selected developer would be open to sell space on their vehicles commercially, and NASA would be nothing more than a commercial customer. Flight demonstrations would begin in 2008. Phase-1 development would concentrate on an uncrewed cargo vehicle, with the option to progress to Phase-2, a vehicle for delivering humans to ISS.

RETURN TO THREE EXPEDITION CREW MEMBERS

Following Discovery's departure, the extended Expedition-13 crew settled down to work. Williams and Reiter installed the ESA experiment rack, which had been delivered by Discovery, in Destiny. They activated the Minus Eighty-degree Laboratory Freezer for ISS (MELFI), which Discovery had delivered and had been set up in Destiny. The freezer was supplied by ESA and contained four compartments offering a total of 300 litres of storage capacity. It would be used to store biological samples

Figure 72. Expedition-13/14: European astronaut Thomas Reiter served with the Expedition-13 and 14 crews. He is shown working with the SWAB experiment.

prior to their return to Earth. On July 19, Vinogradov took three attempts to restart the Elektron unit and succeeded only after the bubbles had been driven out of the system. The crew also completed a check of the oxygen generation system that Discovery had carried to the station. When activated the new system would supplement the Elektron oxygen generator, in anticipation of future Expedition crews consisting of up to six people. Oxygen from the tanks in Progress M-56 was pumped into the station's atmosphere on a daily basis throughout the latter half of July. The crew also began preparations for Williams and Reiter's Stage EVA, by flushing the cooling loops in Quest and the American EMUs. On July 26, Russian controllers fired thrusters on Progress M-56 to raise the station's orbital parameters and place it in the optimum position for the STS-115 launch, in August 2006. The Expedition-14 crew were due for launch in September 2006, on Soyuz TMA-9. The following day, Vinogradov removed the KURS system from Progress M-56 and stored it in Zarya. The last day of July was spent in maintenance of the American Common Cabin Air Assembly in Destiny. The new month began with Vinogradov transferring water from tanks in Progress M-56 to tanks in Progress M-57 and performed other maintenance tasks on the Life Support System in the Russian sector.

Williams and Reiter left Quest wearing American EMUs at 10:04, August 3, 2006, for a Stage EVA that was planned to last 6 hours 20 minutes. Their first task was to install a Floating Potential Probe (FPP) on the S-1 ITS and extend its three sensor arms, to measure the electrical potential of the station as it orbited Earth. They quickly began to get ahead of their planned timeline for the EVA. Their second task was to install two suitcase-size Materials on International Space Station Experiment (MISSE) containers. The MISSE-3 container was installed on one of Quest's high-pressure gas tanks while MISSE-4 was mounted on Quest's outboard end. Following their individual installations the containers were opened to expose the material samples held inside to the space environment. Following these joint tasks, the two men set about individual tasks. Williams installed a controller for a Thermal Radiator Rotary Joint on the S-1 ITS, before installing a starboard jumper and spool positioning device (SPD), also on the S-1 ITS. Meanwhile, Reiter installed a Multiplexer/De-multiplexer, a computer, on the S-1 ITS, replacing one that had failed in 2004, before examining a radiator beam valve module at a site where an SPD was already installed.

He then installed an additional SPD at that site. Finally, he installed an SPD on the port cooling line jumper. The jumpers were designed to assist the flow of ammonia in the radiators once the coolant was installed.

Williams then began the installation of a light to assist future EVA astronauts using the MBS to move along the assembled ITS. Following that work, he removed a malfunctioning GPS antenna. Elsewhere, Reiter tested an infrared camera designed to image the RCC thermal protection on the nose and leading edge of the wings of the Shuttle Orbiters. The camera was designed to show damage by highlighting the difference in temperature within the RCC sections. When he had completed these tasks he installed a vacuum system valve on the exterior of Destiny for use with future scientific experiments. That was the last of the planned EVA tasks, but, as the astronauts were so advanced on their timeline, Houston found a number of "get-ahead" tasks for them to perform. Williams relocated two articulated foot restraints in preparation for the EVAs planned for the visit of STS-115. He then photographed a scratch on the exterior of Quest. Reiter made his way to the exterior of PMA-1 to inspect and retrieve a ball-stack, used to hold equipment during EVAs. With no further tasks the two men returned to Quest and took photographs of each other. In Houston, astronaut Steve Bowen joked, "We will never let this happen again . . . Wait 'til you see next week's schedule." The pair re-entered Quest and closed the hatch at 16:58, bringing their EVA to a close after just 5 hours 54 minutes. For the first time since 2003 a third Expedition crew member, Vinogradov, had been available to remain inside ISS and monitor systems, thereby doing away with the necessity to place the station's systems into un-crewed mode during the EVA.

The day after the EVA, August 4, Reiter broke the ESA endurance record. When his time on Mir and ISS were combined he had broken the previous ESA combined record of 209 days 12 hours 25 minutes, set by French astronaut Jean-Pierre Haigneré. Reiter's position as the first International Partner on an ISS Expedition crew and the first International Partner to make an EVA from the station was seen by many in Europe as a forerunner of European activities following the launch of the Columbus laboratory module, then planned for September 2007. Reiter's flight activities were being run from the Columbus Control Centre in Oberpfaffenhofen, near Munich, Germany. By the time he returned to Earth, at the end of his 5 months on ISS, Reiter would have spent more than a year in space. By that time he would have completed 23 experiments in 6 disciplines in the ESA "Astrolab" programme.

In the same week, the crew prepared for the arrival of STS-115. The Shuttle would deliver the P-3/P-4 ITS, and would see the resumption of construction on ISS. The Expedition-13 crew began packing up items that Atlantis would carry back to Earth. They also performed 2 days of routine maintenance. On August 10, controllers in Houston moved the SSRMS to allow its cameras to view markings on the exterior of ISS as part of the Space Vision System (SVS), which would be used to assist in the correct alignment of the new components that Atlantis would deliver to the station. The following day, Williams walked the SSRMS from the exterior of Destiny on to the recently repaired MT, and then used its cameras to view the exposed end of the P-1 ITS, where the new P-3/P-4 ITS would be mounted.

In the second half of August the Expedition-13 crew continued to pack items for return to Earth on Atlantis. The Carbon Dioxide Removal Assembly was tested in advance of the Shuttle's arrival and the station's orbit was raised by a burn of the Progress M-56's thrusters, on August 23. The crew also continued their experiment programme, routine maintenance, and exercise regimes.

STS-115 was due for launch on August 27, but was delayed, ultimately until September 9. The crew on ISS used the extra time to complete their preparations for the Shuttle's flight. They worked on cosmic ray studies that involved Williams spending a complete orbit in the prone position wearing a helmet with sensors to monitor his brain activity and visual perceptions. Vinogradov spent much of the time maintaining the Elektron unit.

STS-115 DELIVERS THE PORT-3/4 ITS

	STS-115
COMMANDER	Brent Jett
PILOT	Christopher Ferguson
MISSION SPECIALISTS	Joseph Tanner, Daniel Burbank, Heidemarie Stefanyshyn-Piper, Steven MacLean (Canada)

As July turned August 2006, the American press began to fill with stories of STS-115, Atlantis, and how its six-person crew would restart the construction of ISS. The Shuttle would deliver the combined port-3 and port-4 (P-3/P-4) ITS. The inward end of the P-3 ITS would be permanently attached to the exposed end of the P-1 ITS. The P-3 ITS contained a Solar Alpha Rotary Joint (SARJ), which would allow the three outer segments of the port ITS (P-4 and the P-5/P-6) to rotate, in order to keep their SAWs directed towards the Sun. The construction would require three EVAs from Quest.

STS-115 stood at Launch Complex 39 in late August. The weather on August 25 was abysmal, rain; low, dark clouds; and thunder. At one point lightning struck the conductor on the top of the launch structure tower, causing a spike in several electrical systems. When the weather had passed the launch was delayed for 24 hours, from August 27 to August 28, in order to carry out thorough systems checks in the wake of the lightning strike. Even as the checks were completed Hurricane Ernesto was in the Caribbean, approaching Cuba. If it continued on its present course and regained strength over the open sea it might strike KSC with storm winds in excess of speeds in which it was safe for the Shuttle to remain on the launchpad. Two parallel plans were put in place. First, work continued to prepare Atlantis for launch, if the winds abated. Second, work began in preparation to roll Atlantis back to a safe haven constructed in the VAB.

Further difficulties were placed on the launch by the lighting requirements of the cameras that would film the ET during launch and the requirement to jettison the ET on the daylight side of Earth. These requirements had been placed on the first two launches after the loss of STS-107 by the CAIB. Although the current launch window ran until September 13, NASA wanted to launch before September 7, rather than delay the launch of Soyuz TMA-9 carrying the Expedition-14 crew, planned for September 14. The Expedition-13 crew were due to return to Earth in Soyuz TMA-8 on September 24, and any delay in the launch of Soyuz TMA-9 would result in a night-time recovery for Soyuz TMA-8.

Programme manager Mike Suffrendi told the media, "This flight has to occur for the next flight to occur and then the next flight and the next flight ... Even though we say we take them one at a time, this one is a key. This is clearly in the critical path for assembly."

On August 30, NASA began the 12-hour-long roll-back of Atlantis to the VAB. Four hours later, Hurricane Ernesto had altered course, having lost much of its energy over Cuba. The decision was made to stop the roll-back and return Atlantis to the launchpad. Lift-off was rescheduled for September 8.

With the countdown proceeding, an electrical short circuit caused the failure of a coolant pump in one of four power generators in one of Atlantis' three fuel cells. The problem was that the fuel cell might fail in flight, causing Atlantis to return to Earth early, without installing the P-3/P-4 ITS. Despite everything, NASA managers decided that the risk of failure was minimal and the launch preparations continued. The crew were suited up, transported out to the launchpad, and installed in the spacecraft. As the countdown continued, a new problem arose with one of four fuel cut-off sensors within the ET. The sensor was responsible for sensing the amount of propellant in the ET's liquid hydrogen tank and ensuring the three Space Shuttle Main Engines shut down if the main computer failed. There were similar sensors in the liquid oxygen tank. Mission Rules dictated that all the sensors be working before the launch could go ahead. On this occasion flight managers decided that the launch could not continue. The launch attempt was scrubbed and the crew removed from the spacecraft.

The following day, September 9, 2006, the whole procedure began again. Following a near perfect countdown, at 11:15 Atlantis climbed into the blue Florida sky on her first flight since 2002. Although the launch events appeared to go well, review of the numerous video tapes showed that four or five small pieces of foam shed from the

ET. These events all occurred after Atlantis had left the thick lower atmosphere and, therefore, mission managers were sure that the foam did not have enough energy to cause major damage to the orbiter. All launch events passed off as planned and video cameras on the ET captured both the SRB and ET separation. Tanner and MacLean shot hand-held video and still images of the ET as it drifted away. Programme Manager Wayne Hale remarked, "The bottom line is we are looking at nits, nothing of remote consequence. Of course, we will inspect the entire heatshield with a fine-tooth comb." As Atlantis lifted off, ISS was over the Atlantic Ocean, between Iceland and Greenland. The Expedition-13 crew watched the launch on a NASA television link. Once in orbit the crew prepared Atlantis for sustained spaceflight. On the ground, astronaut Mike Fincke told the media, "The logjam behind is huge ... We have a very solid end date for the Space Shuttle, so each mission has to happen one after another in the sequence."

On September 10, Jett and Ferguson began the manoeuvres that would result in rendezvous with ISS. Meanwhile, Ferguson, Burbank, and MacLean used the RMS to lift the OBSS and view Atlantis' right and left-wing leading edges and the carbon–carbon nosecone, before returning the OBSS to its stowed position. Initial reviews of the images obtained showed no damage to Atlantis from the foam shed from the ET during launch. In one place on Atlantis' underside a shim and a tile spacer were seen protruding out from between the TPS tiles. Tanner and Stefanyshyn-Piper prepared the EMUs and EVA tools they would transfer to ISS for the three planned EVAs. On the station, Vinogradov and Williams pressurised PMA-2 in preparation for the Shuttle's arrival.

Rendezvous with ISS took place on September 11, Flight Day 3. Houston joked, "Atlantis is headed your way with a brand new piece of the Space Station in its trunk." As the Shuttle approached the station, Jett and Ferguson performed the r-bar pitch manoeuvre, to allow the station crew to take high-resolution photographs of the Shuttle's underside. In Houston, Pam Melroy was Capcom and remarked, "Station, we see you have visitors. Tell them to give us a wave."

Atlantis docked to PMA-2 at 06:48. Even as the pressure and leak checks between the two vehicles were taking place, Ferguson and Burbank activated the RMS and used it to lift the 17.5-tonne P-3/P-4 ITS out of Atlantis' payload bay and manoeuvred it to the position where it could be handed over to the SSRMS, leaving it in that location. The hatches between the two vehicles were opened at 08:30, and the Shuttle crew transferred to ISS. The visitors were greeted enthusiastically by the three-man Expedition-13 crew before receiving the standard safety briefing. MacLean then joined Williams at the SSRMS station in Destiny and used it to complete the hand-over of the P-3/P-4 ITS from the Shuttle's RMS, at 10:52. The new ITS element was left hanging on the SSRMS overnight to become thermally stabilized in the space environment. The day continued with the EMU's and EVA tools being transferred to Quest. It ended with Tanner and Stefanyshyn-Piper locking themselves in Quest for the program's first use of the new "camp-out" procedures. The two EVA astronauts slept in Quest, with the pressure reduced, in order to reduce the pre-breathing time required to remove the nitrogen from their bloodstream before an EVA.

Figure 73. STS-115 crew (L to R): Heidemarie M. Stefanyshyn-Piper, Brent W. Jett, Jr., Joseph R. Tanner, Daniel C. Burbank, Christopher J. Ferguson, Steven G. MacLean.

Figure 74. STS-115 approaches the ISS carrying the P-3/P-4 Integrated Truss Structure.

Construction of ISS resumed on September 12, 2006. At 05:17, Tanner and Stefanyshyn-Piper transferred their EMUs to internal battery power, thereby officially starting their first EVA. After collecting their tools the two astronauts made their way to the P-3/P-4 , which had previously been moved to its deployed position at the exposed end of the P-1 ITS and held in place by the SSRMS, while motorised bolts were driven home. The pair quickly got ahead of their timeline as they worked their way through a series of tasks including connecting power cables on the ITS, releasing launch restraints on the Solar Array Blanket Box and the Beta Gimbal Assembly. They also configured the Solar Alpha Rotary Joint (SARJ), which would allow the deployed photovoltaic arrays to track the Sun. Having completed their schedule they began a series of "get-ahead" tasks in preparation for the flight's second EVA. They removed the covers that would allow them to access and remove the launch lock bars from the SARJ. As Tanner removed one of the covers a bolt and washer detached from the SARJ and drifted away. The pair returned to the Quest airlock at 11:43, ending the EVA after 6 hours 26 minutes. During the morning, MCC-Houston had informed Jett that additional inspections of Atlantis' heatshield were not required. Meanwhile, the two crews spent the remainder of the day transferring equipment and supplies between the two spacecraft. The day ended with Burbank and MacLean "camping out" in Quest.

Burbank and MacLean commenced their EVA at 05:05 September 13. The entire EVA was devoted to activating the SARJ, with Stefanyshyn-Piper operating the SSRMS in order to use its cameras to video the activities. Both EVA astronauts released the launch locks, but had to overcome several small difficulties, including a stuck bolt and a faulty EMU helmet camera. Next, they prepared the new P-3/P-4 ITS for the MBS, which would be used to transport the SSRMS along the completed ITS. With the major tasks completed, they turned their attention to additional "get-ahead" tasks, including the removal of a keel pin and drag link. They also removed a Space Vision System target, used to align the RMS and SSRMS in order to remove the P-3/P-4 ITS from Atlantis' payload bay. Finally, they installed a temporary rail stop for the CETA.

With Burbank and MacLean back inside at 12:16, after an EVA lasting 7 hours 11 minutes, MCC-Houston began a 4-hour activation and checkout of the SARJ, designed to ensure all systems were working. They engaged Drive Lock Assembly 1 (DLA-1) and rotated the car-sized joint through 180°. When they engaged DLA-2 later in the day they received no talk-back indication to indicate that it had engaged properly. A back-up procedure also failed to result in the required indication of successful engagement. Deploying the P-4 SAWs was delayed while the problem was investigated overnight, and was overcome by a software work-around sent up from the ground in the early hours of the next morning.

At 04:00, September 14, MCC-Houston commanded the SAWs on the P-4 ITS to deploy, a procedure that was not completed until 08:44. For the STS-115 crew the day was a rest-day, while the Expedition-13 crew continued transferring equipment and stores between the two spacecraft. The 66 kilowatts of electrical power produced by the new SAWs was directed to the P-3/P-4 ITS' own systems. It would not be distributed to the remainder of ISS until the system was rewired and the cooling system activated during the flight of STS-116, then planned for December 2006. During the day the SSRMS underwent a "walk-off", from the MBS to the exterior of Destiny.

During preparations for the third EVA a remote power controller tripped out resulting in the loss of power to Quest's depressurisation pump. Tanner and Stephanyshyn-Piper left the airlock and moved to the adjacent Unity module, while still pre-breathing oxygen through face masks. The problem was found to be a momentary power spike and, following checks to ensure there was no short circuit, the breaker was reset, the pump reactivated, and the two astronauts returned to the airlock to continue preparations for their EVA, which was delayed by 45 minutes.

Tanner and Stefanyshyn-Piper began their EVA at 06:00, September 15. Once outside they separated in order to complete individual tasks. Tanner installed bolt retainers on the P-6 Beta Gimbal Assembly, which assisted in the pitch orientation of the SAWs. He also attempted to re-engage a four-bar hinge, which had failed to engage during the flight of STS-97. Meanwhile, Stefanyshyn-Piper recovered the MISSE-5 materials science experiment. Working together, the two astronauts next prepared the Space Radiator on the P-3/P-4 ITS for deployment by removing hardware installed to protect the radiator during launch. With that task completed, they deployed an S-band antenna support assembly on the S-1 ITS. They also deployed a shroud on the failed S-band antenna support assembly, which would be returned to Earth on a later Shuttle flight. Separating again, Stefanyshyn-Piper replaced a base-band signal processor and transponder on S-1, while Tanner installed a heatshield on an antenna group interface tube to help prevent overheating in the area when ISS was placed in particular attitudes in relation to the Sun. Their final tasks included installing an external wireless TV antenna and performing infrared video of Atlantis' wing leading edges. The EVA ended at 12:42, after 6 hours 42 minutes. After their EVA, the two astronauts joked together, "Kind of nice up there on top," Tanner remarked. "Yes, but I didn't get to look around much. At least, I can say I've been there," replied Stefanyshyn-Piper. Tanner added, "Not too many people have."

September 16 was a light day for the STS-115 crew; they had the morning off after their hectic work schedule so far. After lunch they joined the Expedition-13 crew for a joint crew photograph, a press conference, and personal interviews. Jett told the media, "We are off to a good start ... We have a lot of complex missions ahead. We had a few small problems, but the team did a wonderful job of resolving them. I think it bodes well for the future." The two crews also performed the final transfer of equipment between the two spacecraft, as well as moving some of the items of debris, produced during the last EVA, to Progress M-57.

The two crews said their farewells on September 17. The hatches between the two vehicles were closed at 06:27 and, following pressure checks, Atlantis undocked at 08:50. Twenty minutes later Ferguson began the first full fly-around of ISS since 2002. During the fly-around the crew were able to view and photograph the results of their hard work over the past few days. Atlantis manoeuvred away from ISS at 10:30. In Russia, final preparations were underway for the launch of Soyuz TMA-9, with the Expedition-14 crew.

Figure 75. STS-115: Atlantis departs the station with an empty payload bay.

Figure 76. STS-115: construction has resumed. The station displays the new P-3/P-4 Solar Array Wings and ammonia radiator at right. Meanwhile, the Solar Array Wings and ammonia radiator are still deployed on the P-6 Integrated Truss Structure, mounted on the Z-1 Truss.

During the morning of September 18 the Expedition-13 crew, Vinogradov, Williams, and Reiter, were faced with a malfunctioning Elektron oxygen generator in Zvezda, which had been powered off for the visit of STS-115. At a request from Korolev, Vinogradov attempted to restart the Elektron, but it soon shut down again on its own. Following several other attempts the unit was finally restarted. Vinogradov was working on it when it overheated, causing smoking from a melted rubber seal, a strong odour, and the possible release of a small amount of the chemical irritant, potassium hydroxide. The crew were instructed to manually activate a fire alarm to allow the station's software to shut down the air circulation fans. They were also told to don surgical masks, goggles, and gloves. Vinogradov cleared up and bagged a clear liquid that had leaked from the unit. The Elektron was ultimately restarted and the air circulation fans powered on within the hour. On the same day the crew of STS-115 made a final inspection of Atlantis' heatshield, nosecap, and wing leading edges. In the words of Shuttle Programme Manager Wayne Hale, "You can call it anxiety, or you can call it smart. But it's what we do these days. We have no reason not to go look and put every concern to rest."

At 03:00, September 19, Houston linked the crews of ISS, STS-115, and Soyuz TMA-9, which had been launched the previous day, to allow them to talk to each other. Williams told the Soyuz crew, "It's a little crowded in the sky today ... We look forward to having you guys onboard." Jett told the Expedition-13 crew, "We'll see you back on Earth sometime soon." Following the conference Atlantis' crew tested the orbiter's flight control surfaces and RCS thrusters. They spent the day packing up for re-entry, but during the day the flight was extended by one day, to allow for further inspections of the Shuttle's exterior after video cameras showed debris drifting away from the vehicle. The decision to extend was also based on deteriorating weather forecasts for the landing site. TMA-9 continued its approach to rendezvous, while the Expedition-13 crew watched the undocking of Progress M-57 at 20:30. The spacecraft re-entered the atmosphere and burned up just after midnight.

Having left the RMS/OBSS positioned above the payload bay overnight, so that controllers could use its video cameras to inspect the area, it was used to inspect Atlantis' underside on September 20. A shim and a tile spacer seen in an earlier inspection were found to be gone. When no issues with the Shuttle's heat shielding were revealed, flight managers cleared Atlantis for landing in the early hours of the following morning. Meanwhile, Soyuz TMA-9 had docked to Zvezda. The crew entered the station later that morning. Atlantis' crew went to bed at 21:45, waking again at 21:45 to face the final hours of their flight. Atlantis landed at KSC, Florida at 06:21, 21 September, after a flight lasting 11 days 19 hours 6 minutes.

On September 22, Michael Griffin told a press conference, "It's obvious to me we are re-building the kind of momentum that we have had in the past and that we need if we are to finish the Space Station ... We have an awesome task in front of us. I think we will make it."

SOYUZ TMA-9 DELIVERS THE EXPEDITION-14 CREW

SOYUZ TMA-9	
COMMANDER	Michael López-Alegría
FLIGHT ENGINEER	Mikhail Tyurin
ENGINEER	Anousheh Ansari (spaceflight participant)

When the Soyuz TMA-9 crew were originally named, the spaceflight participant involved was Japanese entrepreneur Daisuke Enomoto, but he was grounded for undisclosed medical reasons that caused him to fail his final medical examination. Enomoto was replaced by his back-up Anousheh Ansari, the Iranian-born daughter of an adventure capitalist, supporter of the X-Prize (won by Rutan's SpaceShipOne) and business partner to the Federal Space Agency of Russia in attempts to develop commercial access to space. Following her late allocation to the flight Ansari would perform many of Enomoto's experiments, rather than those with which she had already begun practising, for her own flight.

Ansari flew without an Iranian flag on the sleeve of her Sokol pressure suit, and had promised her American and Russian programme managers that she would make no political speeches while in orbit. Before her launch she had remarked:

> "I believe when you see the Earth without boundaries, without borders, without race, that you can see how important it is for us, everyone, to be more understanding of our neighbours, our friends, other people's beliefs, religions, race and not make issues to start wars ... One thing that I'm hoping is that I'd like for people all over the world to see a different face of an Iranian-born individual, something different than what they see on TV and in the media ... Also, I think my flight has become sort of a ray of hope for young Iranians living in Iran, helping them to look forward to something positive, because everything that they've been hearing is all so very depressing and talks of war and talks of bloodshed."

Soyuz TMA-9 was launched from Baikonur at 00:09, September 18, 2006. Following a standard rendezvous it docked to Zvezda's wake at 01:21, September 20. The hatches were opened at 03:34, allowing López-Alegría, Tyurin, and Ansari to enter ISS. The Expedition-13 crew greeted their reliefs before giving them the standard safety briefing. The intense hand-overs of previous Expedition crews were lightened somewhat by the fact that Reiter had been on the station as part of the Expedition-13 crew since July 2006 and would remain as the third member of the Expedition-14 crew until December 2006.

López-Alegría described his flight:

> "The goals, first of all, are to continue the assembly of the Space Station, which has sort of been on hold for a while since the Columbia accident, so in general our

mission is going to be to receive a couple of Shuttle flights, do some construction while they're there. After they've gone, we're going to be receiving three Progress vehicles, which means a lot of cargo, loading, unloading. In general, the focus is going to be construction and assembly . . . [E]very time something comes, you've got to unpack it, and that takes time. And time is going to be a significant challenge for us. The second thing is, when we unpack that stuff there's got to be some place to put it, and the station is already pretty full of stowage. The challenge is going to be to find where to stow the stuff that's brought up."

During the hand-over period, Ansari performed two ESA experiments, while the two Expedition crews worked together to ensure a smooth transition. Reiter transferred his couch liner and Sokol pressure suit from Soyuz TMA-8 to Soyuz TMA-9, while Ansari took hers in the opposite direction. Throughout the period the Elektron oxygen generator remained powered off and the astronauts burned four SFOG candles each day to produce the required amount of oxygen. Despite now having 120 new SFOGs onboard the crew continued to burn the 40 out-of-date candles in order to use them up. By the end of the hand-over period only four old SFOGs remained. Vinogradov and Tyurin replaced the liquids unit in Elektron and powered it on. It failed shortly thereafter and was designated to be "hard-failed" by Korolev. New spare parts would be delivered by Progress M-58, in October 2006. Throughout the period both Expedition crews participated in ongoing experiments in both sectors of the station. Vinogradov has described the hand-over period saying:

"[T]here also is a bit of a symbolic significance, when the commanders shake hands and say, here, you accept the station and I hand the station over to you and have a successful flight and all. This is quite an important moment associated with a lot of responsibility, and it's quite busy. Then the previous crew leaves and you stay there alone; it's kind of sad, actually. And in a way it's a little bit worrisome. You're there, and you don't have anyone to ask or clarify things. When you part, it's quite an emotional moment. I don't think that there is a single crew that leaves with joy saying, oh fine, it's finally over and I'm going to be home. It's always a sensation that you're leaving your home because always it's that time that evokes a certain amount of sadness. Of course you understand the landing, your family, your kin and friends and all the joys of living on Earth, but it does make you want to go back and it's quite a sad time when you have to leave the station."

Williams explained:

"The plan in the future is to rotate two out of three crewmembers on Soyuz every six months, as we have been doing for the last few years. The third crewmember will rotate on Shuttle; and by default that means that there's a phased rotation. So Thomas gets there after we've been there a little while, and when we get ready to depart, with the next Soyuz, and its crew arrival in the fall, when we depart Thomas will stay on board. He'll be the experience and help with the hand-over of the beginning of . . . Expedition 14."

Figure 77. Expedition-14: American/Iranian spaceflight participant Anousheh Ansari flew to the ISS on Soyuz TMA-9 with the Expedition-14 crew. She returned to Earth with the Expedition-14 crew.

On September 24, Vinogradov and Williams completed systems checks in Soyuz TMA-8 before a full dress rehearsal of the undocking and re-entry procedures, with Ansari the following day. The official hand-over took place on September 27. After a week-long hand-over, followed by fond farewells, Vinogradov, Williams, and Ansari shut themselves in Soyuz TMA-8. Undocking occurred at 17:53, September 28. After a routine re-entry, Soyuz TMA-8 landed at 21:13, the same day. Vinogradov and Williams had been aloft for 182 days 22 hours 43 minutes. Ansari's flight had lasted 10 days 21 hours 5 minutes. Following standard recovery procedures the crew was flown to Kustani for the official welcome home ceremony and then to Zvezdny Gorodok for their 45-day rehabilitation and debriefing. Vinogradov had reviewed his definition of success before launch:

> "First of all I would say it's this internal feeling that you've done everything that you could. The satisfaction with the flight actually comes later, after some evaluations are performed and it's all kind of reviewed and summed up. But the most important thing I would say is the appreciation of your work, of the crew, is when the next crew comes in and can tell you, guys, thanks for doing this and that, it's because of you that we were able to, sort of standing on your shoulders, continue this work. I would say that this is probably the most important assessment of your work on the station that those who are coming after you would be using your experience, would stand on the basis of the results of your work."

By those standards Expedition-13 had been a great success.

On September 26, NASA Administrator Michael Griffin visited Chinese space facilities. He told a press conference:

> "This is an exploratory visit. This is a first date—if you will. I think we need to let it evolve. There are differences between our nations on key points. One of the major points is the control of missile technology. We have been very firm on that in the past, and I believe we will continue to be firm on the importance of appropriately controlling missile technology. China has clearly made great strides in a relatively short period of time ... I have been very impressed with the capabilities, experience and intellectual quality of the people we have met. The facilities we have seen have been first rate."

Griffin made it clear that there were no plans to invite China to join the International Space Station partnership in the near future.

Meanwhile, Russia had increased the cost of their spaceflight participant flights to ISS on a Soyuz taxi flight from $20 million to $21 million. Elsewhere, Space Adventures, the company that markets spaceflight participant flights on behalf of the Russian Federal Space Agency, had also begun talking about an option for their customers to perform a 90-minute EVA whilst docked to ISS. Training for and performing an EVA would cost the customer an additional $15 million on top of the new $21 million fee. Meanwhile, Russian space officials told the media that their country could expect to fall behind "irreparably" if they failed to develop the six-person Kliper spacecraft. Russia and Europe were also considering the development of a new heavy-lift launch vehicle, the Oural programme, as a possible replacement for the Soyuz-2 and Ariane-V launch vehicles at some point in the 2020s.

EXPEDITION-14

López-Alegría, Tyurin, and Reiter quickly settled into a daily routine of performing experiments, housekeeping, maintenance, and exercise. López-Alegría described some of his increment's experiments:

> "We're trying to understand better the effects of long-duration spaceflight on humans, because our goal is to extend our presence not just in low Earth orbit but to go back to the moon with some kind of a longer-term presence, and hopefully on to Mars someday. So, a lot of the science is dedicated to human physiology. We are studying, everything ... on a very small level. There's an experiment called SWAB, which measures bacteria levels on surface[s], water, and air; it's sort of an environmental thing—what actually grows up there and how do we have to worry about reacting with it. Another sort of related idea is something called Epstein–Barr virus. We see how our immune system reacts over time while we're up there. We have a nutrition experiment that will, through taking blood and urine samples, track our intake and how we metabolize the food that we're eating

up there, because, in general, people tend to lose weight in space. In space they lose weight, but when they come back they're usually a little bit lighter and over a lot of time, that obviously can be debilitating. [C]ertainly muscle function, bone loss are very important. We're doing an experiment called TRAC [Test of Reaction and Adaptation Capabilities], which is pretty interesting neuron-reaction time—it's an experiment [where] we've got to have a tracking task in one hand, where you're trying to get a 'pipper' to stay over a target, and in the other hand you're reacting with a keyboard, trying to do things as quickly as possible. Unfortunately because of the assembly and because of the stowage, time challenges, we don't have as much science as we'd like. I guess it's an investment. We're investing the time now to build the Space Station, so that we can have a lot more science time available in the future."

The crew spent their first solo week onboard ISS preparing to move Soyuz TMA-9. They also practised an emergency egress from the station, checking all safety equipment in the process. López-Alegría and Tyurin also completed their first medical experiments. Reiter completed loading unwanted items into Progress M-57 and carried out the standard off-loading of liquid waste from the ISS toilet to the empty water tanks in the Progress. The hatches between the two spacecraft were sealed on October 5. The following day, all three men completed monthly fitness evaluations on the station's stationary bicycle. Tyurin also spent some time during the week troubleshooting the Elektron and replacing components of the instrument panel. The Elektron continued to intermittently malfunction and further repairs were delayed until replacement parts could be delivered on Progress M-58. Meanwhile, the station drew additional oxygen from the tanks mounted on the exterior of Quest. The last of the old design SFOGs had been burned and the igniters in Zvezda were changed out for new ones matching the design of the new SFOG candles now onboard.

CMG-3 began vibrating on October 9 and was shut down, leaving three CMGs to maintain the station's attitude. This was a normal fall-back configuration and had no knock-on effect on the flight. The CMG would be replaced during the visit of STS-118, planned for August 2007. The malfunctioning gyroscope would be stowed on the station before its return to Earth on STS-122. In the meantime the station continued to function normally.

Having set ISS up for unoccupied flight, the three men sealed themselves in Soyuz TMA-9 and undocked from Zvezda's wake at 15:14, October 10, and re-docked to Zarya's nadir at 15:34. They then re-entered the station and reconfigured its systems for human occupation once more. The relocation manoeuvre cleared Zvezda's wake for the arrival of Progress M-58, later in the month. The crew performed routine medical experiments and maintenance while continuing to load Progress M-57 with unwanted items. On October 23, Houston began a 5-day workout of the Thermal Radiator Rotary Joint (TRRJ) on the S-1 and P-1 ITS. NASA explained that the joints would, "... enable the radiators to auto-track, or revolve, when required to dissipate heat from the Truss' avionics equipment ..."

PROGRESS M-58

Progress M-58 was launched from Baikonur at 09:41, October 23, 2006 and climbed into orbit. The spacecraft performed a standard rendezvous before docking to Zvezda's wake at 10:29, October 26. As Progress M-58 approached Zvezda, telemetry failed to confirm that a KURS antenna on the front of the spacecraft had retracted as scheduled. If it had not retracted then it would interfere with the docking latches. Following soft-docking, Russian controllers in Korolev spent 3.5 hours trying to confirm the antenna had retracted. During this period the station's attitude control systems were powered off and the station was allowed to drift so as not to disturb the Progress vehicle and misalign it. In this period the station's drift led to a misalignment of the SAWs and a drop in electrical power. Following routine procedures, the crew powered off non-critical items to conserve power. Finally, the docking probe on the Progress was retracted and the docking latches engaged, securing a hard-dock at 14:00. After hard-docking, the station's attitude control systems were powered on and the station brought back to its correct attitude. The non-critical items were then powered on once more. Among its 2,393 kg of cargo, Progress M-58 carried replacement parts for the Elektron oxygen generator as well as 2.5 tonnes of food, water, propellant, oxygen, spare parts, supplies, and personal items for the crew.

The non-confirmation of the KURS antenna folding on the new Progress continued to worry Russian controllers in Korolev and they began planning a Stage EVA to confirm it was correctly stowed. López-Alegría and Tyurin would make the EVA on November 22. Tyurin would also hit a golf ball away from the Pirs docking compartment in a commercial "experiment" designed to result in the longest golf strike in history. Meanwhile, Reiter began packing material planned to be returned to Earth on STS-116, the Shuttle flight that would bring his 6-month stay on ISS to an end. As the preparations for the next Shuttle launch continued, the Expedition-14 crew continued with their experiments. In preparing for their EVA, López-Alegría and Tyurin attached American EVA lights to the helmets of their Orlan pressure suits. Meanwhile, controllers in Houston continued to test CMG-3, which had shown unplanned vibrations prior to being shut down on October 9. Tyurin repaired the Elektron unit in Zvezda on October 30, after which it was powered on and began supplying oxygen to the station's atmosphere once more.

López-Alegría walked the SSRMS end over end from its position on the exterior of Destiny to the MT, which had been positioned at Workstation 4 on the ITS, on November 1. After manoeuvring the SSRMS so that both end-effectors could be photographed, the MT was moved to the far end of the P-4, where the SSRMS would be used to install the P-5 ITS during the flight of STS-116. Meanwhile, Reiter began preparing for his return to Earth. He would be relieved by American Sunita Williams. Even so, the European experiment programme on ISS continued unabated. The Elektron unit malfunctioned again, developing an internal water leak on November 7. Five days later, the crew began shifting their sleep pattern to match that of the STS-116 crew, then due for launch on November 22. They also closed the hatches between Pirs and Progress M-57.

At 19:17, November 22, López-Alegría and Tyurin exited Pirs wearing Orlan suits. Expedition-14's first Stage EVA began 1 hour late, after Tyurin had to deal with a pinched cooling hose in his suit. He had had to climb out of his suit and reposition the hose in order to release the pinch. Having re-entered the suit, closed the hinged Portable Life Support System behind him, and repressurised the suit he was finally ready to proceed. Once outside, Tyurin mounted a 3-gramme golf ball on a spring mounted tee which he attached to the ladder on the exterior of Pirs. With his booted feet on the ladder he used a gold-plated six-iron to strike the ball away, towards Zvezda's wake. Tyurin remarked, "There it goes, and it went pretty far, I can still see it as a little dot, moving away from us."

Controllers estimated that the ball would re-enter the atmosphere and burn up in three days. Tyurin said he was pleased with the shot and controllers chose not to have him make a second shot. The golf shot was paid for as a commercial venture by a Canadian golf company.

Moving to Zvezda's wake, where Progress M-58 was docked, Tyurin attempted to release a latch on the KURS antenna that had failed to retract. Despite his best efforts using both his gloved hands and a crowbar, alongside commands radioed up from Korolev, the antenna continued to refuse to retract. The two men took a series of digital photographs, which would show that the antenna's dish was stuck behind one of Zvezda's EVA handrails. A tool to remove the antenna would be delivered on STS-116. Moving on, the two men relocated a communications antenna that would be used by the European ATV. By moving the antenna just 0.3 m from its original location it no longer impeded the cover of one of Zvezda's rocket motors. It was this antenna that had prevented an attitude control burn taking place on April 19, 2006, because the cover on the rocket motor could not be opened to its full extent. At Zvezda's ram they installed the BTN-Neutron experiment, designed to record the flux of neutron particles found in low-Earth orbit following solar particle events. Their final task was to jettison a pair of thermal covers for the experiment. The EVA ended after 5 hours 38 minutes.

The last week of November was spent preparing for the arrival of STS-116, Discovery. The Expedition-14 crew prepared Quest for three planned EVAs and they prepared the equipment that would be returned to Earth. On November 29, Tyurin changed out batteries in both Zvezda and Zarya. On the same day, an attempt to re-boost the station using the thrusters on a Progress spacecraft was cut short due to a software problem. Over-sensitive software shut down the thrusters after detecting motion caused by the planned manoeuvre. The planned 18-minute 22-second burn lasted just 3 minutes 16 seconds. The manoeuvre was completed on December 4, and placed ISS in the correct orbit for Discovery's rendezvous.

Tyurin and Reiter disassembled the Matryoshka human torso experiment in Pirs, to remove the 360 internally mounted radiation sensors. They also removed the KURS avionics packages from Progress M-57 and Progress M-58 and stowed them for return to Earth.

As November ended, the US National Research Council released a report criticising NASA for failing to obviously redirect the ISS science programme to meet the needs of Project Constellation, and in particular human flights to Mars, saying,

"The panel saw no evidence of an integrated resource utilization plan for the use of the ISS in support of the Exploration Missions." It continued, "The ISS may prove the only facility with which to conduct critical operation demonstrations needed to reduce risk and certify advanced systems."

STS-116 DELIVERS "PUNY"

STS-116	
COMMANDER	Mark Polansky.
PILOT	William Oefelein.
MISSION SPECIALISTS	Nicholas Patrick, Joan Higginbotham, Robert Curbeam, Christer Fuglesang (ESA)
EXPEDITION-14 & 15 (up)	Sunita Williams
EXPEDITION-13 & 14 (down)	Thomas Reiter

In the wake of a December 7 launch attempt that had been cancelled, due to low cloud, the fact that the ET had been filled with fuel demanded that the next launch attempt could not take place for a further 48 hours. On the second attempt, STS-116 lifted off from Kennedy Space Centre at 20:47, December 9, 2006. In the minutes before launch, Polansky told the launch team, "We look forward to lighting up the night sky." It was the first Shuttle night launch in over four years. He also told the media, "There are always inherent risks when you take people off the planet and try to propel them a couple of hundred miles into space. We try to mitigate the risk as much as possible. If anyone says we can take all the risk out, they are just blowing smoke."

Fuglesang was pragmatic, saying, "I hope this will increase interest in space for Sweden, and actually for science and technology in general. This flight is a small step in the assembly of the Space Station. The station is a small step toward the Moon and Mars."

Meanwhile, Space Station Manager Mike Suffrendi told a press conference, "Many of us consider this the most challenging Space Station flight we've done since we began the assembly effort. When the Shuttle leaves, ISS will not look much different than when Discovery got there, but it will be a dramatically different vehicle inside."

Discovery's seven-person crew would complete three EVAs, during which they would install the port-5 (P-5) ITS and then re-wire the station's electrical system, bringing on-line the SAWs delivered by STS-115. Sunita Williams would join Expedition-14 crew members López-Alegría and Tyurin, while Thomas Reiter would return to Earth in Discovery with the STS-116 crew. The Shuttle also carried a SpaceHab pressurised module full of equipment and supplies for the station. Asked about the mid-term hand-over López-Alegría explained:

> "Well, I think it all stems from the notion that the Russian Space Agency, Roscosmos, would like to have the third seat in the Soyuz available for a paying

passenger . . . and so the third person can't rotate there and he or she will rotate on Shuttle. So, that's . . . the scheme that we have evolved to."

He continued:

"I think there are certain disadvantages, certainly. If you were to build a true Expedition crew, you'd like them to be sort of lockstep with each other all the time. I think this is a little bit different because we do have a fair amount of access to the ground, talking to people, e-mailing friends and family, using the internet protocol phone to be able to converse with people. That probably eases the difficulty in being isolated somewhat that is so specific to certain types of expeditions. I think that the result is that the need to bond really as a single unit is not as stringent as it might be if we were going to live that kind of an experience. However, it does bring up some interesting challenges. I think there's also some advantages because six months with the same two faces all the time, if you don't like one of those two faces it could get old. At least, in this fashion, we will have the opportunity to change those faces once in a while."

Commander of Expedition-15 Fyodor Yurchikhin would express a different point of view, which I quote on p. 277.

On reaching orbit, Discovery was configured for flight and her payload bay doors were opened to expose the vital radiators mounted on their interiors. During the first day in orbit the crew powered up the RMS, and lifted the OBSS from the payload bay door hingeline. They then used the OBSS cameras and lasers to inspect Discovery's heatshield and the leading edges of the wings. Houston confirmed, "There is nothing anyone is excited about so far." The crew also installed the usual selection of rendezvous equipment and cameras and checked the EMUs they would wear during their three EVAs.

Following a standard rendezvous. Williams got her first view of the station and reported excitedly, "Tally-ho on my new home. It's beautiful. The solar arrays are glowing." After the now routine r-bar pitch manoeuvre, Discovery docked to ISS at 17:12, December 11. As usual, the station's bell was rung to welcome the visitors. Following pressure checks, the hatches between the two spacecraft were opened and the Shuttle's crew were welcomed aboard the station at 18:54. As the hatches were swung open López-Alegría joked, "We're having a ball already."

The first order of business was to inspect the tip of Discovery's port wing with the cameras on the RMS, after a vibration was picked up by the wingtip sensor 18 minutes after docking. The additional inspection delayed the crew using the RMS to un-berth the P-5 ITS from its position in Discovery's payload bay. The P-5 ITS was subsequently handed over from the Shuttle's RMS to the SSRMS and was then positioned over Discovery's port wing, where it was left overnight. Williams described the P-5 ITS in the following terms:

"P-5 (and S-5) is a part, that goes between the two major solar array wings. Without them, the two wings would be too close together to actually operate.

Figure 78. STS-116 crew (L to R): Robert L. Curbeam, William A. Oefelein, Nicholas J. M. Patrick, Joan E. Higginbotham, Sunita L. Williams, Mark L. Polansky, Christer Fuglesang.

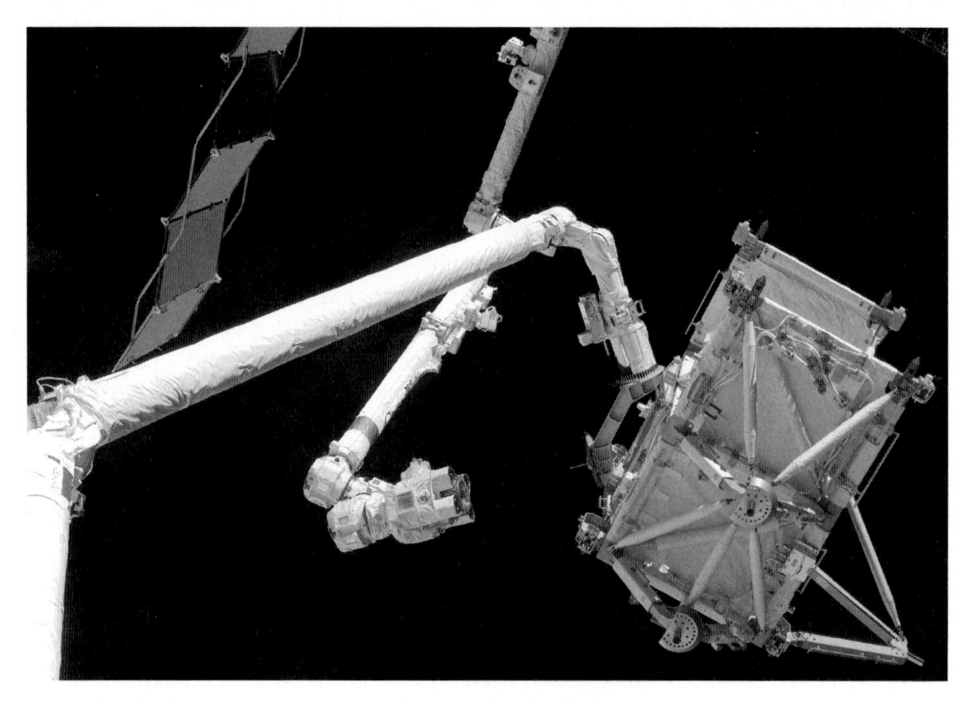

Figure 79. STS-116: astronauts operating the Shuttle's RMS prepare to pass the P-5 Integrated Truss Structure over to the station's SSRMS for installation on the station.

They're a pass-through for all of the thermal, electrical lines going out to the end of the truss and absolutely critical ... to make sure that the two big solar array wings will be able to operate."

At 01:00, Williams installed her couch liner in Soyuz TMA-9 and transferred her Sokol pressure suit from Discovery, thus transferring herself to the Expedition-14 crew. At the same time Reiter transferred his equipment to Discovery and became part of the STS-115 crew. For her first two weeks on the station Williams would spend 1 hour each day, in order to familiarise with the station and the Expedition-14 crew's routines. These hours were unstructured, allowing Williams to concentrate on whatever she felt was necessary to bring her up to speed. Williams has described the advantages of a mid-term crew exchange:

"I think I'm really lucky ... they're going to be there to help me with any ... turnover things that I don't understand. I'm a rookie; never flown before. These two are both experienced space fliers; and them, having lived there for about three months before I get there, I think if I have any questions, they'll be the perfect people to show me the way."

Prior to retiring for the night, the STS-116 crew reviewed the procedures for their first EVA, planned for the following day. Curbeam and Fuglesang sealed themselves in Quest and reduced the pressure in order to "camp out" in the airlock overnight, as part of their EVA pre-breathing regime.

Williams, who would operate the RMS during the P-5 installation, explained:

"The first EVA, which is the P-5 install. Me and Joanie Higginbotham will be operating the robotic workstations. We'll be taking the P-5 Truss from a handoff position from the Shuttle robotic arm and we'll be moving it to the end of P-3/P-4 for the installation. It's a little bit of a tricky installation because the clearances to get the P-5 into its position are pretty tight, about three inches or so. Some of the issues with that is the P-3/P-4 solar array wing is live at the time, so there's going to be some black boxes on the end of P-3/P-4 that are live powered. And so with that clearance, the biggest worry is that you don't hit the box that has the live power on it, 'cause that's going to cause a lot of problems. So, we've practiced this very intensely with the spacewalkers Bob Curbeam and Christer Fuglesang. They'll be out on opposite ends of the P-3/P-4 truss, guiding us in. So this is a very complicated, entire-crew-involved event to try to get this guy installed ... Part of that EVA is also starting up the main bus power switching units, MBSU, and while we're making sure that that's all starting up correctly, the two spacewalkers, Bob Curbeam and Christer Fuglesang, will be moving the CETA carts to the opposite side that they're on, in preparation for the next solar array, which is the S-3/S-4 installation. So, we'll be working with the spacewalkers again as we'll be picking them up and driving them over to the truss, while they'll be grabbing on to the CETA carts. We'll be flying them over to the other side of the MBS."

The first EVA began at 15:31, December 12, when Curbeam and Fuglesang transferred their suits from external power to internal batteries. Exiting Quest, they prepared their tools and then made their way across to the P-5 ITS. The two astronauts guided Higginbotham as she lifted the P-5 ITS into its installation position at 17:45. With the new ITS section in place, the two EVA astronauts bolted it into position, completing the task at 18:21. Moving on, the two astronauts replaced a failed camera, removed the launch restraints and an RMS grapple fixture from the P-5 ITS as well as a cover that would allow the P-6 ITS to be bolted to it when it relocated from its present position on the Z-1 Truss' zenith. With the two astronauts back inside the airlock the EVA ended at 22:07, after 6 hours 36 minutes. During the day mission managers confirmed that Discovery's heatshield was in good condition for re-entry at the end of the flight.

After a night's sleep, phase two of the flight's objectives got underway. The crew spent 6 hours, beginning at 18:17, December 13, sending a series of over 40 commands to retract the port SAW on the P-6 ITS, which had been in place on the station since it was deployed in 2000. The retraction did not go well and the arrays had to be partially retracted, re-deployed, and retracted again. Despite everything, they failed to retract as planned. The guidewires became snagged with only 17 of 31 panels retracted, but it was enough to allow the day's work to continue. At 20:00, the P-4 SAWs began rotating to their operational position. When the ITS was complete and all the SAWs were deployed, they would be able to rotate to track the Sun as ISS orbited Earth. On the evening of December 13, 2006, P-4 became the first SAW to rotate. The manoeuvre was completed without difficulty, and just before 23:00 the valves were opened to allow ammonia to flow into the ITS and the huge radiators mounted there. This was the first stage towards providing permanent cooling for the avionics and electronics on ISS. Inside, the two crews spent the day transferring equipment from Discovery's SpaceHab module and mid-deck to the station.

Meanwhile, in Houston, mission managers met to discuss the various options for completing the retraction of the P-6 SAW. One option was to assign additional EVA tasks, to be performed by the Expedition-14 crew after Discovery's return to Earth, in order to carry out the task manually. The meeting concluded that the partially retracted P-6 port array was in a safe configuration to be left for the remainder of the STS-115 flight and, if necessary, the arrival of the next Progress, planned for launch in January 2007. Despite the difficulties with the P-6 SAW the management team agreed to proceed with the second EVA as planned. As a result, Curbeam and Fuglesang spent a second night camping out in Quest under reduced atmospheric pressure. The two crews began their sleep period leaving the P-4 array rotating as it tracked the Sun and the ammonia flowing through the new cooling system, while they slept. Attempts to transfer orientation of the station back to the CMGs failed, possibly due to increased atmospheric drag as a result of increased solar activity. Discovery remained in control of the combination for the time being.

December 14 began with a planned major power-down of many of the ISS' electrical systems. The systems were powered down because the electrical system that supplied them with power was about to be switched from the P-6 ITS SAWs to the P-4 ITS SAWs. Orientation of the Shuttle/ISS combination was controlled by

Discovery, which maintained orientation by firing the orbiter's thrusters as demanded by its own attitude control system. Curbeam and Fuglesang began their second EVA at 14:41, approximately 30 minutes ahead of schedule. During their 5-hour EVA they worked swapping cable connectors to establish the station's permanent cooling and power systems. By 16:30, one of the external cooling loops was shedding heat into space and the direct-current-to-direct-current converter units were regulating electrical power. By 16:45, controllers were applying power to Channels 2 and 3 for the first time. Additional tasks included relocating one of the CETA handcarts that would run along the ITS. Having caught up a further 30 minutes by performing their tasks more quickly than planned, the two astronauts returned to Quest and ended their EVA at 19:41. Williams and Higginbotham had operated the SSRMS throughout the EVA.

December 15 was a day of internal work. During the first half of the day the crews transferred equipment between the two spacecraft. Following that work they performed two press conferences before taking the remainder of the day off. At 21:04, Williams commanded the stuck P-6 SAW to deploy slightly and then retract by the same amount. The attempt left the SAW in exactly the same position as it was when she started, with just 17 of its 31 panels folded. In Houston, mission managers were still discussing the possibility of a fourth, unplanned EVA to try to complete the folding up of the P-6 SAW. Curbeam and Williams camped out in Quest during the night with the pressure reduced, in preparation for their third EVA the following day.

The flight's third EVA began at 14:25, December 16, but not before controllers in Houston had shut down half of the station's electrical systems: the opposite half to that shut down on December 14. Curbeam and Williams left Quest and prepared their tools. They spent their time swapping electrical connectors once more to bring the station's electrical system to its final configuration. In future, when additional ITS elements were added no new reconfiguration of the electrical and cooling system would be required, except to connect the new ITS elements to the existing system. By 16:18, controllers in Houston were applying power to Channels 1 and 4 for the first time, as they brought the station's electrical and cooling systems back on-line. Their primary task complete, the two astronauts fitted a grapple fixture to the SSRMS and positioned three bundles of radiation shielding for the Russian sector on the exterior of the station. The shielding would be installed in its final locations on a later EVA. Their final task was to position themselves on either side of the partially retracted P-6 photovoltaic array and take turns to shake their respective sides of the SAW while their colleagues inside the station attempted to retract it. Looking at the guidewires on the SAW, Curbeam reported, "It's definitely hanging up." He shook the array and it cleared temporarily allowing it to be further retracted before it snagged again. It was a frustrating procedure that had to be repeated several times.

"This is definitely the right approach. I think we are starting to get there," encouraged López-Alegría from inside the station.

In the control room in Houston, Steve Robinson watched the live television pictures and remarked, "That is an impressive amount of motion and very effective." Curbeam replied, "I'm here to serve."

Curbeam shook the array 19 times and Williams 13 times while the retract command was issued 8 times. At the end of their efforts only 11 panels on the array remained unfolded. The EVA ended at 21:56, after 7 hours 31 minutes.

Whilst the third EVA was underway, mission managers in Houston confirmed that Discovery's flight would be extended by one day to allow Curbeam and Fuglesang to make an unscheduled fourth EVA, in an attempt to finish the retraction of the P-6 SAW.

Working inside the station on December 17, the crews were slightly ahead of schedule in their work to transfer equipment between the two spacecraft. As a result, they spent much of the day preparing for the fourth EVA. Work included positioning the SSRMS and Discovery's RMS to support the EVA. Cameras on the latter would be used to video the astronauts' actions. Discovery's crew also transferred two EMUs to Quest for use during the EVA. Curbeam and Fuglesang spent the night camped out in Quest with the airlock's pressure reduced.

Curbeam and Fuglesang left Quest at 14:12, December 18. Having collected their tools Curbeam mounted the SSRMS and was lifted over to the balky array. Fuglesang made his way manually to the P-6 ITS. Curbeam would attempt to free the stuck guidewires and push on hinges to ensure that they folded the correct way while Fuglesang would stand behind the "blanket box", into which the array was being folded and push it in an attempt to encourage retraction. With the manual work on-site completed controllers in Houston sent commands to retract the SAW one panel at a time. The SAW was finally fully retracted at 18:54, and the two blanket boxes were locked at 19:34. The EVA ended at 20:50, after 6 hours 38 minutes. In making this unscheduled EVA, Curbeam became the first Shuttle crewmember to make four EVAs during a single flight.

During the farewell ceremony held on the station as Discovery's crew prepared to return to Earth, Polansky said, "It's always a goal to leave a place in better shape than it was when you came. I think we have accomplished that."

Williams told Reiter, "I hope Discovery takes you home as smoothly and safely as it brought me here."

With Oefelein at the controls Discovery undocked from ISS at 17:10, December 19, and completed a half-circuit fly-around before finally manoeuvring away. During the fly-around the crew photographed the station in its new configuration. On the following day, Polansky, Oefelein, and Patrick used the RMS-mounted OBSS to survey Discovery's heat protection system. The remainder of the crew began stowing equipment for landing. On the ground, NASA's Phillip Engelhauf remarked, "We are assuming the vehicle is in a 'go' condition for landing unless somebody illuminates an issue out of that data. The assumption is that everything is fine."

At 19:19, December 20, a pair of coffee cup-size Micro-Electromechanical System Based PICOSAT Inspector (MEPSI) satellites were launched from Discovery's payload bay as a single unit, which then separated into its component parts. The technology was designed to allow similar satellites to photograph/film the larger vehicle from which they are launched. A pair of Radar Fence Transponder (RAFT) satellites were launched from the payload bay at 20:58, the same day. They were designed by a group of students at the US Naval Academy to test the American

Figure 80. STS-116: Christer Fuglesang rides the SSRMS to relocate a Crew Equipment Translation Aid (CETA) cart on the Integrated Truss Structure.

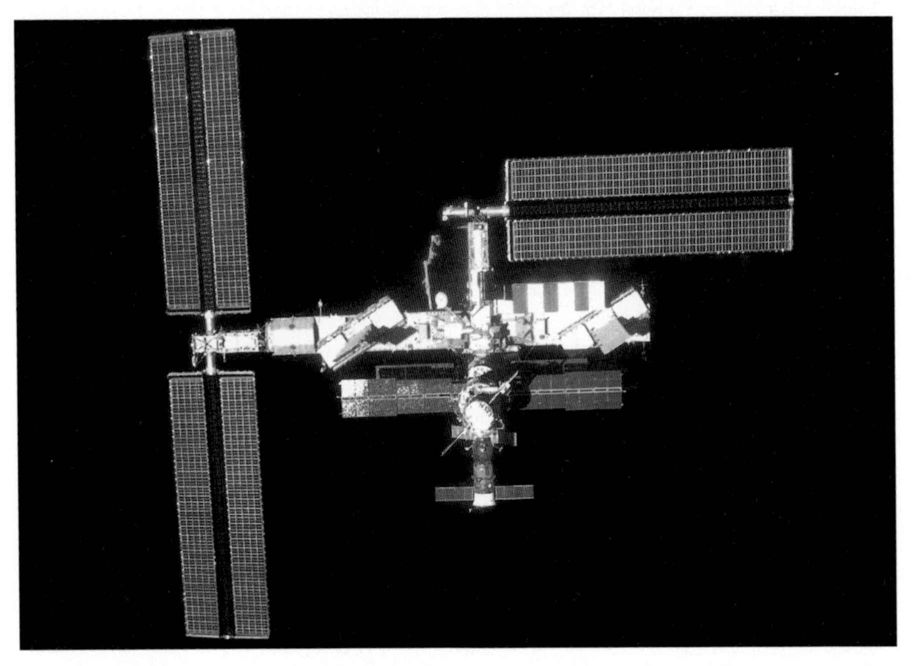

Figure 81. STS-116: as the Shuttle departs the station its lop-sided configuration is obvious. The S-4 Solar Array wings are shown at left and one set of the P-6 Solar Array wings are deployed. The other set of P-6 Solar Array wings were folded and stowed by the STS-116 crew in anticipation of the P-6 Integrated Truss structure's relocation in 2007. The P-5 ITS is partially hidden behind the P-1 ammonia radiators.

Space Surveillance Radar Fence designed to identify hostile objects approaching the continental United States from space.

Preparations for returning to Earth began in earnest on December 21. Oefelein and Curbeam tested Discovery's aerodynamic surfaces and manoeuvring thrusters. Polansky and Oefelein practised simulated landings on a laptop computer. At 12:23, Fuglesang and Higginbotham launched the Atmospheric Neutral Density Experiment (ANDE) micro-satellite from Discovery's cargo bay.

On December 22, the first opportunity to land at KSC was waved off due to the stormy weather conditions there. An opportunity to land at Edwards Air Force Base, California was also waved off due to gusty winds. As the day continued weather in Florida improved and Discovery was able to utilise the second landing opportunity at that site. Polansky glided his spacecraft to a perfect touchdown, just after sunrise, at 05:32, having spent 12 days 20 hours 44 minutes in flight. Reiter had been in space for 171 days. NASA Administrator Michael Griffin was present to greet the crew. He told the assembled crowd,

> "This was a big year ... I've said if we could take the time to get things going properly, we could get back to an operational tempo and finish the station by the time it's necessary to retire the Shuttle ... We have a new understanding in this country that each and every time we do this, it's a minor miracle."

On returning to JSC, Polansky told gathered workers and their families, "It's awesome to see so many people. This is not about us, it's not about this crew. This is about everybody that shares the same dream, the same drive and really believes in what we are doing with human space flight."

CONTINUING THE ROUTINE

With Discovery gone, the Elektron oxygen generator was powered on as ISS was reconfigured for routine operations. It had been powered off on December 10, because Discovery's oxygen supply had been used to support the station during the joint flight. The Expedition crew had a light day on December 20. Monday, December 25, was Christmas Day and the Expedition-14 crew had the day off, before returning to work the following day. They unpacked the material delivered by Discovery, entering it in the computerised inventory, and stowing it around ISS. The crew also resumed their regular schedule of exercise, maintenance, and experiments.

López-Alegría and Williams spent much of the first week of 2007 installing the Oxygen Generation System (OGS) activation kit in Unity. The American system, which would complement the Elektron oxygen generator in the Russian section of the station, was installed in preparation for the intended increase in Expedition crew to six astronauts, following the delivery of extra sleeping quarters in Node-3, by STS-132. The OGS would be activated later in the year. Meanwhile, Tyurin installed

Figure 82. Expedition-14 (L to R): Sunita Williams replaced Thomas Reiter as the third crew member of the Expedition-14 crew. She joined Micheal López-Alegría and Mikhail Tyurin partway through their occupation.

Figure 83. Expedition-15: three SPHERES micro-satellites float in Zvezda during testing.

new fans, vibration isolators, and acoustic shields in the Russian modules in order to upgrade the soundproofing there. During the week the crew installed and ran the first experiments on the Test of Reaction and Adaptation Capabilities (TRAC) experiment, in which they used a joystick to react to movements of a cursor on a computer screen. They also completed the last round of experiments with the European Modular Cultivation System taking the final round of photographs before storing the plants in the freezer for return to Earth.

The crew had a three-day rest period to mark the Russian Orthodox Christmas, before spending the week packing rubbish into Progress M-57, which would be undocked from Pirs at 18:28, January 16, commanded to re-enter the atmosphere several hours later, where it would be heated to destruction. Progress M-59 would replace it at Pirs' nadir. As the week progressed the crew removed the Robotics Onboard Trainer from Zvezda and relocated it to Destiny, Tyurin repaired and tested numerous pieces of equipment in the Russian modules, and Williams performed similar maintenance on American equipment. Automated and hands-on experiments also continued in both sectors of the station.

PROGRESS M-59

Progress M-59 was launched from Baikonur at 21:12, January 17, 2007, and was successfully placed into orbit. The spacecraft's launch shroud carried a painted portrait of Sergei Korolev, the famous Soviet spaceflight pioneer, to mark the 100th anniversary of his birth. Following a standard rendezvous, the unmanned cargo vehicle docked to Pirs at 21:59, January 19. The arrival of 2,561 kg of new supplies was followed by a week of routine exercise, maintenance, and experiments. López-Alegría, Tyurin, and Williams spent time unloading Progress M-59 and also began preparations for a Stage EVA. On January 25, controllers in Houston manoeuvred the SSRMS to the position from which it would support the first EVA, while the crew reviewed their equipment and procedures.

The 50th EVA from ISS, as opposed to from a Shuttle airlock, began at 11:14, January 31, as López-Alegría and Williams left Quest wearing American EMUs. After collecting their tools they made their way to the area between the Z-1 Truss on Unity and S-0 ITS on Destiny, at the centre of the ITS. They worked to de-mate and re-route two electrical connectors running between the Z1 Truss and S0 ITS, to Destiny. During the next EVA the electrical harness would be extended from Destiny to PMA-2. When complete, the Station-to-Shuttle Power Transfer System (SSPTS) would allow docked Shuttles to draw electrical power from the station, thereby extending their flights to 14 days in duration. The SSPTS was due to be used for the first time during the flight of STS-118, then planned for July 2007.

They also redirected four cooling lines, part of the temporary Early External Active Thermal Control System, which had been maintaining the station's temperature since the P-6 ITS had been erected on the Z-1 Truss in 2000, and attached them to connectors for the permanent cooling system, the Low Temperature

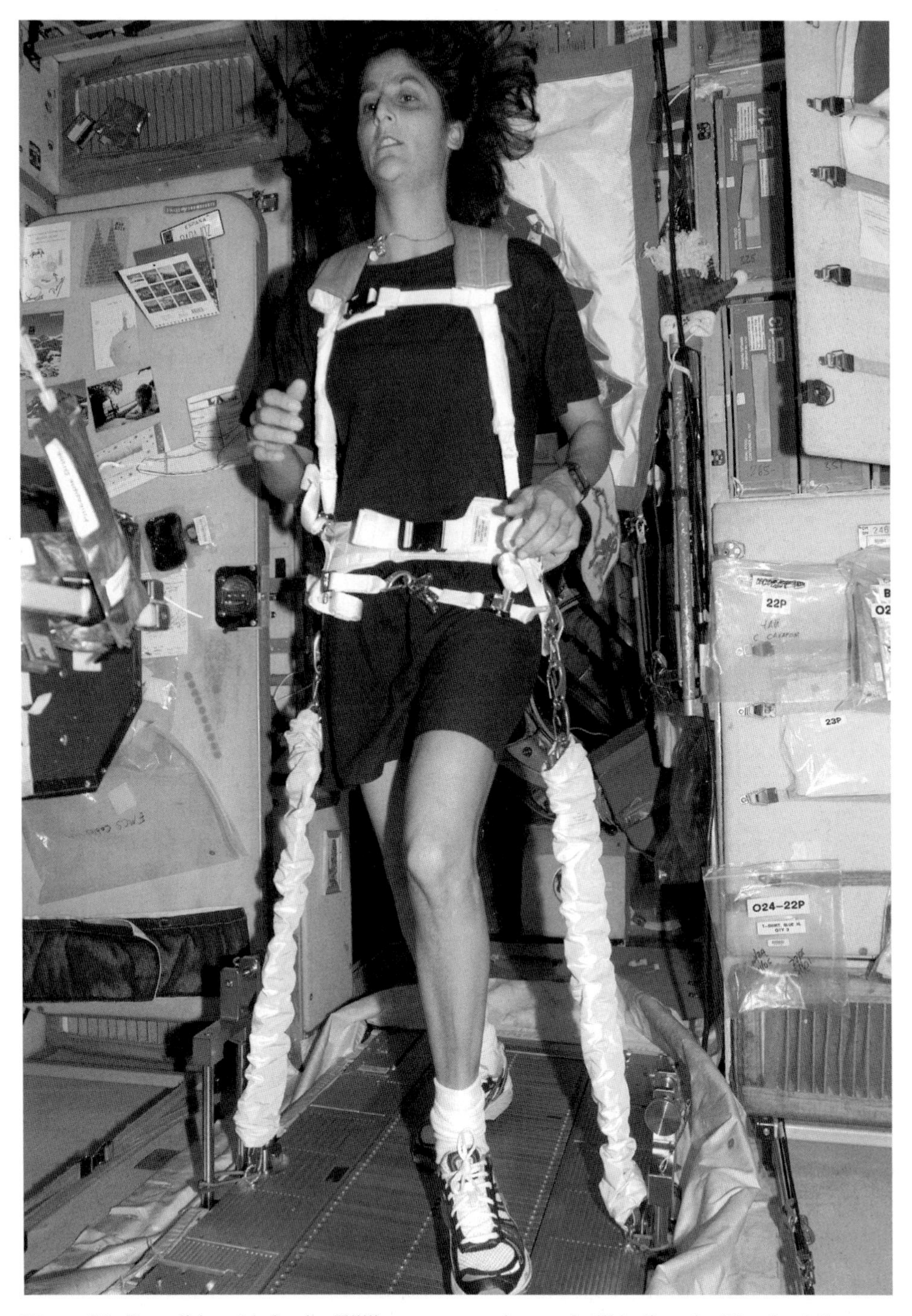

Figure 84. Expedition-14: Sunita Williams runs on the treadmill in Zvezda. The elastic harness keeps her in place in the microgravity environment.

Loop (Loop-A), which connected them to the heat exchangers in Destiny. The Low Temperature Loop carried heat away from the station's environmental systems.

Having completed their work with the SSPTS the two astronauts joined together with controllers in Houston to continue their work on the station's cooling system. Controllers commanded the starboard radiator, one of three, on the P-6 ITS to retract. López-Alegría and Williams secured the retracted radiator in place. The second P-6 radiator would be retracted on the following EVA and the third later in the year, during the flight of STS-118. They covered the radiator to keep it at the correct temperature for the months between its retraction and re-deployment. The astronauts then turned their attention to disconnecting a fluid line to a reservoir, the Early Ammonia Servicer (EAS), on the P-6 ITS, securing it in a storage position. The Expedition-15 crew were to unbolt and jettison the EAS, but in the meantime, by securing the fluid line leading to it, the astronauts were preserving the ability to re-instate the system if needed. The two astronauts returned to Quest at 18:09, after 7 hours 55 minutes.

After two days of rest and a third of preparations, López-Alegría and Williams left Quest again, at 08:38, February 4, 2007. Once again, they made their way to the area between the Z-1 and S-0 ITS, where they had started their previous EVA. There they re-routed a further two electrical and four fluid lines. This time they reconfigured the Moderate Temperature Cooling Loop (Loop-B), which carried heat from the station's avionics and payload racks. Next they joined with controllers in Houston to retract the P-6 aft radiator. The station's orientation in relation to the Sun meant that the aft radiator did not require the installation of a thermal shield to maintain its temperature. With the radiator retracted, the astronauts disconnected and stowed the second EAS ammonia fluid line. López-Alegría, positioned at the base of the P-6 ITS, photographed the starboard SAW and the blanket box into which it would be retracted during the flight of STS-117. With the photographs taken, both astronauts returned to re-routing the electrical system, from the S-0 ITS across the exterior of Destiny, and on to PMA-2, on the laboratory's ram. The cables provided electrical power for the SSPTS. Three of the six cables were connected during this EVA. Lopez-Alegria also removed a sunshade from a data relay box on PMA-1, between Unity and Zarya. The EVA ended at 15:49, after 7 hours 11 minutes, at which time Williams held the record for the total time spent in EVA by a woman.

THE LISA NOWAK AFFAIR

On March 7, NASA dismissed Lisa Nowak from her position as a NASA astronaut. It was the first time that such a thing had happened. Nowak, a US Naval officer had been arrested by police following criminal allegations related to her private life that also involved astronaut William Oefelein and a female US Air Force officer. Nowak and Oefelein were both returned to service with the US Navy. Nowak had flown to ISS on STS-121, in July 2006, and Oefelein had visited ISS on STS-116, in December 2006.

MORE EXTRAVEHICULAR ACTIVITY

The next Expedition-14 EVA began at 08:26, February 8, when López-Alegría and Williams left Quest. They moved to the CETA carts on the ram face of the ITS. Placing their equipment on one cart, they moved it along the rails on the ITS, to the P-3 ITS segment. There, they removed thermal shrouds from the RJMC on P-3. Next, they removed two thermal shrouds from Bay 18 and Bay 20 of the P-3 ITS, to avoid them trapping heat as a result of the station's present orientation to the Sun. Each of the RJMC shrouds was wrapped in one of the bay shrouds and thrown away, towards the station's ram. They then deployed an Unpressurised Cargo Carrier Assembly Attachment System (UCCAAS) on the zenith face of the P-3 ITS. This was done in anticipation of a cargo platform being attached during the flight of STS-118. While López-Alegría worked with the UCCAAS, Williams made her way to the end of the P-5 ITS and removed two launch locks, in preparation for the re-location of the P-6 ITS on to the exposed end of the P-5 ITS. The two astronauts then completed their work, connecting the final four STSPTS electrical cables between Destiny and PMA-2. Whilst in the area they photographed a suspect communications connector on PMA-2 that carried communications between ISS and docked Shuttles while the hatches were closed. Communications at those times had been intermittent on recent Shuttle flights. The EVA ended at 15:06, after 6 hours 40 minutes. López-Alegría completed the EVA as the new American record holder for cumulative EVA time with 61 hours 22 minutes spent in open space.

In the very early hours of February 11, communications were lost between ISS and Houston. A switching unit had suffered a malfunction that caused a circuit breaker to trip, in turn causing a loss of power to the station. All three crew members worked to recover communications and restore power. The difficulties lasted for 90 minutes, but the work to restore the station to its normal routine and return all the affected systems to operation took the remainder of the day. NASA was at pains to point out that, "... the safety of the Expedition-14 crew and the complex was never an issue." The astronauts also began their preparations for their final EVA, when López-Alegría and Tyurin would work out of Pirs wearing Orlan suits. The EVA was planned for February 22, and the two men spent the week beforehand preparing their suits and other equipment, as well as going over their work schedule. Meanwhile, Atlantis was moved to the launchpad in Florida for STS-117. In preparation for that flight, controllers in Houston commanded the MT to move the SSRMS to the starboard side of the ITS.

At 05:27, February 22, López-Alegría and Tyurin exited Pirs to begin their EVA. Tyurin reported that the sublimator, which dumped heat from his suit to the vacuum, had failed to function. As a result, the inside of his faceplate had fogged over. NASA, engineers suggested that the problem was caused by activating the sublimator in the airlock before it was at full vacuum. Tyurin turned the sublimator off and then reactivated it, after which it functioned correctly and cleared his faceplate. Making their way to the stuck KURS antenna on Progress M-58, they cut one of four supporting struts and pulled it back, thus ensuring that it would not impair the spacecraft's undocking. The antenna had become stuck behind one of Zvezda's EVA handrails during docking, but it was now 6 inches clear of that rail.

Their next task was to photograph a Russian satellite navigation antenna, before they changed a Russian materials exposure experiment. They also photographed docking targets and an antenna intended for use by the European ATV when it approached and docked to ISS, then scheduled for later in the year. Photographs were also taken of a German experiment and portions of the Strela-2 crane mounted on the exterior of Pirs. A series of other tasks completed the EVA, which ended at 11:45, after the pair had stowed two foot restraints on the ladder outside Pirs; it had lasted 6 hours 18 minutes, 15 minutes longer than planned. The week following the recordbreaking fifth EVA was spent cleaning up and performing routine experiments and maintenance.

When a thunderstorm passed over KSC on February 26, hail damaged the foam at the top of STS-117's ET as it stood on the launchpad. The Shuttle stack was rolled back to the VAB for inspection and repair. The planned March 15 launch was cancelled and rescheduled for no earlier than May 11, but more likely June. Soyuz TMA-10, carrying the Expedition-15 crew was planned for lift-off on April 7. To raise ISS to the correct orbit to support the rendezvous and docking, two Progress engine burns would be made on March 16 and 28. The Expedition-14 crew's schedule was changed to make the most of the available time before the delayed Shuttle launch. On the last day of the month Williams used a simulation on her laptop to maintain her skills with the SSRMS. She also joined López-Alegría and Tyurin in their experiment programme.

On March 1, the crew was woken up by a caution and warning alarm, when the signals from the RJMC to the Thermal Radiator Rotary Joint (TRRJ) dropped out. The TRRJ, which turned the radiator to the best attitude for heat loss, automatically switched to another command link and operations were not affected. As the month continued, López-Alegría and Williams completed setting up the American OGS in Destiny. Tyurin spent part of the week performing maintenance in the Russian segment. In Zvezda he set up equipment to allow ground controllers to test the satellite navigation system to be used by the European ATV, stowed spare liquids for the Elektron oxygen generator, and installed a new liquid crystal display for the TORU manual docking system for Progress spacecraft. They also completed a series of Russian and American experiments. In Korolev, Russian programme managers agreed to have the crew relocate Soyuz TMA-9 from Zarya's nadir to Zvezda's wake, on March 29. Before that could happen, Progress M-58 would be undocked from Zvezda's wake on March 27. In the meantime, work began to load Progress M-58 with rubbish.

The crew installed a new window with a camera berth in Unity's port-side hatch on March 14. The starboard hatch window had been installed by the Expedition-6 crew, the work being part of the preparation for the relocation of PMA-3 to Unity's nadir, later in the year. A number of water bags had to be relocated to give the crew access to the interior of PMA-3, where they installed upgraded computer cabling. They also cleared everything out from PMA-3, with the exception of a Bearing Motor and Roll Ring Module, which they secured in place, so they would not be lost when the PMA was relocated. The crew also completed packing rubbish into Progress M-58, in preparation for its disposal. As planned, the Progress M-58 thrusters were fired on March 15 to raise the station's orbit.

As the flight progressed, López-Alegría and Williams took part in an experiment to examine how cosmic rays affect brainwaves. For the ALTEA experiment they wore a soft cap with sensors to record brain function and a hard cap with instruments to record cosmic rays passing through the station. It was hoped that the experiment might lead to preventative measures that might be used on long-duration flights to the Moon and Mars. They also worked on a series of medical experiments studying how the human body adapts to spaceflight. With STS-117 delayed, they were able to work on establishing the station's laptop computer network, which would employ new wireless and Ethernet connectivity to avoid cables being deployed between the American and Russian segments of the station. It was estimated that the new network would be up to ten times faster than the present system. During the week, the last propellants were pumped out of Progress M-58's tanks and the last items of rubbish were loaded into its pressurised compartment. Progress M-58 was undocked from Zvezda's wake at 14:11, March 27, to make way for the Soyuz TMA-9 relocation. A few hours later the Progress was commanded to enter Earth's atmosphere, where it burned up.

On March 29, the crew placed ISS in automatic mode and sealed themselves in Soyuz TMA-9. After undocking from Zarya's nadir at 18:30, they flew around the rear of the station and docked at Zvezda's wake at 18:54. After pressure checks they re-entered the station and began the long job of putting it back into occupied operation. The following day was a rest day, to allow the crew to re-adjust their sleep cycle, which had been altered to facilitate the Soyuz relocation. They performed only light duties, routine maintenance, and their daily exercise regimes. The return of Soyuz TMA-9 to Zvezda's wake, which it had only left on October 10, 2006, was to make way for Soyuz TMA-10, at Zarya's nadir.

The crew performed the first SPHERES formation flight inside the station. The 8-inch diameter satellites were battery-powered and each used 12 carbon dioxide thrusters to manoeuvre. They were designed to test automated rendezvous, station-keeping, and docking as an experiment testing possible technologies for use on future spacecraft. The first formation-flying session was considered to be a great success.

As the Expedition-14 occupation approached its end, López-Alegría and Tyurin began preparations for their return to Earth. On April 2, López-Alegría set a new endurance record for an American astronaut on a single flight, when he passed the 196-day record held jointly by Dan Bursch (set in 2001) and Carl Walz (set in 2002). The crew also worked on experiments, repairs, and their daily fitness routines. Experiments included the Lab-on-a-Chip Application Development Portable Test System (LOCAD-PTS), a portable bacteria detector small enough to fit in a compact ice cooler. The experiment would be used five times over the coming weekend science sessions. López-Alegría and Tyurin both tested their hand–eye co-ordination on the TRAC experiment. They also completed a further session with the ALTEA experiment.

OKA-T MODULE

The Russian Central Research and Development Institute of Machine Building (TsNIM) announced that they would develop a free-flying industrial module, designated OKA-T, that would fly alongside ISS and dock to it for servicing, in much the same way as the original European Columbus module had been designed to do. They claimed the module would be launched in 2012.

SOYUZ TMA-10 DELIVERS THE EXPEDITION-15 CREW

SOYUZ TMA-10	
COMMANDER	Fyodor Yurchikhin
FLIGHT ENGINEER	Oleg Kotov
ENGINEER	Charles Simonyi (spaceflight participant)

Two members of the Expedition-15 crew, Fyodor Yurchikhin and Oleg Kotov, along with spaceflight participant Charles Simonyi were successfully launched from Baikonur Cosmodrome onboard Soyuz TMA-10 at 13:31, April 7, 2007. Simonyi, a founder member of Microsoft Corporation, flying under contract to the Russian Federal Space Agency, would return to Earth in Soyuz TMA-9 with López-Alegría and Tyurin, while Williams would remain on ISS, transferring her couch liner and Sokol pressure suit from Soyuz TMA-9 to Soyuz TMA-10, and become the third member of the Expedition-15 crew. Asked, before launch, to describe his role as Commander of Expedition-15 Yurchikhin replied:

> "The main goal of our increment will be to continue the assembly of the station and at the same time we have a lot of people who have a very brief spaceflight experience. I have only one spaceflight, Oleg Kotov has no flight experience, Suni Williams has no flight experience, and astronaut [Clayton] Anderson [who would relieve Williams partway through Expedition-15] has no flight experience. We would like to really prove that we are very good crew members compared to our previous colleagues. We would like to continue their good work. So, all of us are highly motivated to complete our personal goals. My personal goal will be to maintain all the crew members' motivation within the goals of the increment, and to make sure all my crew members are working as a team to achieve their personal goals."

Simonyi had a simpler view of his flight, "I enjoy the whole process of training, and I view the spaceflight as kind of an exclamation point at the end of a very long sentence."

Following a standard approach and a KURS-guided final approach, Soyuz TMA-10 docked to Zarya's nadir at 15:10, April 9. Following pressure checks,

Figure 85. Expedition-14: The Expedition-14 and 15 crews pose together in Zvezda. (rear row) Michael López-Alegría, Sunita Williams, Mikhail Tyurin. (front row) Oleg Kotov, spaceflight participant Charles Simonyi, Fyodor Yurchikhin.

the hatches between the two spacecraft were opened at 16:30 and the three new arrivals transferred to the station, to be received by the Expedition-14 crew. The standard safety brief commenced 10 days of hand-over procedures. High-priority Russian experiment samples were removed from Soyuz and placed in freezers on ISS. Yurchikhin and Kotov would begin working with the samples almost immediately. As with previous spaceflight participants, Simonyi recorded his reactions to spaceflight and took swab samples from the station's inner surfaces in support of ESA experiments. He also recorded radiation readings for the Hungarian Space Agency.

On April 11, Kotov set up the ESA Exhaled Nitric Oxide-2 experiment. It would measure the nitric oxide exhaled by EVA crew members before and after their EVAs. The following day was Cosmonautics Day in Russia, April 12, the anniversary of Yuri Gagarin's flight on Vostok-1. López-Alegría spent much of the day servicing the EMUs that the Expedition-14 crew had used for their three recent EVAs. When the two crews began preparing their main meal together, Simonyi produced a package of gourmet French meat, to add to the day's sense of celebration.

Tyurin and López-Alegría spent part of April 14 in Soyuz TMA-9, running through systems checks and test-firing the thrusters. Tyurin also removed the television cameras and lights from Soyuz TMA-10 and transferred them to Soyuz TMA-9 for return to Earth. Two days later, they returned to their Soyuz and spent 4 hours rehearsing re-entry. They also removed the KURS avionics packages from the orbital compartment of the Soyuz and stored them on ISS, for return to Earth on a later Shuttle. Even as the hand-over continued, both Expedition crews maintained their exercise and experiment programmes. Tyurin worked with the Russian Bio-emulsion experiment, designed to produce micro-organisms for bacterial, fermenting, and medical preparations. Later in the week, he worked with the Pilot experiment, designed to measure changes in his ability to fly a spacecraft following a long-duration spaceflight.

Also on April 16, Williams became the first person to run a full marathon in space. Running on the TVIS treadmill, she officially competed in the Boston Marathon, which was being run on the ground at the same time. As a regular marathon runner, Williams watched live television coverage of the marathon as she ran on the treadmill, held in place with a harness to counteract microgravity. She finished her run in just under 4 hours 24 minutes.

The two crews held their official hand-over ceremony to pass responsibility for ISS to the Expedition-15 crew, Yurchikhin, Kotov, and Williams, in Unity, on April 17. During the day the landing of Soyuz TMA-9 was delayed by 1 day, to April 21. The primary landing site in Kazakhstan was too wet following the spring thaw and flooding after heavy rainfall, precluding recovery operations. The 24-hour delay would allow the Earth to turn beneath the spacecraft, resulting in a landing at a secondary site farther to the south.

The Condensate Feed Unit in the Russian sector of the station, which processed water condensate from the American sector and turned it into potable water, failed at the weekend. Over the next week the amount of potable water on the station decreased considerably, but the station still carried sufficient water to last until the Progress M-60 spacecraft delivered more, in May.

After saying their farewells, the crew of Soyuz TMA-9 sealed themselves inside their spacecraft and undocked from the station at 05:11, April 27, 2007. On ISS, Kotov sounded the station's bell to mark the departure. Simonyi described his feelings at undocking as "bittersweet". Following the standard de-orbit burn, at 07:42, and spacecraft separation, the descent module re-entered the atmosphere and landed safely at 08:31. Tyurin and López-Alegría had completed a flight lasting 215 days 8 hours 48 minutes, a new American endurance record. Simonyi had been in flight for 13 days 19 hours 16 seconds.

EXPEDITION-15

Following the busy hand-over period the Expedition-15 crew began their occupation of ISS with a few days of light workload. While Yurchikhin and Kotov oriented themselves on the station, Williams used her three months of experience to assist them. The new crew participated in drills to maintain medical and emergency skills. Before his launch Yurchikhin had discussed the advantages of having Williams serve with both the Expedition-14 and Expedition-15 crews:

> "[U]p to now, almost all the increments were launched together and landed together. At the initial International Space Station development stage, [Expedi-

> tion crews] always rotated by Shuttle flights, then it moved to Soyuz flights. Now that the Shuttle flights [have] resumed we do combined crew rotation ... The main problem all the increments are facing is the short time allocated for crew hand-over. And to have somebody on board (Suni will have already spent three months on board) that will have several EVAs as well experience working with robotic operations—and experience working on the unit, USOS [United States Operating Segment]. The main problems were, when the previous crew members would close the hatch and would undock and then we realized, 'Oh, I forgot to ask this.' We're not going to have this problem because we're going to have Suni with us."

Contrast this comment with that made by Michael López-Alegría on p. 259.

In Korolev, Russian flight controllers test-fired the two manoeuvring engines on Zvezda on April 25, raising the station's orbit in advance of the Progress M-60 and STS-117 launches. An earlier attempt to make this burn had been prevented when an ATV antenna had prevented the engine covers from opening. That antenna had been moved during an EVA. It was the first time the engines in question had been fired since 2000, when they had been used to help deliver Zvezda into orbit. A second burn took place on April 28, and raised the station's orbit.

Figure 86. Expedition-15: Oleg Kotov and Fyodor Yurchikhin pose in their underwear in Zarya's docking node. They are floating in the hatch leading to Soyuz TMA-10, which is docked at Zarya's nadir.

Williams performed a series of flights with the SPHERES satellites. Yurchikhin and Kotov carried out maintenance work, replacing the water separation unit in the air-conditioning system of the Russian segment. The following week was spent undertaking routine maintenance, in preparation for the arrival of Progress M-60. The crew removed the docking system from Progress M-59, for return to Earth on STS-117, while maintaining their experiment programme. Williams completed a session with the Elastic Memory Composite Hinge experiment, designed to study a new hinge composite in space. She also used a hand-held device intended to identify biological and chemical substances on the station, as part of the crew's health and safety measures. Kotov collected air samples in the Russian modules using the Real-Time Harmful Contaminant Gas Analyser. He also completed maintenance on one of Zarya's battery temperature sensors. The crew also worked together to perform maintenance in Destiny.

On May 8, they tested communications between ISS and Progress M-59. The following day was one of light duties in recognition of Russia's Victory Day, that nation's celebration of the end of the Great War for Independence, World War II. In Houston, flight controllers tested the failed CMG-3 mounted in the Z-1 Truss. By tilting the CMG in different directions they were able to measure the friction involved. At no time was the CMG, which was due to be replaced during the flight of STS-118, spun up to full speed.

PROGRESS M-60

The Progress M-60 cargo vehicle was launched at 23:25, May 11, 2007. After following a standard 2-day rendezvous, automated docking occurred at Zvezda's wake at 01:10, May 15. During the final approach the KURS antennae on the Progress were retracted earlier than on previous Progress flights, allowing the crew on ISS to confirm visually that it had indeed retracted. Following pressure checks, the hatches between the two vehicles were opened later that night, and the unpacking of the 2,561 kg of supplies took place over the next few weeks. On May 27, Williams was informed that she would now return to Earth on STS-117, then targeted for launch on June 8, 2007, rather than STS-118, planned for August 8, 2007, as originally planned. Clayton Anderson, her relief, had been moved forward one Shuttle flight after NASA managers had assured themselves that it would not impact any future operational goals. On receiving the news she replied, "All right, thanks very much. I might see you guys sooner than we all thought. That is pretty good."

Anderson later explained:

> "I had an inkling that it was coming, but at first it was being evaluated ... to make sure that there were no big showstoppers ... [T]he answer came back essentially, "no." ... I tell Suni that I'm her knight in shining armour, I'm going to come up there and I'm going to rescue her from her potential nine-month duration on orbit and I also tell people that it's a clever plot by Michael López-Alegría to keep Suni from breaking his new long-duration endurance record ... It's a very hectic time

for me. We had been scheduled to launch on STS-118 at the end of June, and then, when the hailstorm damage happened to 117's External Tank, STS-118 had moved to August. So ... we breathed a little sigh of relief and I thought, hey, a little extra time to maybe get all this together and relax a little bit. At that point they decided to move me to 117, which now launches earlier than my original date on 118, so, from the perspective of my family getting ready, my guests being ready to go to Florida to watch a launch, all that's a little hectic but it's going to work out."

While commencing preparations for the end of her 6 months in space, Williams made repairs to some exercise equipment and rode the station's stationary bicycle while doctors in Huntsville measured her oxygen intake. She also updated the software in the station's laptop computers. Meanwhile, Yurchikhin and Kotov began preparations for their first EVA, planned for May 30. They checked out Pirs, their Orlan pressure suits, and gathered together and tested the tools that they would use. They also closed the hatch to Progress M-59, docked to Pirs' nadir. On May 23, controllers in Korolev fired Progress M-60's thrusters to place the station in the correct orbit to receive STS-117.

After dealing with an unexpected communications problem, Yurchikhin and Kotov left Pirs, 45 minutes late, at 15:05, May 30, 2007. Having gathered their tools, they moved to the Strela-2 crane on the exterior of Pirs. There, they attached an extension to the Strela boom, to increase its length from 14 m to 18.75 m. Kotov then positioned himself on the end of the extension while Yurchikhin turned the handle to extend the crane until his partner was suspended above PMA-3 on Unity. Using Kotov's verbal instructions, Yurchikhin manoeuvred the crane until its end-effector locked on to a grapple fixture on an adapter stowage rack attached to PMA-3. The rack carried 17 micro-meteoroid debris panels in three bundles, and was referred to by the cosmonauts as the "Christmas Tree". Yurchikhin then manoeuvred the Strela crane, holding Kotov and the "Christmas Tree" to Zvezda's ram, before making his own way to the same location and helping secure the Strela to a grapple fixture. Their first task was nothing to do with the debris panels. It required them to make their way to Zvezda's large conical section where they re-routed a cable for the Global Positioning System which would be used in association with ESA's ATV. With that task complete they returned to the "Christmas Tree", where they removed and opened a bundle of five debris panels each measuring 66 cm × 1 m and installed the panels between Zvezda's large and small diameter sections before returning to Pirs and sealing the hatch at 20:30. The EVA had lasted 5 hours 25 minutes.

The cosmonauts followed the EVA with an easy day, drying their Orlan suits and recharging their batteries. Williams began packing for the end of her flight. May 4 and 5 were spent preparing for the next EVA.

Yurchikhin and Kotov began their second EVA from Pirs at 10:23, June 6. Initially, they installed sample containers on the exterior of Pirs, for a Russian experiment called Biorisk, which was designed to look at the effect of the space environment on micro-organisms. Next, they deployed a length of Ethernet cable on the exterior of Zarya to complete a remote computer network that would allow the

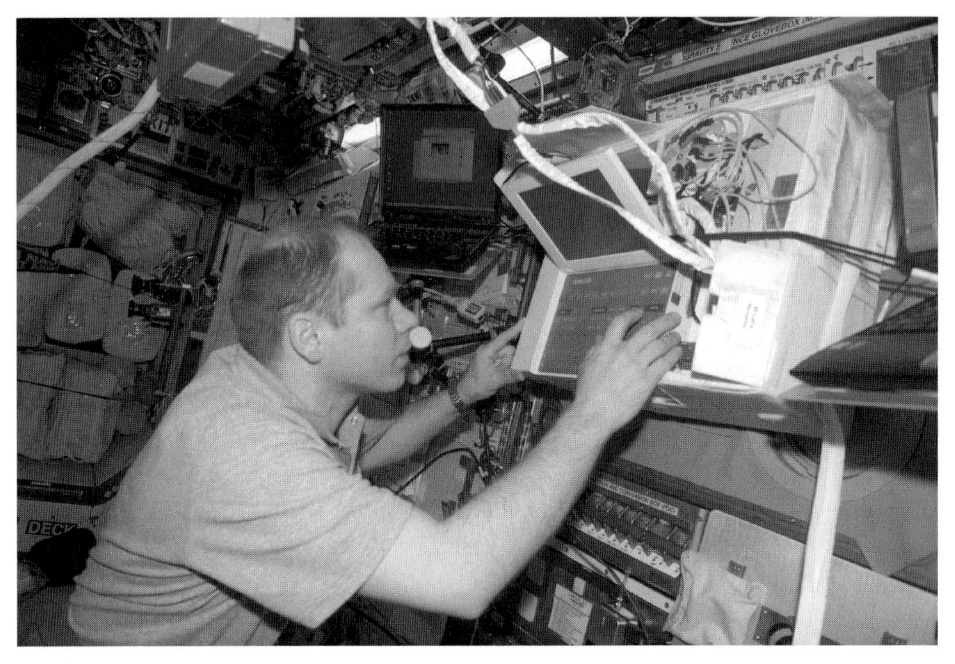

Figure 87. Expedition-15: Oleg Kotov works with the Period Fitness Evaluation experiment keyboard.

Russian modules to be commanded from the US sector if required. As they worked to clamp the cable securely in place they noticed a 6 mm diameter hole in Zarya's insulation. They reported, "This is a dent from a meteorite; it looks like a bullet hole." The remainder of the EVA was taken up with deploying the remaining 12 micro-meteoroid debris panels on the centre section of Zarya. During this work Korolev requested that Yurchikhin return inside Pirs to confirm that the pressurised oxygen bottles were closed correctly: they were. While Yurchikhin returned outside to assist Kotov, Korolev identified that the unexpected reading they were receiving from Pirs was caused by a small amount of oxygen escaping from a fluid umbilical that had improperly sealed when it was disconnected from one of the cosmonauts' Orlan suits. Controllers closed off the flow of oxygen to the hose in order to preserve oxygen. They commanded the flow back on once more, after the two men returned to Pirs; after 5 hours 37 minutes the hatch was closed to end the EVA at 16:00.

NEW SCHEDULES AND NEW FUNDING

In 2004, NASA had announced that it intended to stop funding ISS in 2016, as the Administration turned its attention to the Orion spacecraft, the Ares-1 launch vehicle, and the ultimate return of human astronauts to the lunar surface. Three years later, in April 2007, NASA Administrator Michael Griffin announced that ISS was to be included in NASA's budget requests through 2020. Griffin stated: "The partners have been working for a decade and a half to put in place these four laboratories. I don't think political leaders in 2016 will end their involvement. Assets like the ISS live a lot longer than anticipated. I doubt it will turn into a pumpkin in 2016."

In mid-2007, Roscosmos announced that they had no plans to co-operate with NASA on Project Constellation until at least 2015. The Russians said that their budget had been allocated so as to allow them to support NASA's commitment to ISS until that time, during which they would re-commence lunar exploration with robotic probes while supporting India's and China's robotic lunar exploration programmes. After 2015 ,Russia would be free to reconsider their position regarding Project Constellation. Meanwhile, the Russians continued to talk about building science modules for the ISS, but no new modules had been completed, as yet.

On February 26, STS-117 had been delayed by damage to the ET caused by hailstones while it stood on the launchpad. The delay caused the Shuttle's launch programme to be rescheduled yet again:

STS-117 Atlantis	July 8, 2007	S-3/S-4 Truss structure
STS-118 Endeavour	August 9, 2007	S-5 Truss structure
STS-120 Discovery	October 20, 2007	Harmony
STS-122 Atlantis	December 2007	Columbus

All subsequent flights were similarly delayed.

At the same time NASA announced that they would swap the orbiters assigned to three of those flights in order to ease the pressure in the schedule:

- STS-120 would now use Discovery rather than Atlantis.
- STS-122 would use Atlantis rather than Discovery.
- STS-124 would use Discovery rather than Atlantis.

The rescheduling meant that plans to retire Atlantis in 2008 and cannibalise it to provide spare parts for Discovery and Endeavour had been changed.

NASA also announced that it had added $719 million to its contract with Roscosmos to purchase additional Soyuz TMA flights for American astronauts, to deliver and recover Expedition crews, during the period between the last Shuttle flight in 2010 and the first flight of the Orion/Ares-1 combination to ISS, currently scheduled for 2015. Despite this schedule, continual failure to make the promised annual increase to NASA's budget to support Project Constellation made that date seem highly unlikely.

Asked previously if the Shuttle would continue to fly beyond the announced 2010 deadline set by President Bush, NASA Administrator Michael Griffin had made a one word reply, "No!" That, added to the 2007 schedule delays, meant that NASA would have to concentrate the remaining Shuttle flights on delivering the primary ISS elements into orbit before the Shuttle stopped flying in 2010. Any planned Utility flights, to support ISS operations, might yet prove to be expendable.

STS-117 DELIVERS THE STARBOARD-3/4 ITS

STS-117	
COMMANDER	Rick Sturckow
PILOT	Lee Archambault
MISSION SPECIALISTS	Patrick Forrester, Steven Swanson, John Olivas, James Reilly
EXPEDITION-15/16 (up)	Clayton Anderson
EXPEDITION-14/15 (down)	Sunita Williams

A thunderstorm passed over KSC on February 26, and hailstones "the size of golf balls" struck the foam exposed at the top of STS-117's ET, leaving visible damage. They also damaged approximately 25 tiles on the orbiter's left wing. The Shuttle stack was rolled back to the VAB for inspection and repair. and the planned March 15 launch was cancelled and rescheduled for no earlier than May 11, 2007. That date slipped back to June as NASA and contractor engineers ensured that the ET was safe to use. Shuttle programme manager Wayne Hale told journalists:

> "The speculation that a lot of people have engaged in is that the last flight of the Shuttle to the space station will get pushed out of 2010. That just will not happen due to this problem."

He further stated:

> "There might be some small effect to a couple of the later flights, but by the time we roll around to the end of the year, I expect we would be fully able to catch up."

The Shuttle was rolled back to LC-39A on May 15, with lift-off scheduled for June 8. On that date, Atlantis lifted off into the twilight, at 19:38, beginning the first of four Shuttle flights planned for 2007. In the payload bay was the S-3/S-4 ITS, which would be mounted on the S-1 ITS and deployed during a series of three EVAs. The second set of SAWs on the P-6 ITS would then be retracted into its storage box, in order to allow the S-3/S-4 SAWs to rotate as they tracked the Sun.

Mission Commander Rick Sturckow had previously reviewed the work of earlier crews that had worked with the ITS elements on ISS at a prelaunch press conference, saying, "We're really fortunate that we have those guys to follow. Almost everything went great on those missions, and the things that didn't go so well, we're able to learn from."

To overcome any repeat of the difficulties experienced on STS-115, such as when a bolt locking the SARJ in its launch position had taken much longer than planned to remove Sturckow explained:

> "We have a torque multiplier ... that they didn't have. So if we do encounter the same difficulty with high torques that they had, we'll break out this tool. And we'll apply whatever torque it takes to break the bolt or back it out at the higher torque settings. So I don't have any doubt that we'll be able to remove those launch restraints." He added, "When you're doing assembly operations, everything that you plan to do is contingent on the flight prior to you and the hardware that's already in orbit."

The STS-117 ascent into orbit was flawless, with no obvious signs of foam falling away from the ET in video of the launch. However, while video of the jettisoned ET showed that repairs to the hail damage had remained in place, one piece of foam, approximately 15cm × 8cm was missing. Orbital insertion was followed by the deployment of the payload bay doors and the Ku-band antenna. On ISS, Yurchikhin, Kotov, and Williams watched the launch on a video uplink from Houston.

In the latter part of the day, Forrester and Swanson activated the RMS, to test its function. As they raised the arm from its cradle they noticed that an area of insulation blanket near the port OAMS pod pulled away from the adjacent thermal tiles. Video cameras on the RMS were used to relay views of the 10 cm by 7 cm area to Houston. Similar damage had been identified on Discovery during two earlier flights and both had returned to Earth without incident. Initial reviews of the images of the blanket suggested that the airflow over the OAMS pod during the early phases of the launch had lifted the edge of the blanket and caused it to fold back on itself. Other options included: bad installation of the blanket during processing, or impact damage during launch. The Shuttle crew began their sleep period at 01:38, June 9.

Day 2 began with a wake-up call at 10:10 that morning. Archambault, Forrester, and Swanson activated the RMS, mounted the OBSS in the end-effector, and completed the first inspection of Atlantis' TPS. Prior to placing the OBSS back along the door hingeline, the cameras were used to view the port OAMS pod, with the detached insulation blanket. John Shannon, of the Mission Management team told a press conference, "If we decide this is a problem, we have a lot of capabilities to go address it." He made it clear that Atlantis carried the equipment necessary to repair the blanket, by folding it flat and pinning it in place. While the video inspection was underway, Olivas, Reilly, and Anderson inspected the EMUs that they would use during the flight's three EVAs. During the day the crew also extended the docking ring and installed the centreline video camera that would allow Archambault to see PMA-2 during the final approach to docking.

Following a second sleep period, June 10 began at 09:08. During the morning Olivas used a 400 mm lens on a digital camera to record the lifted corner of the thermal blanket on the port OAMS pod through the flight deck aft windows. The photographs had been requested during the morning briefing given to the crew by controllers in Houston. The written daily briefing was sent up to the crew and

Figure 88. STS-117 crew (L to R): Clayton C. Anderson, James F. Reilly, II, Steven R. Swanson, Frederick W. Sturckow, Lee J. Archambault, Patrick G. Forrester, John D. Olivas.

Figure 89. STS-117 delivers the S-3/S-4 Integrated Truss Structure to ISS.

included the reassurance that, "Although this [damage] does not appear to be a big issue, the teams are discussing several options." On the ground John Shannon told another press conference, "It's not a great deal of concern right now, but there is a great deal of work to be done." He added further detail saying, "There's one option on the table where we just put an astronaut out there on a spacewalk, and they just tuck the fabric right back down. There are other options where they go and try to secure it down with something." A third option was to have the crew use a pressure suit repair kit to sew the blanket back into place using an instrument with a rounded end that looked like a small darning needle.

Rendezvous manoeuvres began at 10:38. Following the r-bar pitch manoeuvre at 14:37, Sturckow moved his spacecraft in for docking with PMA-2 at 15:36. As usual, extensive pressure checks were made to ensure the seal between the two spacecraft before the hatches were opened, at 17:04. Williams rang the station's bell to welcome the new crew aboard ISS at 17:20.

Following the initial greetings and safety briefing Williams transferred her Soyuz couch liner to Atlantis, while Anderson placed his couch liner in Soyuz TMA-10. Williams had been in orbit for 183 days, longer than any other female astronaut. She would now return to Earth in Atlantis, while Anderson began a four-month occupation as part of the Expedition-15 and Expedition-16 crews.

The first task for the STS-117 crew was for Archambault and Forrester to use the RMS to lift the S-3/S-4 ITS out of Atlantis' payload bay and hand it over to the

Figure 90. Expedition-15: Clay Anderson poses with an American Extravehicular Mobility Unit in the Quest airlock.

SSRMS, which was operated by Williams. The hand-over was completed at 20:28, and the S-3/S-4 ITS was left on the SSRMS throughout the crew's sleep period, allowing it to warm in the unfiltered sunlight. Reilly and Olivas spent the night camped out in Quest with the airlock's pressure reduced to purge nitrogen from their bloodstream in advance of their first EVA, planned for the following day.

The crew were up and about at 09:08, but Reilly and Olivas were allowed to sleep in for an extra 30 minutes. After breakfast, Archambault and Forrester used the SSRMS to move the S-3/S-4 ITS towards the exposed end of the S-1 ITS and held it in place. The resulting asymmetry of the new ITS being moved around caused the CMGs in the Z-1 Truss to become saturated and drop off-line and the station began to drift. This had been anticipated by controllers in Houston. Archambault, Forrester, and Kotov commanded the bolts holding the S-3/S-4 ITS in place to close. As a result of the CMG dropout, the EVA began at 16:02, approximately one hour late. Reilly and Olivas made their way to the joint between the S-1 and S-3/S-4 ITS. There, they connected power cables between the two ITS elements and released the launch restraints on the S-3/S-4 ITS SAW blanket boxes and opened them. They also released the launch restraints on the S-3/S-4 radiator, rigidised the four Alpha Joint Interface Structure struts, installed one Drive Lock Assembly, and released the launch locks on the SARJ. The EVA ended at 22:17, after 6 hours 15 minutes. Meanwhile, controllers in Houston activated the two new power channels and deployed the new radiator. Elsewhere on ISS, Williams and Anderson continued their planned hand-over tasks.

During the evening, mission managers extended the flight by 2 days and added a possible fourth, impromptu EVA, to provide time to inspect and repair the lifted thermal blanket on the port OAMS pod. Meanwhile, engineers and astronauts on the ground were trying to establish the best way to make the repair. In the regular end-of-shift press conference at MCC-Houston, Shannon informed the press and media, "We do not want to re-enter until we have done this. I don't want to take a risk of damaging flight hardware, when we have something that looks easy to do, so it's a pretty easy decision to make." On the ground, the various repair methods were being rehearsed and subjected to testing under simulated re-entry conditions. Shannon explained that Shuttle engineers did not think that the re-entry heating on the OAMS pod would be sufficient to burn through the graphite structure beneath the lifted blanket, causing an STS-107 style break-up, but it might cause sufficient damage to require a relatively major repair, thereby throwing the Shuttle launch schedule into total disarray. The 90-minute repair would be carried out by two astronauts riding the Shuttle's RMS. Atlantis would now land on June 21, after a 13-day flight. In orbit the day had gone well, and the crew began their sleep period. Learning from past experience, Houston commanded the new SAWs to extend in a series of small lengths. The first segment was deployed by controllers in Houston while the astronauts slept.

On June 12, the astronauts' day began at 08:08. At 11:43, Sturckow, Archambault, Forrester, Swanson, Olivas, Reilly, and Williams took over the task of deploying the S-3/S-4 SAWs and observing that deployment from inside ISS and Shuttle. Each SAW was deployed separately and in small stages, with regular stops to let the Sun warm the array. The first SAW was fully deployed by 12:29, and the second by 13:58. Reilly told Houston, "We see a good deploy." The new arrays would provide sufficient electricity to power the European and Japanese laboratory modules when they are docked to Harmony.

After dinner the Shuttle crew were given some free time before commencing preparations for the following day's EVA. Throughout the SAW deployment, the station's attitude had been controlled by Atlantis. As the day drew to a close, the station's attitude control was switched back to the station's computers. At that time all three navigation computers and all three command and control computers failed in Zvezda. The computers were built by Daimler-Benz in Germany, under an ESA contract, and one of their tasks was to activate Zvezda's thrusters if ISS attitude manoeuvres were beyond the capabilities of the CMG. Controllers elected to let the CMGs continue to manage the station's attitude, but Atlantis' thrusters would be used for large manoeuvres, rather than Zvezda's thrusters. The day ended with the pre-positioning of the MBS on the ITS and that night Forrester and Swanson camped out in Quest; both were preparations for the EVA planned for the following day. While the astronauts slept, controllers in Houston began the retraction of the remaining SAW on the P-6 ITS. They succeeded in retracting 7.5 of the 31.5 panels of the array.

The day started at the usual time and the EVA began at 14:03. Forrester mounted the RMS end-effector and was lifted to the P-6 ITS, mounted on the Z-1 Truss, while Swanson made his own way to the location. Once in place, they oversaw and assisted with the retraction of a further 5.5 panels of the 2B SAW, as commanded from inside the station. Moving back to the S-3/S-4 ITS, they removed the remaining locks holding the SARJ. Although they had originally been planned to remove the launch restraints, they left them for the third EVA. When the restraints were finally removed the joint would be free to rotate, as the SAWs tracked the Sun. They also installed a second drive-lock assembly. and that was where their problems arose. Commands sent to the second drive-lock assembly were received by the unit installed during the first EVA. Controllers in Houston confirmed that the first unit was in the "safe" condition and had to confirm that the second unit was similarly configured. The SAW retraction would continue during the following day. The EVA ended at 20:33, after 7 hours 16 minutes. Once again Anderson spent the day completing hand-over tasks in preparation for his 5-month stay on ISS, as well as assisting Expedition-15 crewmates Yurchikhin and Kotov to transfer supplies from Atlantis to ISS. During the day mission managers confirmed that at least part of the third EVA would be spent repairing the port OAMS pod thermal blanket.

On June 14, Houston awoke the crew officially at 08:39. In fact, they had been woken up at 07:23, when a fire alarm sounded in Zarya. It was a false alarm set off by the loss of three Russian command and control computers, affecting the life support system and causing power outages throughout the Russian sector of the station. During the day controllers in Korolev temporarily rebooted the navigation computer and then turned it off again to continue work on the original problem. By 11:38, Sturckow, Lee, Archambault, Swanson, Williams, and Anderson resumed the attempt to retract the P-6 SAW. Meanwhile, Forrester, Reilly, and Olivas reviewed the procedures for their third EVA. The three of them practised the plan to staple the

Figure 91. STS-117: Patrick Forrester works removing launch restraints from the S-4 Solar Alpha Rotary Joint.

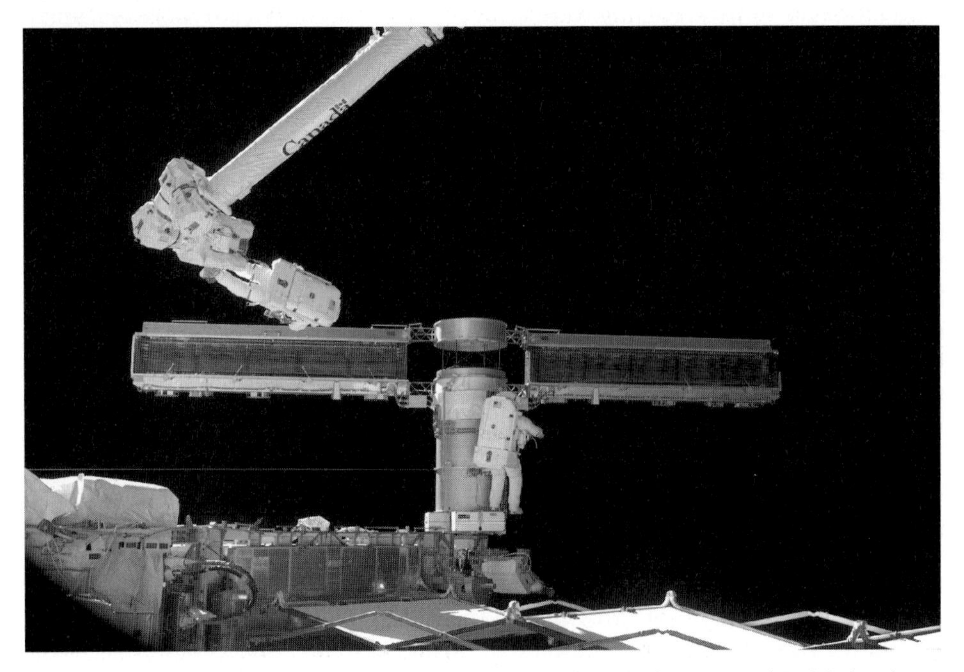

Figure 92. STS-117: James Reilly and John Olivas work with the retracted P-6 Solar Array Wings.

two sections of thermal blanket on Atlantis together and pin it to an adjacent thermal tile. Sturckow told Houston, "When we first saw it, we were not too concerned. We're still not. This is just the right thing to do, the conservative thing to do. We appreciate everyone taking a look to make sure we have the right configuration for re-entry."

In Moscow, Russian engineers continued to work with their American counterparts on the computer problem throughout the day. The leading theory as to the cause of the computer problem was a bad electrical power feed between the American and Russian sectors of ISS, as the computers now drew their power from the ITS. NASA's Mike Suffredini explained, "A power line has a certain magnetic field around it, and that can affect systems near it." Plans included disconnecting power cables between the two sectors of the station, rebooting the computers, and then reconnecting the power cables. If the problem recurred, the computers in the Russian sector could receive electrical power from the photovoltaic arrays on the Russian modules. Sturckow was objective, "These challenges, they come up when you bring new pieces of hardware or when computers are improved. This is to be expected. Things aren't always going to go well." Meanwhile, NASA Associate Administrator for space operations Bill Gerstenmaier was positive, telling a press conference, "This is a complex station. This failure is not easy to understand. It's some combination between Russian systems and our systems. It's just going to take a little bit of time to get this worked out." He added, "We're still a long way from where we would have to de-man the station."

NASA was at pains to point out that the station had 2 months of oxygen supply if the Russian oxygen generator could not be brought to full operation. Also the American oxygen system was nearly completely installed. Carbon dioxide removal systems were also available in both the Russian and American sectors. In Russia, the difficulties led to discussions as to whether or not to advance the next Progress launch by two weeks, to July 23, and use that launch to deliver new computer parts to the station.

Before going to bed, Williams and Anderson checked power lines and circuits connected to the new S-3/S-4 ITS that supply electricity to the Russian modules with a number of diagnostic instruments, but found nothing that could account for the computer difficulties. Meanwhile, the STS-117 crew had been instructed to power down some of Atlantis' systems, just in case the Shuttle mission needed to be extended by a further day, to continue supplying back-up attitude control. A NASA spokesman told the press, "I expect we will have the computers back in the next several days. It's not an urgent situation, but we clearly need to get this resolved." Asked if the station would be evacuated if the computer problem persisted NASA's Mike Suffredini insisted, "The best thing we can do is keep the crew onboard to keep working this problem until we sort it out. That is what our plan is." Meanwhile, John Shannon confirmed that Atlantis would remain docked to ISS until June 20. Reilly and Olivas spent the night camped out in Quest breathing oxygen at a lowered pressure in preparation for the crew's third EVA.

As the new day started Russian controllers disconnected the Russian modules from the new electrical power supply, returning them to the supply provided by their own photovoltaic arrays. The STS-117 crew's wake-up call came at 08:41, June 15,

Reilly and Olivas began preparations for their EVA straight after breakfast. The EVA began at 13:38. After collecting their tools they prepared to repair Atlantis' thermal blanket. Olivas mounted the Shuttle's RMS and was manoeuvred to the port OAMS pod, where he pushed the folded thermal blanket back into the correct position. Using a stapler from the Shuttle's medical kit he fixed the offending blanket to the blanket next to it. Finally, he drove a metal pin through the blanket, securing it to the adjacent thermal tiles. The repair took the full 2 hours that had been allocated for it. As he completed the task Olivas told controllers in Houston, "Hopefully it's going to be good, good enough." Sturckow, the Shuttle's Commander added, "He's done an absolutely wonderful job."

While Olivas repaired the thermal blanket, Reilly installed a hydrogen vent in the forward face of Destiny. The vent would be used by the new American oxygen generation system. The new system would separate water into oxygen as the Russian Elektron oxygen generator did, for the crew's life support system, and hydrogen which would be vented overboard through the new vent. With the repair and the valve installation complete, both men moved to the P-6 ITS, where they assisted in the retraction of the last 15 bays of the 2B SAW. While the retraction was commanded from inside the station, the two EVA astronauts assisted by helping to fold the SAW and ensuring that they were correctly stored in the blanket box. Finally, they secured the lid of the blanket box itself. The retraction was completed at 20:40. The EVA ended at 21:36, after 7 hours 58 minutes.

While the Americans concentrated on their EVA, the Russians continued to work on the failed computers. All of the computers were taken off-line at 06:00 and left off-line throughout the day. Yurchikhin and Kotov used a jumper cable to bypass a power switch, thus allowing them to get both C&C computers partially running. It was decided that the one processor in each computer that did not re-boot would be replaced using spares to be delivered by the next Progress. The computers, which only required one processor each to perform their role on ISS, were left running overnight. A telemetry downlink over a Russian ground station allowed controllers in Korolev to monitor the computers' performance over the test period. NASA spokeswoman Lynette Madison told the press, "They're up and operational and this is good news for all." Mike Suffredini made a similar comment, "We feel like the computers are stable and back to normal."

Whilst the two crews were asleep, Sunita Williams became the most experienced female astronaut in history. At 01:47, June 16, she passed the female endurance record of 188 days 4 hours set by her NASA colleague, Shannon Lucid. Later in the day Williams remarked:

> "I feel like a lot of this was just sort of being in the right place at the right time. It just sort of happened. It's just an honour to be up here. Even when the station has little problems, it's just a beautiful, wonderful place to live. I'm just happy to be part of history that provides a steppingstone for the next generation of explorers and women to come up here and do that. To me, it's no big deal."

Williams admitted:

"My biggest desire is to go for a walk on the beach. I grew up near the beach in New England, and I love going to the beach."

Lucid, who had worked in MCC-Houston during Williams' flight, told a press conference, "She's done an absolutely wonderful job. I think it's really great because it shows the space programme is getting more mature when you have more and more people stay in space for longer periods of time." She added, "I said [to Williams], 'Enjoy your last few days because all too soon you will be back to bills, dirty dishes and laundry'."

The Shuttle crew's wake-up call came at 08:38, June 16. They spent the day transferring supplies from Atlantis to ISS and preparing for the flight's fourth EVA. Yurchikhin and Kotov used a second external cable to redirect the power supply, allowing them to bring the final two computer processors on-line. With Zvezda's computers performing well, controllers in Korolev began assigning them some of their usual control tasks. The computers were now talking to the equivalent C&C computers in the American sector of ISS, something they had not been doing for the past 3 days. By this time most people did not believe the S-3/S-4 power supply had caused the computer problem. Flight Director Holly Ridings stated, "In the last 24 hours, we've had a lot of successes." ISS programme manager Mike Suffredini, summed up his feelings succinctly, "Spaceflight is a challenging business and these are the things you occasionally have to deal with. We can all go home and not do it, or we can choose to explore. We choose to explore." He told the media, "We're having a great day on orbit." As the day continued NASA suggested that Atlantis would land on June 21. On hearing the news, Sturckow replied, "That's great news."

June 17 began at 07:38. Following breakfast everyone began preparations for the final EVA. Forrester and Swanson had spent the night camping out in Quest. Their EMUs were transferred to internal battery power, commencing the EVA at 12:25. Kotov shadowed Reilly as intravehicular crew member, in preparation for his assuming that role during an up-coming Expedition-15 Stage EVA. Having collected their tools, Forrester and Swanson set to work. Their first task was to retrieve a camera and its stand from a mounting on the exterior of Quest and move it to the S-3 ITS. While on the S-3/S-4 ITS, they confirmed the Drive Lock Assembly-2 configuration and then removed the final six SARJ launch restraints, leaving the SAWs free to rotate and track the Sun as ISS orbited Earth. Reilly told them, "Great job, Guys," Swanson replied, "That's what we came here to do."

Still on the ITS, they moved the temporary stops installed on the MBS rails, leaving the MBS free to travel along its rails on the new length of ITS. They also removed additional equipment that had held the S-3/S-4 ITS in the Shuttle's payload bay. The task was the final one scheduled for STS-117 and was completed by 16:17. The remaining activities were "get-ahead" tasks. First they installed a computer network cable on the outside of Unity and then moved to open the newly installed hydrogen vent valve on Destiny. Finally, they tethered two debris shield panels on Zvezda. The EVA ended at 18:54, after 6 hours 29 minutes. Inside, the Russian computers had been returned to controlling the station's systems, and even the

Elektron oxygen generator was powered on, but not configured to produce oxygen. The computers remained stable.

During an evening press conference, Anderson was asked how he was adapting to life on ISS. He replied, "I think I'm hanging in there. It kind of reminds me of my first swimming lesson. I just got tossed in the water and told to survive." Questioned on the subject of Zvezda's computers, Yurchikhin remarked cautiously, "We're slowly moving back toward a normal mode of operations."

While the two crews slept, American controllers in Houston activated the SARJ on the S-3/S-4 ITS and tested its rotation. At 20:00, the SARJ was placed in auto-track mode. The ISS now had a new symmetrical shape, with a pair of SAWs at either end of the ITS and the P-6 SAWs fully retracted, although the P-6 ITS was still attached to the Z-1 Truss, with one of its huge radiators still deployed.

The final day of docked Shuttle operations began with the crew rising early, at 07:08, June 18. The crew of Atlantis had the morning off after the hectic pace of the past few days. They completed the final transfer of equipment to Atlantis during the afternoon. At 10:28, the Shuttle's thrusters were used to manoeuvre the station into the correct position for a combined potable water and waste water dump and then manoeuvre it back to the original position. Once the combination was stable, after the second Shuttle manoeuvre, at 10:34, attitude control was passed to the Russian terminal computer. The computer activated the thrusters on the Russian modules to maintain the station's attitude. At 12:09, attitude control was handed back to the American computers and the CMGs mounted in the Z-1 Truss. The test was summed up by NASA's Phil Engelhauf, "There was absolutely nothing anomalous out of the testing. Everything performed exactly as it should have." Only after the test of the Russian computers was successfully completed did NASA managers confirm that Atlantis would undock at 10:42, June 20.

After saying their goodbyes to the Expedition-15 crew, Sturckow led his crew back to Atlantis, securing the hatches between the two vehicles for the final time at 18:51. Sunita Williams was included in that crew, returning to Earth after almost 6 months in space. Williams remarked, "It's sad to say goodbye, but it means that progress is being made." Yurchikhin told controllers, "We had a really great time with Suni up here." Her place on the Expedition-15 crew had been taken by Clayton Anderson who remarked, "I hope I can carry on, and do half as well as she did."

Atlantis undocked at 10:42, June 20, with Archambault at the controls. The Pilot manoeuvred the orbiter around the station so that the crew could complete a photographic and video survey, before performing the separation manoeuvre. Sturckow told the Expedition-15 crew, "Have a great rest of your mission." Yurchikhin replied, "Godspeed, and thanks for everything."

Following the separation burn, the crew used the RMS-mounted OBSS to carry out further scans of the Shuttle's nosecap and wing leading edges. The images were down-linked to Houston before the astronauts began their evening meal and sleep period. Following an early morning start, the crew spent what should have been their last full day in space carrying out all of the routine preparations for re-entry. These included a test of the aerodynamic surface and a test-firing of each of the Shuttle's manoeuvring thrusters. While the crew prepared to come home, the weather over

Figure 93. STS-117: image shows the protruding corner of a thermal blanket on the orbiter's OAMS pod.

Figure 94. STS-117 departs ISS. Note the station's new symmetry and the retracted P-6 Solar Array Wings.

Florida threatened the plans for returning to that location. Sturckow told controllers in Houston, "Get us some good weather for Thursday, if you can. It doesn't have to be good, just good enough."

On June 21, the final preparations were made. Following breakfast the Ku-band antenna was stowed, and the payload bay doors were closed at 10:05. Retrofire was planned for 12:50. That attempt was cancelled due to bad weather. The day's final landing opportunity demanded retrofire be completed at 14:25, but that attempt was cancelled at 13:38, as the weather in Florida showed no signs of clearing. Sturckow was informed, "The rain showers and cloud ceilings will keep us from making it into Florida today. We are going to try again tomorrow."

Atlantis' payload bay doors were re-opened and a manoeuvre was performed to adjust the orbit to support the five available landing attempts (two at KSC in Florida and three at Edwards Air Force Base in California) on June 22. The crew spent the extra day and night in orbit, before beginning re-entry preparations again.

On June 22, Sturckow was informed, "Our mindset is we're going to land you somewhere safely today." The payload bay doors were closed at 09:32. The first attempt to land in Florida was cancelled. Landing finally occurred on the dry lakebed at Edwards Air Force Base, California, at 16:49. The flight of STS-117 had lasted 13 days 20 hours 11 minutes. Williams had been in space for 195 days. At the post-landing press conference NASA Associate Administrator for space operations Bill Gerstenmaier told the media, "My hat's off to the team that really pulled off an awesome mission." Returning Atlantis to Florida on the back of a Boeing 747, NASA carrier aircraft would require a week's work and cost $900,000.

In the weeks following Atlantis' recovery, pressure suit engineers discovered a small cut in the outside layer of one of Curbeam's EMU gloves. As a result, NASA introduced a new rule that required astronauts on an EVA to examine their gloves approximately every 60 minutes while they were in a vacuum.

PLANNING THE END

Even as STS-117 continued the construction of ISS, NASA was thinking about its end. Under the original plan for the present ISS configuration, as defined in the Memorandum of Agreement signed by all ISS partners, the station was intended to be operational until 2016, at which time it would be de-orbited in such a way that any items that survived re-entry would fall harmlessly into the Pacific Ocean. To achieve that end, NASA had originally planned to dock a Shuttle to the station and use its thrusters to fire the de-orbit burn. As the operational portion of the programme was continually delayed the possibility of operating ISS beyond 2016 came under discussion, but the 2016 end date remained as a NASA budget planning tool. Recently, NASA had been openly talking of funding and operating ISS beyond 2016 (which had always been a distinct possibility), but now the Shuttle was due to stop flying in 2010, so it would no longer be available to de-orbit the ISS at the end of its operational life.

In mid-2007, one possibility that NASA was considering to replace the Shuttle in the vital ISS de-orbiting role was the ESA ATV. In the plan NASA would purchase an ATV and its Ariane-V launch vehicle from ESA. When the final flight was required ESA would launch the ATV for NASA and fly it to a docking with Zvezda's wake. The ATV's thrusters would then be used to perform the de-orbit burn, bringing ISS down as planned, with any surviving debris falling in the Pacific Ocean.

ESA's original contract was for six ATVs, but only five ATV launches were currently shown on the ISS launch manifest up to 2016. If the sixth vehicle was not built, ESA would suffer financial penalties. The NASA de-orbit plan, which offered a possible customer for the sixth ATV, was raised during discussions between NASA and ESA regarding ATV launch schedules which had been subject to numerous delays.

As the negotiations continued, the first ATV launch was due in January 2008. Jules Verne, the first ATV, was undergoing pre-flight review in Noordwijk, Holland, to clear it for shipping to the Ariane-V launch site in Kourou, French Guiana.

One major problem with the ATV de-orbit plan lay in the fact that NASA was now budgeting for America's continued use of ISS, using the proposed Orion spacecraft, through 2020. ESA could not promise to fund the ATV and Ariane-V programmes, with their associated infrastructure through 2020, if the fifth and final ATV cargo delivery flight was scheduled for 2016. Meanwhile, ESA was reviewing the computers to be used on the ATV, which were the same as those that ESA had supplied for Zvezda.

MORE COMPUTER TESTS

Yurchikhin described the activities following the undocking of a visiting Shuttle in the following terms:

> "What happens after the Shuttle goes back? First of all, we are always sorry to see them go because we are used to working together by that time. But after the Shuttle goes back to Earth ... we're going to have a lot of cargo items left [on ISS] ... First of all, when they're unloaded into the station, we place them in temporary stowage locations. After that we're going to open all the containers, pull out all the equipment, and then place the equipment in the allocated stowage areas ... Then the next stage will be to activate this equipment ..."

With Atlantis gone, Yurchikhin, Kotov, and Anderson turned their attention back to the computers in the Russian sector of ISS. Two computers in each set of three were working through a jumper cable, but one computer in each set was working in a back-up capacity. On June 21, Yurchikhin powered up the two back-up computers which were also using a jumper cable to bypass a secondary power switch. As the station passed over a Russian ground station, he turned off the computers, removed the jumper cable, and attempted to re-boot the computers.

They failed to re-boot and were left powered off, with the jumper cable removed while troubleshooting continued at Korolev. During the day the Elektron oxygen generator began to produce oxygen for the first time since the computer crash.

Meanwhile, Anderson began his experiment programme, and completed his first *Saturday Morning Science* presentation for younger television viewers. He explained how Newton's laws of motion applied to sports activities. As the new week began he collected blood and urine samples for a nutrition experiment that also required him to record all of the food and drink that he consumed. Anderson and Kotov also participated in a medical emergency exercise while Yurchikhin replaced an antenna in the Russian Regul communication system.

During the week they all performed regular housekeeping, including an inspection of the windows in the Russian modules. They also worked on a number of Russian experiments. Kotov spent further time working with Korolev in an attempt to finally overcome the problem with the Russian computers in Zvezda. Anderson wore an "acoustic dosimeter" to record noise levels as he moved around the station, and his attempts to locate a leak in the MSG were unsuccessful. On June 28, Yurchikhin and Kotov worked with the Russian Profilaktika experiment, studying the long-term effects of microgravity, and Yurchikhin also worked with the Matryoshka radiation detection experiment, while Kotov inventoried medical supplies on the station. Propellant and oxygen was transferred to the station from Progress M-59 on July 11. Two days later Elektron was deliberately shut down. Meanwhile, the American Oxygen Generation System in Destiny was tested over the period July 11–14. The installation of new software in the Russian computers led to a successful re-boot of all Zvezda's major computers on July 16. The computers were used to command the firing of Progress M-60's thrusters, to raise the station's orbit on July 21.

As July had begun, Yurchikhin and Anderson prepared the EMUs and Quest for their first Stage EVA together, scheduled for July 23. Yurchikhin and Kotov also continued to work with Korolev in an attempt to finally overcome Zvezda's computer problems. Their painstaking inspections of the electrical system supplying the computers showed that one relay on the secondary power system was showing a lower voltage than expected. Inspection of the power-monitoring box that had been bypassed revealed moisture, in the form of condensation, inside. One connector was also found to have some corrosion on it and a second was discoloured.

When the reports on the Zvezda computer failure were finally released they pinpointed this corrosion as the cause of the entire problem. In the event of a major electrical power spike the system had been designed to shut down to prevent that spike crippling the system. Water vapour from the dehumidifier above the power-monitoring box had led to a build-up of condensation in the box. The moisture in the power-monitoring box and the corrosion in the connector pins had caused a power line to fail and short-circuit. That short-circuit had tripped the safety system and shut down the entire computer control system in the Russian sector. When the box was replaced by a new unit, delivered from ESA, condensation and microbial growth was discovered on the module's wall behind the original box. When the new unit was installed, an old book was placed between the box and the spacecraft wall, to act as an insulation layer. With the new hardware in place the jumper cables were removed and the Russian C&C computer system was back in place.

Software upgrades were also completed on the computers in the American sector of the station, thus allowing them to support the addition of Node-2, Harmony, Columbus, and Kibo, as the station was expanded in the coming year.

Prior to the Stage EVA, the station was yawed through 180° so that PMA-2, on what had been Destiny's ram now became the station's wake. This was to provide a wider space for the two large items that would be jettisoned towards the station's wake during the EVA. The EVA itself began at 07:25, July 23, when Yurchikhin and Anderson left Quest and prepared their tools. Removing the stanchion for a television camera from the storage pallet on the station's exterior, they moved to the P-1 ITS and installed the stanchion on the nadir face. They then split up for their next tasks. Anderson reconfigured the power supply for an S-band antenna assembly before placing the foot restraint in the open end of the SSRMS and climbing on to it. Meanwhile, Yurchikhin replaced a Remote Power Controller Module (RPCM) to ensure a redundant supply of electrical power to the Mobile Transporter on the wake face of the ITS.

Working together once more, they removed a Flight Releasable Attachment Module and other equipment, which Anderson jettisoned from the end of the extended SSRMS, which was controlled by Kotov from within Destiny. Kotov was the first Russian cosmonaut to be qualified to operate the SSRMS. Yurchikhin then made his way to the Z-1 Truss, where he disconnected and stowed cabling connected to the EAS, mounted on the P-6 ITS. The EAS had been installed by the STS-105 crew in 2001 and had contained an emergency supply of ammonia for use in case of a leak in the station's cooling system. With the station's permanent cooling system now active, the EAS was no longer required and had to be removed before the P-6 ITS was relocated. No leak had ever developed and the emergency ammonia supply had not been used. With Anderson working from the end of the SSRMS and Yurchikhin now on the P-6 ITS, they worked together to sever the final connections and remove the EAS from the P-6 structure. Anderson then held the EAS while Kotov manoeuvred him until he was below the ITS, where he pushed the EAS away from his body towards the station's wake. NASA expected the tank to orbit Earth for up to 11 months before burning up in Earth's atmosphere. Their final scheduled task was to clean Unity's nadir CBM in preparation for the temporary relocation of PMA-3 during the flight of STS-120, then scheduled for October 2007. Finding themselves ahead of their scheduled timeline the two men were also able to complete three "get-ahead" tasks. They relocated an auxiliary equipment bag from the P-6 ITS to the Z-1 Truss, removed a malfunctioning GPS antenna from the S-0 ITS, and released bolts on two fluid trays, also mounted on the S-0 ITS. In the future, the trays would be re-located to the exterior of Harmony, following its delivery on STS-120. Yurchikhin and Anderson returned to the airlock, ending the EVA at 15:06, after 7 hours 41 minutes. Following the EVA, ISS was rotated through 180°, to return PMA-2 to the station's ram. The thrusters on Progress M-60 were used to boost the station's orbit and prepare for the arrival of Progress M-61 and STS-118 later that evening.

Following pressure checks, Progress M-59 undocked from Pirs at 11:07, August 1. The spacecraft did not perform a separation burn, and Yurchikhin had to use the TORU manual control system to fire the de-orbit burn at 14:42. The Progress was destroyed as it re-entered the atmosphere.

PROGRESS M-61

Progress M-61 was launched from Baikonur at 13:33, August 2, 2007. After a standard rendezvous it docked automatically to Pirs' nadir at 14:40, August 5. The new Progress delivered a 2,569 kg cargo of propellants, water, oxygen, dry cargo, and personal items for the crew.

STS-118 DELIVERS "STUBBY"

STS-118	
COMMANDER	Scott Kelly
PILOT	Charles Hobaugh
MISSION SPECIALISTS	Tracy Caldwell, Rick Mastracchio, Barbara Morgan, Benjamin Drew Dafydd Williams (Canada)

STS-118 was Endeavour's first flight in 4.5 years, during which the orbiter had undergone extensive modifications. With Endeavour on the launchpad in mid-July, NASA engineers displayed the usual "Go Endeavour" banner on the gate across the bottom of the crawlerway ramp, only the banner in question read "Go Endeavor," the American spelling, while the Shuttle's name is spelt the British way, having been named after the Royal Navy ship sailed by the British 18th-century explorer Captain James Cook. When the mistake was pointed out the banner was quickly removed and replaced by one with the correct spelling. It was a small thing with no effect on the launch preparations.

In the run-up to launch, much of the media coverage centred on Barbara Morgan, who had first been selected to fly the Shuttle into space in the early 1980s, when she was one of two high school teachers selected for President Ronald Reagan's "Teacher in Space" programme. Ultimately, Morgan was selected as the back-up and she watched the launch of STS-51L from the roof of the VAB, at KSC, on January 28, 1986. From there she saw the Shuttle explode, just 75 seconds into its flight, killing Commander Richard Scobee, Pilot Mike Smith, Mission Specialists Judy Resnik, Ellison Onizuka, Ronald McNair, Payload Specialist Gregory Jarvis, and high school teacher Christa McAuliffe. Morgan had returned to teaching before applying to become a full-time professional astronaut. She was subsequently selected as part of the 1998 astronaut group. When President George W. Bush announced his Educator Astronaut Project, a plan to select one professional educator in each subsequent annual astronaut group, Morgan was named as the first Educator Astronaut in 2002 and began training as a Mission Specialist on STS-118. The original Teacher in Space and the Educator Astronaut Project had the same objective, to encourage America's schoolchildren and college population to study mathematics, science, and engineering and to be generally inspired by the act of spaceflight itself. In her pre-launch interview Morgan remarked:

> "I'm really excited about going up and doing our jobs and doing them well. I'm excited about experiencing the whole spaceflight, seeing Earth from space for the very first time and experiencing weightlessness and what that's all about. I'm excited about seeing what it's like living and working onboard the International Space Station."

To some journalists it seemed appropriate that Christa McAuliffe's back-up should make the first Educator Astronaut flight on Endeavour, the orbiter that had been constructed to replace Challenger. On the subject of STS-51L, Morgan said, "The legacy of Christa and the Challenger crew is open-ended. I see this as a continuation. The great thing about it is that people will be thinking about Challenger and thinking about all the hard work lots of folks over many years have done to continue their mission."

NASA Administrator Michael Griffin was more succinct, stating, "Every time we fly I know that we can lose a crew. That occupies a large portion of my thoughts. Unless we're going to get out of the manned spaceflight business, that thought is going to be with me every time we fly."

Morgan's Educator Astronaut tasks were secondary to her other mission tasks; she would operate Endeavour's RMS to provide images of the installation of the S-5 ITS, which would be manoeuvred into place by the SSRMS, mounted on the MT. The S-5 was a spacer, used to provide sufficient room to allow the S-6 SAWs to function correctly when they were installed. A similar unit, the P-5 ITS, was already in place. The S-5 and P-5 spacers, which were each the size of a compact car, also included all of the plumbing and electronics needed to ensure that the S-6 and P-6 ITS elements could function as part of the overall ISS power and cooling system. It would require three EVAs to complete all of the necessary connections to make the S-5 ITS an integral part of the ITS. Anderson explained:

> "We add this little spacer, S-5, which we call 'Stubby'—P-5 was 'Puny,' so you had 'Puny' on the left and 'Stubby' on the right—and you add that piece so that when STS-120 comes to bring me home ... we're going to take the P-6 module that's been sitting up on top of the station since its arrival, they're going to take that off, move it outboard on the P[ort] side and stick it on. So now we'll have three sets of solar arrays ..."

Lead Shuttle Flight Director Matt Abbott described the flight in the following terms, "This mission has lots of angles. There's a little bit of assembly, there's some re-supply, there's some repairs, and there's some high-visibility education and public affairs events. It's a little bit of everything."

As well as the S-5 ITS, STS-118 would also carry the last SpaceHab logistics module to ISS. SpaceHab was carried rather than an MPLM because the S-5 ITS left insufficient room in the orbiter's payload bay for the latter. The crew would also replace the CMG, in the Z-1 Truss, that had failed in October 2006, using procedures similar to those employed by the STS-114 crew when they replaced a malfunctioning CMG in 2005. In its basic form, STS-118 was a standard 11-day flight to ISS. However, Endeavour would be the first orbiter to carry the Station–Shuttle Power Transfer System (SSPTS). As the name suggested, the SSPTS allowed electrical power generated by the station's ITS SAWs to be transferred to a docked Shuttle orbiter, powering its systems. If the SSPTS worked correctly, flight managers would have the potential to extend the stay on ISS by up to 3 days, and even to add a fourth EVA to the flight. Future Shuttle flights might be extended by up to 6 days by using the SSPTS.

During the countdown, a faulty valve caused a pressure leak in Endeavour's crew compartment. It was replaced by a similar valve, removed from Atlantis. The leaking valve, together with thunder storms that caused delays in the pre-launch work on the launchpad, led to the August 7, 2007 launch being delayed by 24 hours on August 3. The delay gave NASA the opportunity to launch the Phoenix probe, which only had a 3-week launch window if it was going to reach its proposed landing site on Mars.

STS-118 finally lifted off at 18:36, August 8, 2007. Representatives of some of the STS-51L crew members' families were at KSC to see the launch. Ironically, given the media's concentration on Morgan, members of the McAuliffe family were not among them. As Endeavour left the launchpad, ISS was over the Atlantic Ocean, southeast of Nova Scotia. Onboard cameras showed that nine pieces of foam separated from the ET and at least three appeared to strike Endeavour. The first was seen at $T+24$ seconds, and looked as though it struck the body flap at the orbiter's rear. The second piece was seen at $T+58$ seconds, and resulted in a spray and decolourisation on the right wing. The third and subsequent pieces departed from the ET after STS-118 was above the sensible atmosphere and were therefore not thought to have had sufficient energy to damage the orbiter. A few minutes later, Endeavour was in orbit. In Houston, astronaut Rob Narvis announced, "... Class is in session." Approximately 90 minutes after launch, a fragment of a Delta launch vehicle launched in 1975 passed within 2.4 km of Endeavour. Flight Day 1 ended at 00:36, as the crew began their first sleep period in orbit.

Day 2 began at 08:37. During the first full day in space, Hobaugh, Caldwell, Mastracchio, and Morgan used the RMS to lift the OBSS from the opposite side of the payload bay at 11:20, and spent the next 6 hours using its cameras and laser sensors to view the orbiter's TPS. Elsewhere, the crew checked their rendezvous equipment, installed the docking system centreline camera, and extended Endeavour's docking ring. During a television broadcast, Morgan told the audience, "Hey, its great being up here. We've been working really hard, but it's a really good fun kind of work." The day ended at 23:36.

Figure 95. STS-118 crew (L to R): Richard A. Mastracchio, Barbara R. Morgan, Charles O. Hobaugh, Scott J. Kelly, Tracy E. Caldwell, Dafydd R. Williams, Alvin Drew, Jr.

Figure 96. STS-118 approaches ISS, the payload bay holds External Stowage Platform 3, the S-5 Integrated Truss Structure and a SpaceHab module full of supplies for the station.

August 10 began at 07:36. After eating their breakfast the crew commenced preparations for rendezvous and docking with ISS. Final rendezvous manoeuvres began at 11:14, when Kelly performed the Terminal Initiation burn. He told the Expedition-15 crew, "We're about 30,000 feet away. You're looking very good." Anderson replied, "All right, man, keep up the good work. We're waiting for you."

Closing to within 200 m of the station, Kelly manoeuvred Endeavour through its r-bar pitch manoeuvre, allowing Yurchikhin and Kotov to expose a series of high-definition digital images of the orbiter's underside, which were down-linked to Houston. When experts on the ground examined the photographs they identified a 7.5 cm × 7.5 cm gouge in the underside of Endeavour's starboard wing, close to the main landing gear door. When he was told of the damage, Kelly replied, "Thanks for the update." It was decided to carry out an OBSS inspection of the area with the orbiter docked to ISS.

Docking to PMA-2 on Destiny's ram occurred at 14:02. The hatches between the two spacecraft were opened at 16:04, and Endeavour's crew entered Destiny, where they were greeted by Yurchikhin, Kotov, and Anderson. After a short ceremony, the new visitors received the usual safety brief from their hosts. At 16:17 the crew activated the SSPTS and electrical power produced by the ISS' SAWs flowed into a docked Shuttle for the first time. One of the first activities following docking was Mastracchio using the RMS to lift the S-5 ITS out of the payload bay and hand it over to the SSRMS, operated by Hobaugh and Anderson. The S-5 ITS was then left on the end of the SSRMS overnight to acclimatise to the space environment. Work began at 18:00 to unload the first of the supplies from Endeavour's SpaceHab module. Meanwhile, Mastracchio and Williams spent the night camped out in Quest with the air pressure reduced, in preparation for their first EVA the following day.

In Houston, engineers had inspected the video from the various cameras mounted on the Shuttle during lift-off. They showed that the gouge on the underside of the right wing had been caused by a grapefruit-sized piece of foam that had shed from the ET and struck a bracket on the fuel line at the rear of the tank as it fell away. The impact had changed its course, leading to the impact with the underside of the right wing. Although the planned OBSS inspection would go ahead, NASA now felt sure that Endeavour could re-enter Earth's atmosphere without the need for an EVA to repair the TPS. Mission Manager John Shannon stated, "It's a little bit of a concern to us because this seems to be something that has happened frequently."

During the night, Endeavour's crew was woken up by an audible alarm, which had activated because the use of the SSPTS had allowed one of the Shuttle's fuel cells to cool down more than on previous flights. The alarm's tolerance level was adjusted to take the lower temperature into account.

Endeavour's crew was officially woken up at 07:38; they then ate their breakfast. After supplying electrical power to Endeavour throughout the night, the SSPTS was powered off prior to the EVA, which began at 12:28, August 11, after Hobaugh had used the SSRMS to position the S-5 ITS at the exposed end of the S-4 ITS. Mastracchio and Williams left Quest and made their way to the area, where they gave verbal instructions to assist in the final positioning of the new ITS element. The tolerances between the S-5 ITS and other parts of the station were at times only a few centimetres, and with no cameras in the area, verbal instructions were vital. They then set about bolting the new ITS in place and completing the electrical and plumbing connections between the S4 and S-5 ITS. The S-5 ITS was considered to be officially "in place" at 14:26.

At 15:52, while the EVA was underway, the primary American Command & Control computer, in Destiny, shut down without warning. The back-up computer immediately assumed the primary role and the third, stand-by computer assumed the back-up role, exactly as they had been programmed to do. The EVA was not affected and ISS was never in danger. Controllers in Houston began troubleshooting the problem.

After installing the S-5 ITS, Mastracchio and Williams made their way to the P-6 ITS, where they retracted the forward radiator. This was the final task to prepare the P-6 ITS for its removal from the Z-1 Truss and relocation to the exposed end of the P-5 ITS. That task would be completed by the crew of STS-120, after they had temporarily installed Harmony on Unity. The STS-118 EVA ended at 18:45, after 6 hours 17 minutes. The SSPTS was powered on after the EVA was complete and began supplying power to Endeavour once more.

August 12 began on Endeavour at 07:07 and was the day that the Shuttle crew inspected the damage caused to their vehicle by foam and ice shedding from the ET during lift-off. Hobaugh and Anderson used the SSRMS to lift the OBSS from its storage position, and handed it off to Endeavour's own RMS at 09:45. Kelly, Caldwell, and Morgan then spent 3 hours operating the RMS to view five damaged areas of TPS on the underside of Endeavour using the cameras and laser instruments on the OBSS to form a three-dimensional image of the damaged areas. Four of the five areas inspected offered no threat to re-entry. The fifth and by far the largest area was the 7.5 cm × 7.5 cm gouge seen earlier in the flight. Engineers in Houston would reconstruct the large gouge and test it under simulated re-entry heating temperatures. In the event that the gouge was found to threaten Endeavour during re-entry, the crew had three options for repair, all of which would involve an EVA:

- Apply a thermal paint to the exposed surface of the gouge.
- Install a cover plate of TPS material.
- Fill the gouge with a caulking material.

Throughout the day, Williams, Hobaugh, Mastricchio, and Drew spent their time transferring equipment and logistics from SpaceHab to ISS. In Houston, flight managers reviewed the operation of the SSPTS and extended the flight of STS-118, pushing Endeavour's undocking from the station back to August 20, with landing 2 days later. The extra time would be used for a fourth EVA, by Williams and Anderson, to install a berth on the exterior of ISS to hold the OBSS. Just after 21:00, Mastracchio and Williams began their second overnight camp-out in Quest, in preparation for their second EVA.

That EVA began at 11:32, August 13, when the two astronauts left Quest and made their way to the Z-1 Truss, using the SSRMS operated by Hobaugh and Anderson. Once in location they removed the CMG that had failed in October

2006. The failed CMG was carried to a temporary stowage location, while they installed a new CMG, which had been carried into orbit in Endeavour's payload bay, mounted on External Stowage Platform 3. With the EVA progressing well, Caldwell told Mastracchio and Williams, "You guys rock." The final task was to move the failed CMG from its temporary location to External Stowage Platform 2 on the exterior of ISS, where they secured it to await recovery and return to Earth on a later Shuttle flight. The EVA ended at 18:00, after 6 hours 28 minutes.

Throughout the EVA, Drew continued to transfer equipment from SpaceHab to ISS, a task that other crew members assisted with once the EVA was complete. In Zvezda, Yurchikhin and Kotov continued searching Zvezda for the cause of the computer failure that had occurred during the visit of STS-117. On removing some wall panels in the Russian module they discovered that condensation had collected behind them.

In Houston, mission managers announced that the OBSS survey had shown that the largest gouge on Endeavour's underside passed right through the TPS tiles and had exposed the felt that was laid between the TPS tiles and the orbiter's aluminium skin. Discussions were underway as to whether or not the crew should make a repair before leaving ISS, but Mission Manager John Shannon told a press conference, "This is not a catastrophic loss-of-orbiter case at all. This is a case where you want to do the prudent thing for the vehicle." He added, "we have really prepared for exactly this case, since Columbia [STS-107]. We have spent a lot of money and a lot of people's efforts to be ready to handle exactly this case."

Shannon explained that it might be better to add a complicated TPS repair EVA to the flight, rather than risk more serious damage being caused during re-entry that would require long repairs and throw the already tight launch schedule into further disarray. He also assured journalists that mission managers had ruled out the second repair option, screwing a pre-prepared plate of TPS material over the damage, as the damage did not warrant such drastic measures. The choice now was between the thermal protection paint and the caulking material. On the subject of what this foam-shedding event meant to the launch schedule, Shannon would not be drawn, saying only, "We have a lot of discussion to have before we fly the next [External] Tank."

Endeavour's crew began their sleep period at 22:06.

Two events were marked during the August 14 working day. Endeavour's crew were woken up at 06:07, by a recording of Caldwell's nieces and nephews singing, "Happy Birthday, dear Tracy," to mark the astronaut's birthday. Also, at 11:15, Zarya, the first American-financed ISS module to be launched, completed its 50,000th orbit.

Caldwell and Morgan used Endeavour's RMS to lift External Stowage Platform 3 out of the Shuttle's payload bay, and hand it over to the SSRMS. Hobaugh and Anderson then moved it into position. The 4 m × 2 m platform, which held a second, spare CMG, a nitrogen tank assembly, and a battery charger/discharger unit, was attached to the P-3 ITS at 12:18. The previous two ESPs had been installed on Destiny and Quest by astronauts making EVAs.

During the day, Kelly, Caldwell, and Morgan were interviewed by news organisations, and Kelly answered a question on the damage sustained by Endeavour during launch, saying, "My understanding is that the tile damage is not an issue for the safety of the crew. We may still choose to repair, but I'm not concerned with our safety."

Morgan then joined Anderson, Williams, and Drew in answering questions from children at the Discovery Centre in Boise, Idaho. One child asked Morgan how being an astronaut compared with being a teacher. She replied:

> "Astronauts and teachers actually do the same thing, we explore, we discover and we share. And the great thing about being a teacher is you get to do it with students, and the great thing about being an astronaut is you get to do it in space and those are absolutely wonderful jobs."

The video conference continued with the students asking questions on a wide range of subjects and receiving answers from the astronauts on Endeavour. Asked about what it felt like to fly into space, Williams replied:

> "As soon as the engines stop, you float forward in your seat, your arms rise up, and it's an incredible sense of freedom. The first thing we like to do is go up to the window and look at the Earth. It's an amazing sight."

On the ground, mission managers were still awaiting heating tests before deciding whether or not to repair the gouge in Endeavour's underside. They continually made it clear during the day's press conferences that any repairs undertaken would be to prevent prolonged repairs after flight and were not a matter of preventing the loss of Endeavour during re-entry. Mastracchio and Anderson spent the day preparing for their third EVA, the following day. They spent the night "camped out" in Quest. The crew's day ended at 22:06.

August 15's work day began at 06:07. Mastracchio and Anderson left Quest at 10:38 to begin EVA-3. The two men worked together to relocate an S-band Antenna Sub-Assembly from the P-6 ITS to the P-1 ITS. They also installed a new transponder on the P-1 ITS, before removing the transponder on the P-1 ITS. Both men watched while Hobaugh and Kotov used the SSRMS to relocate the two CETA carts from the track on the port side of the MBS to its starboard side. In so doing, they cleared the track to the port side for the MBS to complete the transfer of the P-6 ITS, from the Z-1 Truss to the outer edge of the P-5 ITS, during the flight of STS-120. Throughout the movement of the CETA carts, Morgan used the cameras on Endeavour's RMS to provide live images of what was happening.

At 14:54, Mastracchio carried out one of the new periodic checks of his EMU gloves and discovered a cut in the thumb of his left glove that had passed through the outer two layers of the glove's five layers of construction. The cut did not puncture the gas bladder, there was no leak, and Mastracchio was in no danger. Even so, he was instructed to return to Quest and return his Extravehicular Mobility Unit to the electrical power supply inside the airlock. In Houston, NASA spokesman Kyle Herring told a press conference, "The suit is perfectly fine. This is just a precaution." Anderson completed the final task before returning to Quest at 16:05, after 5 hours

Figure 97. STS-188: Canadian astronaut Dave Williams deploys a new Control Moment Gyroscope for installation in the Z-1 Truss. The old, failed CMG was temporarily stored on the exterior of the station. Williams is standing in a foot restraint held by the SSRMS.

Figure 98. STS-118: Astronaut Richard Mastracchio at work on the Integrated Truss Structure during the flight's third EVA.

28 minutes. Plans to recover the MISSE-3 and MISSE-4 experiments were delayed until a later EVA.

Inside ISS, the remainder of the two crews spent the rest of the day on the transfer of equipment from SpaceHab to the station, which was half-complete. In Houston, mission managers delayed the fourth EVA from August 17 to August 18, but had not yet decided if it would be devoted to repairing Endeavour's underside or to "get-ahead" tasks. Either way, on August 17 the crew would perform preparations for a repair, until managers told them that it was not required. If the repair went ahead, Mastracchio and Williams would ride on the end of the OBSS, which would be held in the end-effector of Endeavour's RMS. They would be moved to the underside of the orbiter, where they would apply heat-resistant black paint to the gouge in the tiles then they would apply the caulking material to the gouge.

At 08:06, August 16, Morgan and Drew spent time talking to students at the Challenger Centre for Space Science Education, Alexandria, Virginia. The centre had been set up by the families of the STS-51L crew. June Scobee-Rogers, the wife of STS-51L Commander Richard Scobee at the time of the STS-51L flight, was there to oversee the session. She began the session by greeting Morgan, "Barbara, we have been standing by, waiting for your signals from space for twenty-one years ..." The questions that followed were many and varied. One student asked Morgan if she had had any special teachers in her own life. She replied that the seven people on STS-51L were mentors, "... that have meant more than anything to me. They were my teachers and I believe they are teaching us today, still." As the broadcast ended, Morgan held an STS-51L mission patch up to the camera. The question-and-answer session was followed by interviews with a number of television and radio stations. Later in the day, Morgan also answered questions from the education district where she had been a high school teacher, during an amateur radio session.

The two crews spent most of the day transferring equipment between the two spacecraft, while Mastracchio, Williams, and Drew spent their time preparing for the fourth EVA and the repairs that they might have to make to Endeavour's TPS. Just before the crew signed off for the day, Houston told them that the Mission Management Team (MMT) had decided that they would not carry out a repair to Endeavour's underside. On receiving the news, Kelly replied, "Pass along our thanks for all the hard work the MMT and everyone down there is doing to support our flight."

In Houston, John Shannon told a press conference that the decision, "... was not unanimous, but pretty overwhelming."

The fourth EVA would now be dedicated to installing two antennae and a berth to hold the OBSS on the exterior of ISS between Shuttle flights. They would also recover the two MISSE trays that they had not recovered during the previous EVA. The day ended with both crews being given some off-duty time.

August 17 was another quiet day, with Mastracchio, Williams, and Drew preparing for the EVA now planned for the following day. The majority of the day was spent transferring equipment between the two spacecraft and talking with reporters in Houston and the Canadian Space Agency Headquarters in Montreal. Mission managers also had a new problem to consider, Hurricane Dean was approaching Houston, where the Shuttle control room was located. Plans were made to shorten the fourth EVA and even to undock a day early if the hurricane threatened the control centre. Mastracchio and Williams spent the night "camped out" in Quest when the crew began their sleep period at 23:06.

The new day, August 18, began with a wake-up call at 05:03. The crew were informed that due to the threat offered to Houston by the approaching hurricane the planned 6.5-hour EVA had been shortened by 2 hours. The shorter EVA would allow for the hatches between Destiny and Endeavour to be closed at the end of the day, preserving the option to undock one day earlier than planned. If required, mission control would move to a back-up facility at KSC, in Florida.

The final EVA began at 09:17. Williams and Anderson installed the External Wireless Instrumentation System antenna, part of a system to measure stresses in the ISS structure. Next, they installed a stand for the OBSS, before recovering the two MISSE packages deferred from their previous EVA. Plans to secure micro-meteoroid shielding on Zvezda and Zarya and moving an external toolbox had been removed from the plan to facilitate the shorter EVA. During the EVA, ISS passed directly over Hurricane Dean, as it travelled across the Caribbean. Williams remarked, "Wow! Man, can't miss that." Anderson added, "Holy smoke, that's impressive!" Both men returned to Quest at 14:19, after an EVA lasting 5 hours 2 minutes.

Questioned during the crew's joint press conference with the Expedition-15 crew about NASA's decision not to repair Endeavour's TPS, Kelly told journalists:

> "I think it was absolutely the right decision to forego the repair ... I think they took the appropriate amount of time to come to that conclusion."

He added,

> "We have had Shuttles land with worse damage than this. We gave this a very thorough look; there will be no extra concern in my mind due to this damage."

Talking about the proposed repair, Mastracchio said:

> "We were not looking forward to doing it, only because there was a lot of risk involved and a lot of long hard hours involved getting it all prepared ... We felt comfortable we could go and accomplish it."

Williams had the final say:

> "We [NASA] do not take chances; we manage risk. We are in the business of mitigating risk, and that is a data-driven process. To analyse all the appropriate data took time. They made the right decision. Going beneath the belly of the orbiter is something that has its own risk."

By 17:10, the two crews had completed their goodbyes, during which Yurchikhin and Kelly embraced before the Russian told his American colleague, "Have a good

Figure 99. STS-118: the decision was made not to repair this tile damage on Endeavour's underside. The damage to the orbiter's heat protection system had no effect on re-entry, and Endeavour was recovered successfully.

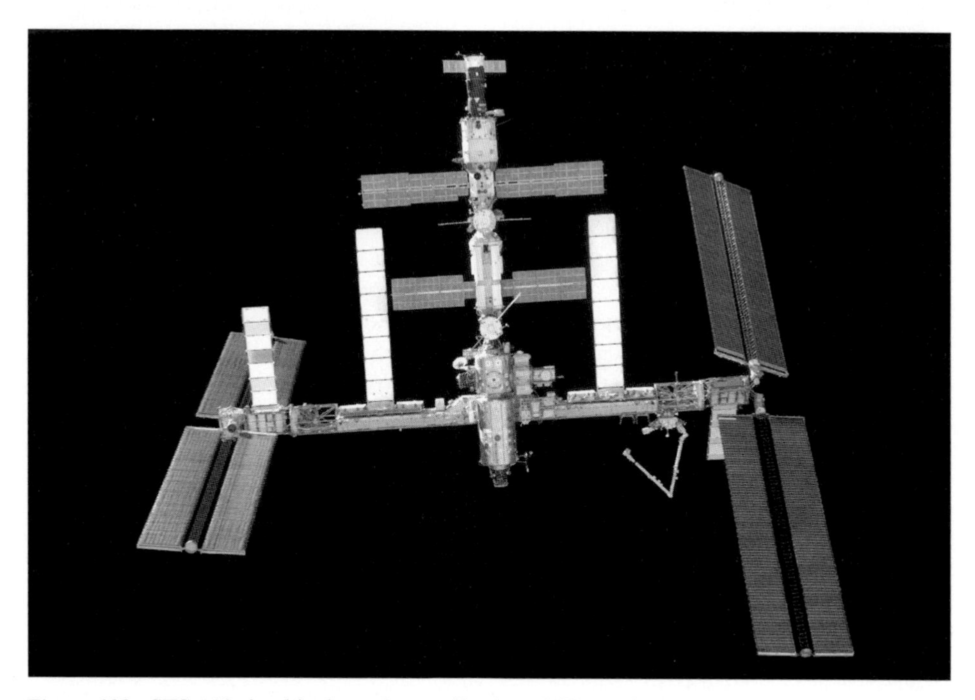

Figure 100. STS-118: in this departing nadir view, ISS's new symmetry is less than obvious because the S-4 and P-4 Solar Array Wings are positioned at 180° to each other (note the position of the outer pair of radiators that rotate with the SAWs).

trip back to Earth." Kelly led his crew back to Endeavour. The hatches between the two spacecraft were sealed and both crews settled down to their respective sleep periods.

Endeavour's wake-up call came at 04:37, August 19. After final checks, Endeavour undocked at 07:56 and backed away from Destiny's ram. As the two spacecraft separated, Yurchikhin rang the ship's bell on the station and remarked, "Endeavour departed." Anderson added, "Thanks for everything Scott, and Endeavour crew, Godspeed." Kelly replied, "We couldn't have gotten everything accomplished without you guys. We look forward to seeing you back on planet Earth."

After a partial fly-around of ISS, the Shuttle was manoeuvred clear at 08:23. Following a second burn, at 09:30, the crew used the RMS to lift the OBSS and use it to inspect Endeavour's nosecap and the leading edges of both wings. They berthed the OBS back along the payload bay door hingeline at 14:37. At 16:35 the crew began four hours of free time in advance of their sleep period, beginning at 20:36.

The last full day in space began at 04:37, and the crew spent the day stowing their equipment prior to re-entry. Kelly, Hobaugh, and Mastracchio tested Endeavour's thrusters and aerodynamic surfaces. Kelly, Morgan, and Williams took time to talk to schoolchildren in Canada, before closing out the SpaceHab module. The day ended at 20:36.

Overnight, Hurricane Dean had hit land in Jamaica, in the process losing most of its power. It was downgraded from a Category 5 hurricane to a Category 1 tropical storm. Endeavour's crew were woken up at 04:36, August 21, and at 07:26 began their final preparations for re-entry. The payload bay doors were closed at 08:45 and retrofire occurred at 11:25, resulting in a landing at KSC on the first of two opportunities that day. Kelly landed his spacecraft in Florida at 12:33, after a flight lasting 13 days 17 hours 56 minutes. As Endeavour rolled out to wheel-stop, Capcom in Houston told the crew, "Welcome home. You've given 'Higher Education' a new meaning."

After the toxic fuels had been pumped from the spacecraft the crew were allowed to leave. On reaching the ground Kelly walked underneath Endeavour to look at the gouge in the TPS. He later described it as, "Somewhat underwhelming ... it looked rather small." Small or not, that one gouge had been the subject of some 4,000 hours of computer time and NASA Administrator Michael Griffin praised the engineers who had carried out those studies. He later reminded journalists at a press conference, "This is very much an experimental vehicle. Anyone who doesn't believe that just doesn't get it."

As to the cause of the gouge, NASA engineers believed that ice had formed on the ET, causing the foam to break away from one of the brackets holding the fuel line. That foam had then struck a second bracket and changed direction, striking the TPS on the underside of the orbiter. The bracket had already been redesigned to minimise foam loss, but tanks with the new brackets were not scheduled to come on-line until spring 2008, with the flight of STS-124. In the meantime, engineers were considering removing some of the foam from the bracket on the ETs of the three Shuttle flights before STS-124.

Predictably, throughout the flight of STS-118 much of the media coverage had concentrated on Barbara Morgan. There had been endless words written and spoken about her role as the back-up Teacher in Space for STS-51L, and how "Challenger exploded in January 1986, killing high school teacher Christa McAuliffe and six other crew members." Regretfully, 21 years after the event, most of the journalists had seemed unable to name the "six other crew members", or thought it was unnecessary to do so in their reports.

ALONE AGAIN

With the Shuttle gone, the Expedition-15 crew was alone on ISS once more. They quickly adjusted their sleep pattern and returned to their daily routine of exercise, maintenance, and experiments. Anderson has described some of the experiments in which he would take part:

> "You know, some of them are quite simple. For example, scientists on the ground want to know what I eat, how much I eat and drink, and how often I eat and drink, and they want to know my vitamin-D content. Well, vitamin-D is something we get from sunlight on Earth, primarily, but we need it for strong bones and muscles, and so folks, elderly folks that have trouble with bones and osteoporosis and those kind of problems can benefit by the scientists that are doing experiments and gathering data on how a crew member in space that doesn't get sunlight anymore and has to supplement his vitamin-D with either drugs—or vitamin tablets—and what he eats and what he drinks, and that's a key experiment that's pretty simple. They just have to gather the data. Another experiment that's quite interesting to me is I'm going to wear a special watch, for the entire increment, and it's called an actiwatch, and it's a sleep watch: it knows when you move, it knows when you go to sleep, it knows when the lights go on and when the lights go out, and scientists will gather that data that we'll downlink periodically through the mission. What they're trying to do is try to figure out ways to benefit people on Earth that do shift work or that have trouble sleeping or that sleep too much, and ways to work with your circadian rhythm and your body and try to help you not go through these periods when you can't stay awake or when you can't go to sleep. So that's quite neat. And those are simple experiments. The more complex experiments include things, we're going to look at combustion on the station through an experiment that has ... several samples that just rotate through a chamber and they look at the flammability and take the data and then they'll evaluate it on the ground. We'll grow some plants, we'll grow some worms, and the key there, of course, is when you go on long duration, can you grow plants, can you eat those plants, how do physical things like worms adapt to zero-g in a long-duration mission, such that we can apply that to humans."

As to his spare time, Anderson had plans of his own:

"From a personal standpoint, I like to write music and I'm going to try to write a song when I'm in space. Now I don't know how much time I'll have, I don't know how successful I will be. The other thing I'd like to do is there's a guitar on board and I've always wanted to learn to play guitar and hopefully I'll have enough time and there's some software on our computers that will try to guide me through the learning process to learn how to play the guitar. I think what I want to do the most, though, is I want to try to absorb as much as I can, all that I experience and see while I'm there, and try to take as much of that with me as a memory either through video media or computer e-mails or what have you, but I want to try to take as much of that away as possible so I can relate it to people back here on Earth."

PMA-3 had been delivered to ISS by STS-92 in October 2000 and installed on Unity's nadir. In December of the same year, STS-97 had docked to it in that location in order to install the P-6 ITS on the Z-1 Truss. STS-98 had also docked to PMA-3, in order to install Destiny, in February 2001. During the visit of STS-102, in March 2001, PMA-3 was moved from Unity's nadir to Unity's port CBM, thus allowing Unity's nadir CBM to be used for the berthing of MPLMs carried by Shuttles docked to PMA-2 on Destiny's ram. On August 30, 2007, the Expedition-15 crew relocated PMA-3 in preparation for the arrival of STS-120. Anderson operated the SSRMS from inside Destiny, while Yurchikhin operated the relevant CBM docking mechanisms on Unity. Kotov backed up both of his colleagues. After latching the end-effector of the SSRMS onto PMA-3, Yurchikhin commanded the docking system on Unity's port CBM to release and Anderson removed the PMA at 09:18. During the undocking a fault alarm sounded when one of the latching bolts registered zero load. The work was stopped to study the situation before continuing. During the relocation fault, alarms sounded intermittently on three securing bolts and work was stopped a second time to further review the situation. Finally, Houston gave the command to continue. After Anderson had manoeuvred PMA-3 next to Unity's nadir CBM, Yurchikhin commanded Unity's docking mechanism to close, holding the PMA in place at its new location. The move was completed at 10:07.

The relocation of PMA-3 was required so that Harmony could be temporarily docked to Unity's port CBM. This temporary installation on Unity was necessary because the SSRMS could not reach to install a PMA on Harmony's ram if the new module was docked directly to Destiny's ram when it was delivered by STS-120. When STS-120 had departed, the SSRMS would be used to undock PMA-2 from Destiny's ram and move it to Harmony's port. Next, the SSRMS would be used to move Harmony, with PMA-2 on one end, and docking, via its exposed CBM on the other end, to Destiny's ram, leaving PMA-2 exposed on Harmony's ram to receive visiting Shuttles. Zvezda's thrusters were fired on September 24, to adjust the station's orbit in advance of the launches of Soyuz TMA-11 and STS-120, planned for October 10 and October 23, respectively.

Anderson updated the software in the American navigation systems and installed new American computer hardware, and Kotov tested and upgraded the Russian computers before the two computer systems were integrated as one system.

The station was re-oriented on September 11, to reduce drag as it passed through the upper atmosphere. It was estimated that the manoeuvre would save the equivalent of the total amount of manoeuvring propellant for Zvezda's thrusters delivered on two Progress flights. Onboard, both loops of the Russian thermal control system suffered a single pump failure, leaving both loops operating on just one pump each. Repairs made by Kotov on October 3 returned one loop to full operation, the second loop would have to await the delivery of a new pump on Soyuz TMA-11. Both men worked to replace Russian EVA support equipment in Zvezda and Pirs on September 19–20. The original equipment had passed its use-by date.

Even as preparations for the arrival of STS-120 continued, Yurchikhin and Kotov were also preparing for the end of the stay on ISS. Progress M-60 was undocked at 19:37, September 18, but was not commanded to re-enter immediately. Rather, it was commanded to perform six manoeuvres as part of the "Plasma–Progress" experiment. Progress M-60 finally re-entered on September 25.

As the Expedition-15 crew brought their experiment programmes to an end, they also began packing to go home. Before that could happen they had to relocate Soyuz TMA-10. On September 27, having prepared the station for unoccupied flight, all three men donned their Sokol pressure suits and sealed themselves in the Soyuz, which was docked to Zarya's nadir. Yurchikhin undocked his spacecraft at 14:20 and manoeuvred clear of the station before flying along its length and manoeuvring to dock at Zvezda's wake. Docking occurred at 14:47 and the crew returned to the station after leak checks. They then began the task of reconfiguring the station for occupation once more. The move cleared Zarya's nadir for the docking of Soyuz TMA-11, flown by the Experdition-16 crew. That launch was planned for October 10.

The following day, September 28, Zarya's starboard photovoltaic array was retracted. This was required to prevent impact with the starboard radiator, mounted on the ITS, that would be deployed during the visit of STS-120. Zarya's port photovoltaic array was retracted on September 29, to prevent impact with the port radiator, which would be deployed after STS-120 had left the station. Meanwhile, on September 30, STS-120 was transferred from the VAB to LC-39A, in preparation for its launch on October 23.

In Destiny, Anderson activated the American Oxygen Generation System and measured the sound levels that the machinery produced on October 2. The new oxygen generator was set at 50% and left running, while Houston monitored its performance. When running at full power the American system would provide sufficient oxygen to fill the entire station at "Core Complete" plus the International Partners' modules. It would be capable of supporting the entire station, even when operated by six Expedition crew members.

The MT/MBS combination was moved to the port side of ISS on October 3, in preparation for the arrival of STS-120 and the relocation of the P-6 ITS. Moving the P-6 ITS from the Z-1 Truss and relocating it to the far end of the port side of the ITS, and the installation of the S-6 ITS on STS-119, represented the ultimate tasks that the SSRMS would be called upon to perform during the construction of ISS. The SSRMS would be operating fully extended and at the extreme limit of its reach.

On the same day, the crew reopened the hatch to Progress M-61, which they had sealed prior to relocating Soyuz TMA-10.

October 4, 2007 was the 50th anniversary of the launch of Sputnik-1, the world's first artificial satellite. Asked about the significance of the anniversary in a pre-launch interview, Expedition-16 crew member Yuri Malenchenko said:

> "I believe we have achieved a considerable progress over such a short time period. We learned to live in space just a short 50 years ago, but didn't live in space. We weren't even thinking, or rather we were thinking, but weren't sure if it is possible, to live in space constantly. Currently we have a continuous presence of humans in space, not only living in space but performing complicated activities and tasks, performing science experiments, and it has been going on for years. Of course, space exploration is unique. All steps, all achievements, regardless of where, which country and when, have been completed, are important, and each step is an important stage for subsequent steps."

The Expedition-15 crew marked the anniversary onboard ISS, but also had to continue with their own work as well as their preparations for the arrival of Soyuz TMA-11 and their own return to Earth. Meanwhile, Anderson worked with Kotov to use the Oxygen Uptake Measurement equipment to collect data as he exercised on the stationary bicycle. The Oxygen Generation System in Destiny continued to operate at 50%, waiting for the system's water supply to be depleted. During the week before the launch of Soyuz TMA-11, the crew mounted the centreline camera in Unity's port CBM, where it would be used in support of the initial docking of Harmony during the flight of STS-120. The three men also completed the medical experiments and extra exercise that all long-duration crews perform as their flight approaches its end. They also made room in Zvezda where the Soyuz TMA-11 spaceflight participant would perform his experiments during his short visit to the station.

On October 6 the TVIS treadmill in Zvezda failed during Kotov's exercise period. The crew worked with engineers in Korolev to replace three roller bearings and return the vital unit to use. As they prepared for their return to Earth, Yurchikhin and Kotov continued to work with both Russian and American experiments.

SOYUZ TMA-11 DELIVERS THE EXPEDITION-16 CREW

SOYUZ TMA-11	
COMMANDER	Peggy Whitson
FLIGHT ENGINEER	Yuri Malenchenko
ENGINEER	Sheikh Muszapher Shukor (Malaysia) (spaceflight participant)

When Soyuz TMA-11 launched towards ISS at 09:21, October 10, 2007, it contained an extremely experienced crew. Soyuz Commander Yuri Malenchenko served 126 days on Mir and had commanded the 185-day occupation of ISS as part of the Expedition-7 crew. He had also visited ISS during the 12-day flight of STS-106. Whitson had made two Shuttle flights, serving as Pilot. Whitson had also served on ISS for 185 days as part of the Expedition-5 crew. Following the Soyuz TMA-11 crew's transfer from Soyuz to the station, she would become the first woman to command an ISS Expedition crew. Sheikh Muszapher Shukor was a commercial customer of the Russian Federal Space Agency on his first flight. On launch day, Russian engineers presented Whitson with a ceremonial Kazakh riding whip and suggested that she might use it to keep her male colleagues in line while on the station. Asked about how she viewed being the first female Commander of ISS, Whitson replied:

> "I think being a woman doesn't really play too much into that. I think it's special that I get the opportunity to play that role, but I think it's also special to have an opportunity to demonstrate how many other women also work at NASA. So I'd like to be able to do that as well."

In answer to a different question she explained:

> "Actually it's going to be kind of exciting. During STS-120 Pam Melroy will be commanding that Shuttle mission; my lead flight director is Holly Ridings. I also have Lead Flight Directors for two different Shuttle missions during those phases, Dana Weigel and Sally Davis. And so we have a big team, which is consistent with any mission, but it happens this time around we have a number of females in the leadership roles. So I think it's exciting."

Soyuz TMA-11 lifted off from Baikonur at 09:22, October 10, 2007. As was the standard procedure, during the launch and solo flight, Malenchenko served as Soyuz Commander. Following a standard 2-day rendezvous, Soyuz TMA-11 docked to Zarya's nadir at 10:50, October 12. After leak and pressure checks the hatches between the two spacecraft were opened at 12:22 and Whitson led the Expedition-16 crew on to ISS. After the standard safety brief, Shukor and Anderson moved their seat liners and Sokol pressure suits between the two Soyuz spacecraft, Anderson thus becoming a member of the Expedition-16 crew and Shukor preparing for his return to Earth in Soyuz TMA-10 with the Expedition-15 crew. The next 9 days were spent in joint experiment programmes while the new crew, who were both ISS veterans, also took time to re-associate themselves with the station. Shukor performed his experiment programme in Zvezda. That programme consisted of five Malaysian experiments and three ESA experiments. In a Malayan press release, he was identified as the country's first *angkasawan* (astronaut). Meanwhile, Anderson replaced a failed audio terminal unit in Quest on October 11. The new unit would lock up during the Soyuz TMA-11 hand-over period. Houston began an investigation.

On October 16, China expressed an interest in getting involved with ISS. Li Xueyong, a Chinese minister of science and technology, stated:

> "We hope to take part in activities related to the International Space Station. If I am not mistaken, this programme has 16 countries currently involved and we hope to be the 17th partner ... The Chinese government has always pursued a foreign policy of peace and consistently worked for the peaceful use of outer space."

In Whitson's first in-flight press conference, the subject of Russian cosmonauts' attitude towards their female colleagues was discussed. The new station commander remarked, "Russian cosmonauts are very professional. Having worked and trained with them for years before we got to this point makes it better." Yurchikhin added, "It's not a problem, women running operations. The problem is whether we are professional or not. We are professionals. She is our friend and colleague." On the same subject, Anderson joked, "I'm a little concerned about this whip. I'm kind of waiting for her to take it out and put me in line sometimes."

Figure 101. Expedition-16 crew (L to R): Sheikh Muszaphar Shukor, Yuri Malenchenko, Peggy Whitson.

After a week of shared maintenance, experiments, and daily exercise, the official hand-over of command took place on October 19, at which time Whitson told Yurchikhin and Kotov, "It's been a very impressive mission, and you guys have performed exceptionally." Yurchinkhin, Kotov, and Shukor said their farewells the following day and sealed themselves in Soyuz TMA-10. They undocked from the station around 03:14, October 21. Following the standard retrofire manoeuvre, Soyuz separated into its three parts. The descent module re-entered the atmosphere but soon deviated from its planned trajectory and followed a much steeper, ballistic trajectory. The course change had been commanded by the onboard computer. After a re-entry in which the crew pulled higher *g*-forces than planned, the spacecraft's parachutes deployed and lowered Soyuz TMA-10 to a safe landing in Kazakhstan, at 06:37. The Expedition-15 crew had been in flight for 196 days 17 hours 5 minutes. Shukor's flight had lasted 10 days 21 hours 14 minutes. Although the landing was 338 km south of the target, the recovery forces had tracked its descent and recovery helicopter crews had the descent module in sight as it descended on its parachute. After removal from the module by the recovery forces, the cosmonauts used the satellite phone, added to the Soyuz spacecraft after a similar re-entry trajectory switch by Soyuz TMA-1, to speak to Korolev. Talking about the re-entry afterwards, Yurchikhin stated:

> "The overload was really powerful, but nobody fainted ... I remember the overload going to 8.5 or 8.6 g."

Shukor was more descriptive, stating:

> "I was not really scared, it happened so fast ... It felt like an elephant pressing on my chest, but the Russians trained us very well."

On the subject of his flight he remained optimistic, saying:

> "I hope to go back and inspire a generation of Malaysian youth ... I hope other Muslims would be united, stay away from war and be peaceful."

STS-120 DELIVERS HARMONY, "THE PIECE THAT MAKES THE REST POSSIBLE"

STS-120	
COMMANDER	Pam Melroy
PILOT	George Zamka
MISSION SPECIALISTS	Scott Parazynski, Douglas Wheelock, Stephanie Wilson, Paolo Nespoli (ESA)
EXPEDITION-16/17 (up)	Daniel Tani
EXPEDITION-15/16 (down)	Clayton Anderson

In the days leading up to the launch of STS-120, a degradation of the outer protective coating was observed on the leading edge of both of Discovery's wings. One RCC panel on one wing and two on the other showed the degradation, which had been present for the past three of Discovery's flights. Launch managers decided that the problem fell within the limits of acceptable risk and decided not to roll STS-120 back to the VAB and replace the panels in question. A roll-back would have caused the flight to be delayed by a minimum of two weeks.

Rain threatened on the morning of launch, but in the end the weather held back. One technical problem that had threatened to delay the launch was a build-up of ice on a propellant line under the left wing. By the time the countdown reached its final stages, the ice had melted sufficiently to offer little threat to the Shuttle.

Discovery lifted off on time at 11:38, October 23, 2007, after what Launch Director Michael Leinbach described as, "One of the cleanest countdowns we've had since I've been a Launch Director." Discovery passed through a succession of cloud layers as it sped towards orbit. Melroy reported that several pieces of ice struck the orbiter's forward windows during launch, but did no damage. All observed ice shedding from the ET took place after the critical first 2.5 minutes of flight, by which time Discovery was beyond the thick lower atmosphere, where the supersonic slipstream might slam the ice into the orbiter, causing damage.

On NASA's website, Commander Pam Melroy had shared her enthusiasm for the flight, saying:

> "STS-120 is such a cool mission. Node-2 is the expansion of the Space Station's capability to bring international laboratories up. It's the expansion of our capability to carry additional people. It has additional life support equipment that will allow us to expand out beyond a three-person crew. It's this big boost in capability which is really exciting."

With lift-off behind them, the crew opened the payload bay doors to deploy their vital radiators and deployed the Ku-band antenna before spending several hours configuring their spacecraft for orbital operations, before settling down to their first eat and sleep period at 17:30.

On ISS, Whitson and Anderson worked on the TVIS treadmill, while Malenchenko serviced the toilet in Zvezda. The following day, Malenchenko serviced the KOB-1 and KOB-2 Thermal Control Loops performing a major plumbing overhaul that returned both systems to partial operation.

Up again at 01:30, October 24, the STS-120 crew's first full day in space was occupied by using the OBSS on the end of the RMS to inspect Discovery's Thermal Protection System, including the RCC panels on the leading edges of both wings. An initial review of the data showed no immediate problems for re-entry at the end of the flight. They also prepared the EMUs stored in Discovery's airlock as well as the equipment they would use during the rendezvous and docking the following day, including installing the centreline camera and extending the docking ring. Inside the orbiter, a high-speed computer modem presented the one difficulty of the day. The modem was due to be used to download the crew's digital photographs to MCC-

Figure 102. STS-120 crew (L to R): Scott E. Parazynski, Douglas H. Wheelock, Stephanie D. Wilson, George D. Zamka, Pamela A. Melroy, Daniel M. Tani, Paolo A. Nespoli.

Figure 103. STS-120 approaches ISS with Node 2, Harmony, in the payload bay.

Houston. On ISS, Anderson was approaching the end of his 4.5-month occupation, and was undergoing an increased daily exercise regime, in preparation for his return to Earth. Anderson and Malenchenko also prepared the cameras they would use to photograph Discovery's underside during the now standard r-bar pitch manoeuvre prior to docking. Whitson performed pressure leaks in PMA-2 in advance of Discovery's docking. The Shuttle crew's day ended at 17:38.

Awake once more at 01:39, October 25, Melroy's crew ate breakfast together before making the final preparations for rendezvous with the station. Melroy began the rendezvous manoeuvres just before 03:00. Two hours from docking Anderson told Discovery's crew, "We can't wait to see you. We welcome you with arms open. The towels are clean and laid out." At 07:32, at a range of 200 metres below the station, Melroy had Discovery perform a nose-over-tail pitch manoeuvre so that Anderson and Malenchenko could photograph the TPS on the Shuttle's underside. Those digital images were sent to MCC-Houston, so that specialists could search them for evidence of any damage caused by the ice or foam shed from the ET during launch.

Docking, with Melroy at the controls, occurred at 08:40, off the coast of North Carolina, and was greeted with cheers from both crews. On ISS, Whitson rang the ship's bell and announced, "Discovery arriving."

Parazynski remarked, "Everyone here is ecstatic. We are so fired up to be here." As usual, docking was followed by pressure and leak checks, before the hatches between the two spacecraft were opened 2 hours later. As the hatches opened, Whitson, the first female commander of ISS, greeted Melroy, only the second female commander of a Shuttle flight. Before launch Melroy had talked about this moment, saying, "The most important thing to me is the picture we take when our hands first meet across the hatches." In Russia, before her own launch, Whitson had sounded less enthusiastic about the meaning of that handshake, saying, "I look forward to their arrival ... She thinks it will be a special moment."

In reality, Whitson embraced Melroy as she entered Destiny. The media-hyped meeting, as the female commanders in charge of two separate spacecraft, was pure coincidence, caused by the delays in past Shuttle launches. Originally, STS-120 had been scheduled to launch before Whitson took command of Expedition-16.

The remainder of Melroy's crew were greeted with handshakes and hugs. After the formal greeting onboard the station and the standard safety brief, Discovery's crew began moving spacewalking equipment into the Quest airlock. At 12:12, Tani installed his couch liner and Sokol launch and re-entry suit in Soyuz TMA-11, becoming part of the Experdition-16 crew, while Anderson moved his equipment into Discovery, transferring him to STS-120. As the day ended, Melroy's crew were told that initial inspection showed no damage to Discovery's Thermal Protection System. Melroy replied to the news, saying, "Oh, man. That is fantastic news. Obviously, that was a question that has been on our minds." Parazynski and Wheelock spent the night "camped out" in Quest, in preparation for the first EVA the following day.

October 26 began at 01:39. After breakfast, Discovery's crew commenced preparations for their EVA. Wheelock and Parazynski exited Quest at 06:02, half an hour earlier than planned, at the beginning of a planned 6.5-hour excursion. Italian astronaut Nespoli choreographed the EVA from inside Discovery. As the preparations came to an end and Quest was depressurised, Whitson joked, "We'll open the hatch so you guys can go out and play."

Parazynski replied, "They call it work, but there is no better job, is there?" As the outer hatch swung open, Parazynski was awed by the view of Earth and remarked to Wheelock, "You're not going to believe this." Their first task was to remove a malfunctioning S-band antenna from its position on the Z-1 Truss and store it in Discovery's payload bay for return to Earth. They also disconnected the final umbilicals running between the Z-1 Truss and the P-6 ITS, in preparation of the latter's relocation later in the flight. Parazynski was subjected to a small ammonia leak while disconnecting the umbilicals and had to undergo cleaning procedures after returning to Quest at the end of the EVA. As they passed over the Gulf Coast, Wheelock remarked enthusiastically, "Oh, boy, look at that; Hello, Houston."

Returning to the payload bay, they put in place a payload and data grapple fixture that could not be mounted on Harmony during launch, due to lack of room within the closed payload bay doors. Their next task was to disconnect the umbilicals supplying electrical power and cooling fluids to Harmony. Tani, Anderson, and Wilson then grappled the new module with the SSRMS, lifted it out of the payload bay, and manoeuvred it to its temporary location on Unity's port CBM. It was the first new pressurised module to be added to ISS in six years.

As the EVA drew to a close, Parazynski remarked, "Great day in outer space." The ammonia decontamination procedures were first used on STS-98 and consisted of partially pressurising Quest, venting the airlock to vacuum once more, in an attempt to remove any residue ammonia crystals, before pressurising Quest to allow the other astronauts to briefly open the internal hatch, pass in wet towels, and close the hatch once more. The two EVA astronauts then wiped down the exterior of each other's EMUs, before bagging the towels and finally leaving the airlock to return to the station. The EVA ended at 12:16, after 6 hours 14 minutes.

Flight Director Dereck Hassaman described Harmony in the following terms:

> "It's the gateway to the International Partners. As the station is configured today, there's nowhere to put the International Partner modules until we deliver and activate Node-2. That's the piece that makes the rest possible."

Flight Director Rick LaBode added:

> "We're going to put it on the left side of Node-1 [Unity], and then, after the mission undocks, we'll robotically remove the port the Shuttle docks to [PMA-2] from the end of the lab [Destiny] and put it on Node-2 [Harmony]. And then we're going to take the Node-2 [with PMA-2] and put it on the end of the lab."

In orbit, with the job of delivering Harmony already completed, Parazynski stated, "Now the crews that are hot on our heels have a place to come."

As the day continued the Mission Management team in Houston decided to add an unplanned task to the second EVA, planned for October 28. The starboard SARJ had been experiencing increased friction over the previous 6 weeks. Parazynski and Wheelock would remove the thermal covers and make a 360° inspection of the joint.

Meanwhile, October 27 began at 01:39. After breakfast, Whitson and Nespoli worked together to prepare Unity's port CBM, before the crew opened the hatch giving access to the interior Harmony. That happened at 08:24, when the hatch was swung back allowing Whitson and Nespoli to become the first people to enter the new module. All crew members wore surgical masks during their first visit to the new module in case there was any loose debris floating around that might be inhaled. With Harmony only in a temporary location, their task was not to power up the module before preparing it for the arrival of Columbus and Kibo. Rather, they applied minimal electrical power and installed a temporary ventilation line to circulate air into Harmony's interior. Later in the day the two crews used the new Node to host a press conference, during which Whitson remarked, "We think Harmony is a very good name for this module because it represents the culmination of a lot of International Partner work and will allow International Partner modules to be added on." Melroy added, "This is a really special moment for the station. This kicks off the international science portion of the Space Station's life cycle." Flight Director Rick LaBrode told the media, "It's beautiful; bright shiny. The report from the crew is that it's as clean as can be. Perfect shape!" Melroy also praised the work of the other members of both crews during the previous day, saying, "I just sat around and made lunch for everyone, and watched them do a totally fantastic job."

During the morning Discovery's OBSS was returned to its storage position along the orbiter's payload bay hingeline. The second inspection of Discovery's Thermal Protection System, planned for that morning, had been cancelled the previous day. During the remainder of the day, Tani reviewed the plan for him to inspect the SARJ during the second EVA, and he also spent time with Anderson, working on hand-over procedures. At 15:23, as the crew's day ended, Parazynski and Tani were locked inside Quest and the pressure was dropped, to allow them to "camp out" overnight, in preparation for their EVA the following day.

The crew's wake-up call on October 28 came at 01:09. After breakfast, Parazynski and Tani donned their EMUs inside Quest, while Wilson and Wheelock manoeuvred the SSRMS to grasp the P-6 ITS, mounted on the Z-1 Truss. Exiting the airlock at 05:32, Parazynski remarked, "It's a beautiful day," and Tani replied, "Awesome." After collecting their tools they made their way to the base of the P-6 ITS, where they disconnected the final electric cables and the bolts that held the structure in place. Wilson and Wheelock then lifted the P-6 ITS away from the Z-1 Truss. The 15 m long P-6 was left hanging overnight on the end of the SSRMS. Meanwhile, the two EVA astronauts set about performing separate tasks. Parazynski moved to the exterior of Harmony, where he installed EVA handrails. Tani made his way to the starboard ITS, where he checked the CETA cart for sharp edges on its handrails and then moved on to the SARJ, where he removed the thermal covers and inspected the joint for friction points. He discovered the joint was covered in a black dust, which included metal shavings, and there was friction wear on the race ring, then he replaced the covers. Station managers decided to limit the amount of rotation that the joint was subjected to while the investigation into problem continued. Tani also reconfigured connections on the S-1 ITS that would allow Houston to deploy the S-1 cooling radiator at a later date. For their final task they worked together to install a second PDGF on the exterior of Harmony, by which it would be held during its transfer from Unity's port side to Destiny's ram. They also removed launch covers from the exterior of Harmony. The EVA ended at 12:05, after 6 hours 33 minutes. After the EVA, NASA's Mike Suffredini commented to a press conference regarding the port SARJ, "I really don't think we are in any situation we can't recover from. It's just a matter of time. We have an obligation to try and get our partners to orbit as quickly as we can."

October 29 was a day of robotic work, with astronauts inside ISS and Discovery moving the P-6 ITS around outside the station. The day began at 01:39 and after breakfast the two crews set about their individual tasks. Parazynski and Wheelock had a relatively quiet day preparing Quest and the station's EMUs for the third EVA, with Nespoli's assistance. Meanwhile, the remainder of the two crews separated into their own work teams. Wilson and Zamka operated Discovery's RMS while Anderson and Tani operated the SSRMS. At 04:08 the RMS was manoeuvred to grapple the P-6 ITS, after which SSRMS was commanded to release it. Discovery's RMS held on to the P-6 ITS while the MBS holding the SSRMS was commanded to travel to the far end of the port ITS, from where it would still be stretched to its limits to install the P-6 ITS in its final location. The MBS translation along the port ITS took 90 minutes.

Tani has described his activities during the 3 days of work required to relocate the P-6 ITS:

> "Conceptually it's not that difficult: It's four bolts—very big bolts but four bolts—it's about a dozen electrical connectors and, and some fluid connectors. During the first couple of EVAs we will disconnect the electrical connectors; on the second EVA I will help unbolt the actual element. We'll have Doug Wheelock inside running the arm. He will initially move the P-6 out and away from the station. Now the difficulty here is that the arm is not long enough to take it from its initial position and move it out to its final position. So we have to do a juggling act. We move it out to the side of the Space Shuttle and I believe Stephanie [Wilson] or George [Zamka] will then grab the P-6 Truss so that the Space Station arm can let it go, and then we utilize the Mobile Transporter, which is this little rail car that's on the truss, and they'll drive this little rail car, with the Space Station arm on it, all the way out to the end of the truss as far as they can go. Then the next day, I'll run that Space Station arm to go pick up the P-6 Truss again and hand it off from the Shuttle. On the next EVA, I'll run the arm and we'll do a final install during the EVA with Doug and Scott outside, to do its final install and bolt it to the end of the truss and then redo those electrical and fluid connectors ... right now the P-6 solar arrays have been fully retracted. It's a big element but at least it doesn't have these huge wings hanging off them. It's hard to think of an analogy, but we are adding a huge source of power to the Space Station, or we're

moving it, and the power reconfiguration to protect all the circuitry, once you hook that up, is very extreme. In fact, we'll have to power down half the Space Station while we do this because you don't want to do what we call a 'hot mate'. You don't want power in one connector and have arcing across these connectors. So we will be powering down half the Space Station while we do this. We mate the P-6 to the P-5 and then, as soon as we can, once the electrical connectors are made, the folks on the ground will start powering those channels back up and we will start attempting to deploy these solar arrays."

During the MBS translation, Whitson and Tani worked inside Harmony, installing avionics racks. The remaining crew members spent most of the day transferring items from Discovery to the station. During the day, Houston informed them that an additional day had been added to the flight plan, giving them a day of additional light workload between EVA-4 and EVA-5. Also plans were added to try and clean the starboard SARJ during EVA-4. As a result of the last point an inspection of the port SARJ was added to EVA-3 in order to provide data with which to compare the descriptions of the starboard SARJ obtained during EVA-2. No attempt would be made to repair the port SARJ on this flight. The changes meant that plans to test a space age caulking gun, designed to be used to repair gouges in the Shuttle's Thermal Protection System caused by foam or ice impacts during launch would be abandoned and moved to a later Shuttle flight. As the day ended, Parazynski and Wheelock were shut inside Quest and the pressure reduced, in preparation for EVA-3 the following day. Meanwhile the Mission Management Team had studied the effect of stopping the continuous rotation of the starboard SARJ, which was a reduction in electrical power production. The reduced electricity supply would be sufficient to support the launch of Columbus, then scheduled for December 2007, but might not support the addition of Kibo, due to be launched in early 2008. Work would continue to resolve the problem.

October 30 began at 00:38. After breakfast, Parazynski and Wheelock began dressing for their EVA while the RMS and SSRMS teams began their own preparations for the hand-off of the P-6 ITS and its re-installation on the far end of the P-5 ITS. The two astronauts left Quest at 05:45. The P-6 ITS was offered up on the end of Discovery's RMS towards the SSRMS, now positioned on the end of the port ITS. After the hand-off to the SSRMS, the P-6 ITS was manoeuvred and then offered up to the exposed end of the S-5 ITS. With few cameras in the area, Parazynski and Wheelock were there to give verbal instructions. Following a successful re-mounting, the two astronauts drove home the four bolts and completed the connections with the P-5 ITS and the station's power system. This move and the similar installation of the S-6 ITS represented the design limits of the SSRMS, even so the two teams on ISS made the task look simple. As all astronauts are pleased to acknowledge, this was all down to the highly professional nature of their training and the dedication of their training teams.

Parazynski then moved to the port SARJ and removed the thermal covers. He described the joint as "pristine". The EVA ended at 12:53, after 7 hours 8 minutes, but on getting out of his EMU, Parazynski discovered a small hole in the outer layer of the thumb on his right-hand glove.

As the EVA reached its final moments, controllers in Houston commanded the first SAW to deploy on the P-6 ITS. When it was fully extended, Discovery's crew commanded the second SAW to deploy. When it had deployed to approximately 30 m, 80% of its full length, Melroy called a stop, "We've detected something that appears to be a wrap-around or some damage." Houston replied, "We see it."

The live television pictures in the control room showed a tear in the SAW. Programme Manager Mike Suffredini later told a press conference, "This will take time and needs to be worked, but my personal opinion is we've got the time to work this issue, so we can be methodical about it, and we will." The remainder of the day was spent discussing the new problem, transferring items from Discovery and talking to the press.

October 31 began at 00:38. As breakfast ended, Parazynski and Wheelock began configuring a spare EMU to replace Parazynski's original suit, which had suffered from cooling problems during the third EVA. Together with Nespoli, they would spend the day preparing for the fourth EVA, which was now planned for November 1, and would be dedicated to a thorough inspection of the starboard SARJ and sampling of the debris seen in the joint during the second EVA, as well as trying to identify the root cause of the friction. Meanwhile, Whitson and Tani worked inside Harmony, removing launch restraints and deploying the Zero Gravity Stowage Rack. On the subject of Harmony, Whitson explained:

> "Node-2, Harmony, like Node-1 [Unity], has six different ports that we can add modules on to, to build the station. So it's, it's our next big connecting piece in our puzzle of putting this huge station together on orbit. Node-2 is required to power and provide the thermal heat rejection for the science laboratory modules that'll be coming up, the one built by the European Space Agency and the one built by the Japanese Space Agency. So it's a pretty key module for us, for the continued development of the station."

As the day proceeded, the priorities for STS-120 changed. Although the ripped SAW on the P-6 ITS was producing 98% of the electricity that it would if fully deployed, Houston decided to make it the priority for the remainder of Discovery's flight. The fourth EVA would be slipped back 24 hours, to November 2, or even November 3, if more time was required for preparation, and would now concentrate on repairing the ripped P-6 ITS SAW before the damage got any worse. The Mission Management Team decided that the priority was to fully deploy the SAW and thus hopefully prevent further damage. Initial plans called for Parazynski to carry out the repair while riding the end of the OBSS mounted on the SSRMS, while Wheelock provided verbal instructions for Wilson and Tani operating the SSRMS. The repair itself would consist of threading wire through holes in the SAW blanket on either side of the tear and using an aluminium strip to support it from beneath, thus closing and supporting the tear in much the same way as a cuff-link works on a man's shirt sleeve. Meanwhile, in Houston, Suffredini was blunt:

> "I need this array. We believe over time we could tear the blanket further. If we do enough damage, we could potentially get into a configuration where we could not stabilise the array. If we can't, we have to figure out what to do. We don't have a lot of options, and the most likely option is that we would have to jettison it."

He continued:

> "The station is a robust vehicle. We have many options with how to deal with the problems. It's not a situation where anyone is particularly panicked. But on the other hand, we want to get this fixed to a point where we can continue with the assembly the way we planned . . . This is not about style points. It doesn't have to look good. It just has to produce power."

Suffredini also paid a compliment to the team of engineers who had been working on the problem since the torn SAW was first identified:

> "We give this team a little time to start thinking about creative solutions, and it doesn't take them long to blow you away with what they come up with."

During the afternoon press conference with the crew, the President of Italy congratulated Nespoli on his flight, but predictably the conversation returned to the damaged SAW. Melroy described what she had seen as the second SAW deployed:

> "It was a tough situation. The Sun was shining directly into our camera views. At one point, we did stop because we were concerned we had lost our big picture. We can second guess ourselves, and there may have been something we could have done, but I think we certainly aborted as soon as we saw something that was not right."

Parazynski added:

> "My initial take was the guide-wires that became frayed earlier may have been the culprit. However, it looks to our eyes, via the binoculars and photos, like the guide-wires may be intact."

During the day the fourth EVA was pushed back to November 3, to give ground teams more time to come up with a work schedule and to give the astronauts additional time to prepare. Even so, Whitson, Commander of ISS, remained confident stating, "If there is a way to do this, we will figure out a smart way to come up with whatever workaround we need to make it happen."

The crew spent the remainder of the day making the hinge stabilisers that they would install when they repaired the SAW and preparing their EMUs. At one point Ex-President George Bush Senior and his wife Barbara visited the control room in

Houston and were able to talk to the crew. Talking to Melroy, he told her, "Good luck to you. Pam, we want to wish you well and all of your team. We're so proud of your team ... Barbara and I."

November 2 began at 01:38 and was another day of preparation. During the morning, controllers moved the MBS back from the far end of the P-5 ITS to the centre of the ITS. There it was used to take hold of the OBSS and remove it from Discovery's payload bay hingeline. The OBSS was then handed to Discovery's own RMS, where it would stay overnight, while the MBS moved back to the far end of the port ITS.

In Houston, NASA made the media aware of some of the risks involved in making repairs to a SAW that was still actively producing electricity. Astronaut David Wolf, head of the EVA branch of the Astronaut Office, said, "We are faced with a difficult situation. At some point, we have to execute the plan we've got, as long as it's very safe, instead of having a perfect plan and having it be too late to execute." He added, "It's a real test of the adaptability of this team, of our baseline knowledge of how to work in space ... We have some risks here."

The two rips, one just under 1 m long and one 0.3 m long, would be repaired using five bracing straps made from 12-gauge wire with a 10 cm long aluminium strip at each end. The aluminium strips would be fed through existing holes in the SAW to hold the damaged areas together along a 5 m length. The straps, which the team that developed them had begun calling "cuff-links", varied in length from 1 m to 2 m. To prevent an electrical discharge and possible injury to Parazynski, Kapton tape, an insulating material, had been wrapped around each of the straps, as well as the tools that would be used and the exposed metal parts on the outside of Parazynski's EMU. The panel was "live", with up to 100 volts of electricity passing through it, and could not be turned off. As a result, the two EVA astronauts had been instructed in which parts of the P-6 ITS represented shock hazards. At the end of the day, Parazynski and Wheelock "camped out" in Quest under reduced pressure.

November 3 began at 01:38 and breakfast was followed by the hand-off of the OBSS from Discovery's RMS to the SSRMS. The EVA started at 06:03. Melroy encouraged her two crew members as they left the airlock with the call, "Go out there and fix that thing." Parazynski replied, "We will." Even so, Houston warned, "Time is of the essence."

Having mounted the OBSS, Parazynski spent 90 minutes being swept through 180° of open space, taking him from the centre of the ITS to the worksite 30 m above Quest and 50 m out to the port side of the station. As he watched Earth sweep by below him, he told Houston, "This is just indescribable. Words just can't do it justice. At least, not mine."

On arrival at the damaged area, he found that the guide wires used during the SAW's deployment were damaged, but the wires carrying electrical current were not damaged. His helmet camera showed a view of the deployment guide wires that he described, "It appears severely frayed."

Melroy viewed the area with binoculars from Discovery and described it as a "furball". She added, "I'm sure that is causing shudders on the ground somewhere." Tani told Parazynski, "You are a dot to us."

Figure 104. STS-120: damage to the P-6 photovoltaic array was stabilised with loops of wire referred to as "cufflinks" by the crew.

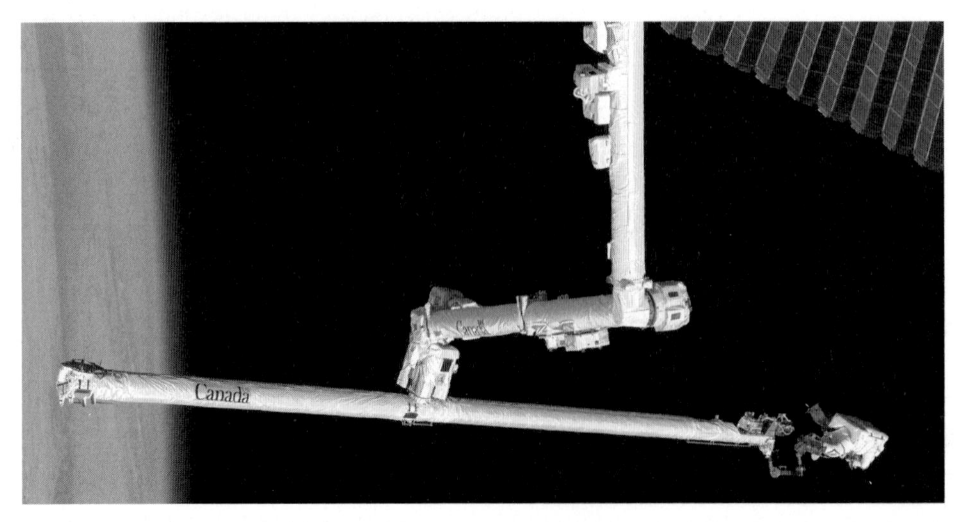

Figure 105. STS-120: Scott Parazynski rides the OBSS held in the SSRMS during the fourth EVA. During the EVA he installed six wire loops to stabilise damage to the P-6 photovoltaic array (see above).

Parazynski cut one of the guide wires with an insulated tool, and Wheelock, working at the base of the SAW, used a pair of pliers to feed it into the relevant take-up reel. Parazynski then installed the five cuff-links, poking the aluminium straps through existing holes in the SAW. As the EVA ended, Parazynski sighed, "What an accomplishment?" Whitson complimented them, "Excellent work guys. Excellent."

With the repair complete, just past 11:00 controllers in Houston began commanding the SAW to complete its deployment. Fifteen minutes and 13 computer commands later the SAW was deployed to its full extent. Meanwhile, it took an hour to sweep Parazynski back through open space to the centre of the ITS, from where the two men made their way back to Quest, closing the hatch at 13:22, after an EVA lasting 7 hours 19 minutes.

Lead Station Flight Director Derek Hassman called it, "One of the most satisfying days that I've ever had in Mission Control." Suffredini was equally enthusiastic, "We are in great shape, fixing the array lets us get on with the assembly ... This was just a fabulous effort. Our baby is still beautiful to us."

The remainder of the day was spent clearing up after the EVA and transferring equipment. While electricity from the P-6 ITS 2B SAW was integrated into the station's main power supply, that from the repaired 4B SAW remained isolated while testing of the repaired SAW continued.

November 4 began at 02:08, before the clocks were put back an hour for the change from EDT to EST. During the morning briefing, Houston told the two astronauts, "This will go down as one of the biggest successes in EVA history. Words cannot express how proud you made everyone with the execution by the entire team." After breakfast the two crews completed the final transfer of items between the two spacecraft before beginning to get ready for Discovery's undocking. 992 kg of new supplies were now on ISS, in addition to Harmony, while 916 kg of scientific samples and other items would be returned to Earth in Discovery. Anderson's occupation of ISS was at an end, but like everyone else before him he was in two minds about how he viewed the prospect of leaving the station:

> "I have a lot of blood, sweat and tears left aboard the International Space Station. What we are doing here is very important for all of human kind. It's worth the risk. It's worth the cost ... Five months ago I was on my back preparing to launch and wondering what the heck I had gotten myself in to. Now, I'm poised to return to Earth after having served very proudly ... Part of me is ready to go and part of me wants to stay."

After saying their formal farewells, the two crews locked arms and swayed back and forth to music, laughing together and some of them even shed a few tears. At 12:28 Melroy led her crew, including Anderson, back to the Shuttle, closing the hatch between the two spacecraft at 12:03. Whitson, Malenchenko, and Tani remained on the station to continue the Expedition-16 occupation.

A 02:38 wake-up call on Discovery on November 5 was followed by a quick breakfast and final preparations for undocking. Zamka backed Discovery away from PMA-2 at 05:32. In Destiny, Whitson rang the station's bell to mark their departure.

Figure 106. STS-120 departs ISS. The RMS holds the OBSS, which lies across the empty payload bay.

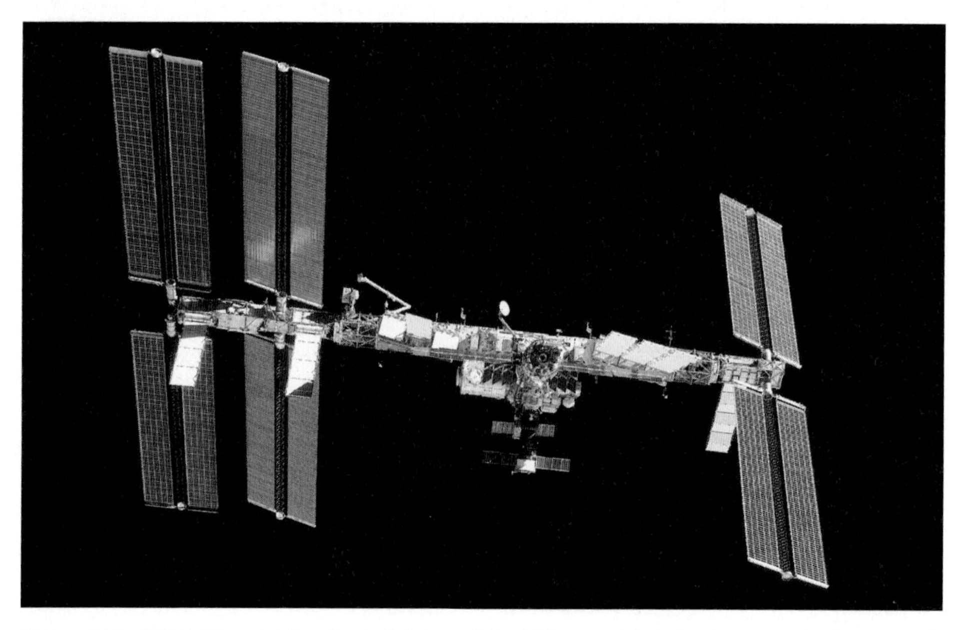

Figure 107. STS-120: a nadir view of ISS as STS-120 completed its fly-around. Harmony is shown docked to Unity's port CBM, opposite the Quest airlock. The P-6 ITS has been relocated from the Z-1 Truss to outboard of the P-5 ITS.

She told them simply, "Thanks, guys." Zamka performed a full 360° fly-around of the station while the crew photographed and videoed its new configuration from all angles. Back opposite PMA-2, Zamka performed the separation burn at 07:15. Discovery's crew spent the day using the OBSS mounted on the end of the RMS to inspect the orbiter's wing leading edges and nosecap for damage sustained while in flight. They found none, and Discovery's TPS was cleared for re-entry. Anderson spent the day exercising in the mid-deck, in advance of his return to Earth.

Discovery's last full day in space, November 6, began at the 02:38. Melroy and Zamka undertook the standard test of the orbiter's flight surfaces and thrusters. The remainder of the crew spent the day packing for re-entry. During the afternoon Anderson's recumbent chair was set up on the mid-deck and the Ku-band antenna was stowed. As Discovery passed over KSC, Melroy noted, "We can see the runway from orbit. So, the weather is looking pretty good." Discovery had been planned for a night landing, but Melroy had asked for the flight plan to be changed in favour of a daylight landing, due to the unintended length and complexity of the flight.

During the day, Melroy talked to the media about how she felt "extremely concerned" for Parazynski's wellbeing during the EVA to repair the 4B SAW.

Parzynski recalled, "It was a phenomenal personal experience to be out on the end of the boom."

Anderson discussed his mental preparations for his return to Earth, "I've enjoyed my time up there immensely, and it's kind of a bittersweet time for me to come home, but I'm ready."

The final day of STS-120 began at 02:38 November 7. Following breakfast, preparations for re-entry began at 08:03. Discovery's payload bay doors were closed at 09:20. Melroy and Zamka began preparing for the de-orbit burn, turning Discovery to a tail-first attitude before igniting the rocket motors at 11:59. When the burn was complete, Melroy turned the orbiter so that its flat underside faced the on-coming atmosphere. Following the standard radio blackout caused by the sheath of ionised air surrounding the vehicle, Melroy flew a series of large, sweeping S-turns in the sky to bleed off energy. Finally, approaching Florida she flew the spacecraft across the state and headed out over the ocean as Discovery turned around the heading alignment circle to line it up with the end of the Shuttle Landing Facility at KSC. Melroy put the rear undercarriage on the runway at 13:01, after a flight lasting 15 days 2 hours 23 minutes. Anderson had been in space for 152 days.

After the flight, Melroy described the mission emotionally, saying, "What you saw is who we are at NASA." NASA's Administrator Michael Griffin watched the landing from alongside Runway 33, KSC. He took a similar tone when he described the flight to reporters as, "NASA at its best."

PREPARING FOR THE INTERNATIONAL PARTNER MODULES

With Discovery gone and the P-6 ITS relocated, Whitson, Malenchenko, and Tani settled down to the remainder of their occupation. They continued their daily routine of experiments, maintenance, and exercise, but beyond that they would oversee the

Figure 108. Expedition-16: Daniel Tani poses in his sleeping bag mounted between two EMUs inside the Quest airlock.

transformation of ISS into a truly International Space Station. The Expedition-16 crew had a quiet day on November 5, in the wake of Discovery's departure and in advance of a busy period during which Harmony would be moved to Destiny's ram. That work began on November 8, when the crew spent the day preparing their EMUs and the Quest airlock for a Stage EVA.

At 04:54, November 9, Whitson and Malenchenko exited Quest to carry out work that should have been completed by the STS-120 crew, but had been rescheduled because of the urgent need to repair the P-6 ITS SAW. Making their way to Destiny's ram, their first task was to disconnect the SSPTS cables from PMA-2, before disconnecting eight other cables between Destiny and the PMA. Whitson also removed a CETA light on Destiny, to clear the area for equipment trays to be installed at a later date. Their third task was to disconnect the rigid umbilicals on the side of Destiny. Both astronauts covered the receptacles left open by the de-mated umbilicals with dust caps as they worked. Separating, Whitson completed connections for the PDGF that would be used when Harmony was relocated. Malenchenko moved up to the Z-1 Truss' wake face to remove and replace a failed RPCM. Working together once more, they made their way back to Harmony, on Unity's port CBM. On the new module's exposed end, they removed a dust cover that had protected the CBM in that area. As they removed the dust cover, Tani observed from inside the station. Looking through the window here, all I can see is a big aluminium foil. It looks like turkey cooking in the oven." Whitson and Malenchenko recovered the dust cover for disposal on a Progress spacecraft. Malenchenko's next task was to re-route an electrical cable at the wake of the Z-1 Truss, while Whitson moved to the "rats' nest", the area between the Z-1 Truss and the S-0 ITS, where she made changes to the electrical connections in that region. Next, Whitson recovered a base-band signal processor and returned to the airlock with it. It would be returned to Earth and refurbished. Finally, Malenchenko redistributed EVA tools between two storage bags and then moved one of those bags to the S-0 ITS. The EVA ended when they returned to Quest, at 10:49, after 6 hours 55 minutes exposed to vacuum. Even as Whitson and Malenchenko completed their EVA, STS-121 Atlantis was moving out to the launchpad where the European Columbus Science Laboratory was already waiting in its payload container.

In orbit, Whitson and her crew began preparations on November 13, for the arrival of Columbus. On that date, Tani commanded the SSRMS from inside Destiny and used it to grapple the PDGF on PMA-2. At 04:35 Whitson commanded the first of four mechanical bolts holding the PMA in place to unwind. The final bolt was released at 05:02, and Tani moved the PMA away from Destiny's ram 10 minutes later. The SSRMS was used to manoeuvre the PMA to a position below Destiny where the station's cameras were used to inspect its mating surfaces. When the survey was complete, Tani moved the PMA to its new position on Harmony's outboard CBM, at which time Whitson commanded the four bolts to secure it in place. The final bolt was secured at 06:29. Later that same day, Houston placed a ban on EVAs. A ground test of an EMU on Earth had resulted in a smell of smoke. Subsequent testing of the suit revealed no signs of burning.

The following day, Tani and Whitson repeated their roles, using the SSRMS to grapple Harmony and release the CBM bolts holding it in place. Whitson released the first bolt at 03:58 and the last at 04:21.Tani then used the SSRMS to move Harmony from its temporary position on the side of Unity and relocated it on Destiny's ram. The relocation manoeuvre was completed at 05:45, much earlier than planned. Capcom Kevin Ford told them, "You guys are really cooking with gas." During the manoeuvre the station had passed over the Atlantic Ocean; Whitson looked out of Destiny's window and remarked, "It's amazing. I love my job!"

On November 15, the P-1 radiator was deployed, increasing the area available to the station's ammonia cooling system. On the same day, NASA cleared the EMUs on the station for future EVA work. NASA's Lynett Madison stated, "There is no indication of combustion or an electrical event. We've been cleared to conduct spacewalks." The smoke odour detected in the suit test earlier in the week was thought to have been caused by a canister of metal oxide used during ground tests of the suit.

Whitson described the two EVAs that she and Tani had originally been expecting to make to outfit Harmony, in the following terms:

> "The EVAs that have to be conducted between the arrival of Node-2 [Harmony] and before arrival of Columbus are critical. We can't accept the new module without the completion of those EVAs ... [T]he two EVAs that Dan and I will conduct actually will lay what we call the umbilical trays, and they are the fluid lines that will connect the Thermal Control System that's based in the truss. We have to run them along the laboratory module and then connect [them] to the Node-2. [T]he reason that's important is the Node-2 has six different heat exchangers; some of those will be providing the thermal heat rejection for each of the new modules that come up later. So it's got a big thermal job, and we have to connect all those lines that will allow it to happen. Obviously we also have to do the electrical and the data connections as well, so that we'll be able to transmit data and receive telemetry back and forth throughout not only the Node-2 module but then later, through the laboratory modules on Columbus and the JEM ... We do some mating on the inside: the internal Thermal Control System's mated on the inside. We also have power and data connections that are done on the inside."

The first of those two EVAs began at 05:10, November 20, 2007, when Whitson and Tani left the Quest airlock wearing American EMUs. Exiting the airlock as the station passed over the Atlantic Ocean, Tani remarked, "A nice day at the office here."

After preparing their tools, they set about individual tasks to maximise their time outside. Whitson removed an ammonia jumper, part of a temporary cooling system, on the outside of the station, vented it, and then stowed it securely in place. The jumper's removal allowed for the establishment of the new Loop-A, one of two loops in the permanent cooling system. As she worked Whitson reported that frozen ammonia crystals were escaping from the open end of the system, "They appear

Figure 109. Expedition-16: Peggy Whitson makes a Stage EVA following the departure of STS-120. In the background Harmony has been relocated to Destiny's ram, and PMA-2 is on Harmony's ram.

frozen and just bouncing off me." Houston replied, "Not a problem at this time. We're ready to press on."

At the same time, Tani retrieved a bag of tools left outside the station during the EVA on November 9. He then removed two fluid caps, as part of the preparation of the permanent cooling loop. His next task was to reconfigure an electrical circuit that was used to fire a pyrotechnic during the deployment of the P-1 cooling radiator on November 15. Both astronauts then made their way to the centre of the S-0 ITS where they co-operated to unbolt the 6.5 m long Loop-A fluid tray from its storage position. In order to move the tray, they took it in turns to move ahead of the tray and secure lines to ensure that it did not drift away if they lost control of it. The tray was then moved forward and the next set of lines attached to it before the previous set of lines were released. In that manner they moved the tray to the exterior of Harmony, where they secured it in place. Next, they secured six fluid connections, two at the tray, two on the S-0 ITS, and two inbetween those two locations. Tani's final planned task was on the port side of Harmony, where he mated 11 avionics lines, meanwhile Whitson configured heating cables and connected electrical harnesses linking PMA-2 and Harmony. With time to spare they were also able to complete a number of "get-ahead" tasks. Tani connected five avionics lines on Harmony's starboard side, before joining Whitson to connect a series of redundant umbilicals and connect the SSPTS cables to PMA-2 in its new location. The EVA ended at 12:26, after 7 hours 16 minutes.

Whitson and Tani's next EVA took place on November 24 and was for all intents and purposes a mirror image of the EVA completed four days earlier. Where the earlier EVA had set up Harmony's primary cooling loop (Loop-A), the second EVA would establish the back-up cooling loop (Loop-B). Ammonia, circulated through the umbilicals installed during these two EVAs, would take up the heat produced by Harmony's electrical equipment and transport it to the large radiators on the ITS, where the heat would be radiated to space and the ammonia recirculated. The EVA began at 04:50, with the crew exiting from Quest wearing American EMUs. They worked together to prepare their tools, before Whitson removed, vented, and stowed the ammonia lines associated with the original, temporary cooling loop. Tani disconnected two fluid caps in preparation for the establishment of Loop-B of the permanent cooling loop. His next task was to relocate an articulated portable foot restraint from its location on the port side of Harmony, to its new position on the lower portion of the module's ram endcone. The two astronauts then joined together to move the Loop-B cooling tray from the S-0 ITS to its permanent location on the port avionics tray on Destiny's zenith, where they bolted it in place. They used the same method to move the fluid tray as they had during the previous EVA. With the Loop-B fluid tray in place, they made the same six connections that they had made on the Loop-A fluid lines: two on the fluid tray, two on the S-0 ITS, and two inbetween. Whitson then made her way to Harmony's starboard side where she removed the launch restraints from the petals on the CBM that would provide soft-docking for Columbus when STS-122 delivered it. That delivery was planned for December 2007. At the same time, Tani made his way to the starboard SARJ, removed one of the thermal covers, allowing him to photograph the joint and recover samples of the metal shavings contaminating the joint. It was a repeat of the work he had carried out during the visit of STS-120. During the inspection, Tani reported, "I see the same damage that I saw before ... I would say there is more damage than I saw before." Tani took the thermal cover back to Quest, leaving the joint open to the video cameras on the SSRMS. The video survey would be completed after the visit of STS-122 and would include at least one full rotation of the SARJ. The EVA ended at 11:54, after 7 hours 4 minutes. The crew had light-duty days on November 25 and 26 following their week of hard work.

On November 28, NASA announced that they feared Harmony may have developed a pressure leak, although the overall pressure leakage rate for the whole station had not increased. (All pressurised modules leak. The rate of leakage is included in the module's design stage and confirmed during manufacture and pre-launch testing. Under normal operations the gases used to pressurise the module are supplied at a rate that will maintain the correct internal pressure in addition to the known leakage rate.) That evening, Whitson was instructed to secure the area between Harmony and Unity's hatches, so that the internal pressure could be monitored. The fact that the overall pressure leak rate had not increased suggested that the problem might actually lie in one of the measuring instruments and might not be a leak at all. The test was repeated and again showed no loss of pressure in the space between the two hatches. As a result, preparations went ahead on the station for the arrival of STS-121, in early December, while Houston continued to monitor the "pressure leak" problem.

With Harmony now on Destiny's ram and PMA-2 on Harmony's ram, ISS was finally configured to receive the next few Shuttle flights, which would deliver the European and Japanese modules to the station. The astronauts from those two nations would begin flying to the station in greater numbers and with increasing regularity. Following the delivery of Node-3, with its extra sleeping facilities, the station's crew would be increased to six people, increasing its capacity to perform first-class orbital science. The last two items of American ISS hardware, the S-6 ITS and the Cupola, would also be launched and installed. In time the European ATV and the Japanese HTV would begin delivering consumables to the station alongside the Russian Progress spacecraft.

As the STS-122 launch was delayed in November 2007, the future schedule for ISS through the end of the Shuttle programme was mapped out:

STS-122	DISCOVERY: Columbus
STS-123	ENDEAVOUR: JEM ELM-PS (placed in temporary position) and Canadian Dextère robotics system. Four EVAs to install equipment
Soyuz TMA-12	Expedition-17 crew up.
STS-124	ATLANTIS: Kibo, two EVAs to install lab and Japanese RMS. Relocate JEM ELM PS to permanent position
STS-128	ENDEAVOUR: MPLM. Establish six-person Expedition crew
H-IIA	ATV-1
STS-119	ENDEAVOUR: S-6 Truss
Soyuz TMA-13	Expedition-18 crew up.
STS-126	MPLM
STS-127	DISCOVERY: JEM-ES and JEM-EF
STS-129	DISCOVERY: EXPRESS Logistics Carrier 1 & 2
STS-130	MPLM
STS-131	DISCOVERY: EXPRESS Carrier 3 & 4
STS-123	ENDEAVOUR: Node-3 and Cupola
STS-132	ENDEAVOUR: EXPRESS Logistics Carrier 5 & 6

As the Shuttle approaches the end of its career, the Russian Soyuz will become the principal vehicle for crew delivery and recovery including the astronauts from all of the ISS International Partners. Given the support of Congress and the new President (the Presidential election is in 2008) the American Project Constellation spacecraft, Orion, and its Ares-1 launch vehicle will be developed and flight-tested. As 2007 drew to a close, only Presidential candidate Hillary Clinton had made positive statements on Orion during her campaign. Clinton's spokesperson, Isaac Baker, had stated, "Senator Clinton does not support delaying the Constellation Programme and intends to maintain American leadership in space exploration." Meanwhile, Senator Barack Obama had called for Project Constellation to be delayed for 5 years and the money spent on education and social programmes.

If they are built, Orion and Ares-1 will assume the role of American crew delivery and recovery in the ISS programme, but flying to ISS is not the principal role for which Orion is being built.

As America prepares to return to the Moon, hopefully taking their International Partners with them, what role does that leave for ISS? During the pre-launch interview for his Expedition-11 flight, Sergei Krikalev voiced his view of the importance of the ISS programme to Project Constellation and the future of human spaceflight in general:

> "[The International Space] Station is not the ultimate goal. It's an intermediate goal. That may be the significance of this Station. This is an intermediate step you have to make before you go any further. Life science experiments can be conducted on the Station to understand how far we can go with the configuration we have right now and what else we need to do to provide more efficiency of human beings on this long-duration mission, and long-distance mission. We continue to conduct technological experiments to see how materials change and how they behave inside, and outside, the Station, to know how to build new vehicles. We are even learning how micro-organisms change inside the Station, and some of these organisms might be a biological hazard for materials inside. Certain micro-organisms can destroy insulation on wires and create big trouble. We have to be prepared especially if we are to go on long-distance missions. On these long-distance missions (not only long-duration missions, as we are flying on the Station right now) you have to be much more autonomous. Even small things that people don't think about very often can change the quality of our development. Being [a] participant on Mir flights and now [on an] ISS flight I see that [the] experience of people, on the ground, operational experience, is very important. Unless we gain this experience, unless we do this step, we will never be able to move any farther from the Earth. It needs to be done on the Station before we can make any further steps."

6

Project Constellation

As the Shuttle returned to flight following the loss of STS-107, initial definition was well under way on the new Project Constellation space vehicles, intended to fulfil President Bush Junior's vision of returning humans to the lunar surface and then moving on to a human landing on Mars. The Constellation hardware consisted of two launch vehicles and two spacecraft. The Crew Exploration Vehicle (CEV), later named the "Orion" spacecraft, would be launched by the Crew Launch Vehicle (CLV), renamed the "Ares-I". The Lunar Surface Access Module (LSAM), "Altair", and its heavy-lift launch vehicle, called "Ares-V" are currently of no relevance to the future ISS flight programme and are therefore not reviewed in this volume.

ARES-I CREW LAUNCH VEHICLE

The Crew Launch Vehicle (CLV) had been named Ares-I. Ares being the Greek god of Mars, and the "I" designation being given in recognition of the Saturn-I/IB, America's first heavy-lift launch vehicles developed specifically for spaceflight. Ares-I will consist of two stages.

First stage

The Ares-I first stage will be constructed at Lockheed's Michoud facility, where the Orion spacecraft will also be constructed. It is derived from a single five-segment SRB similar to those used on the Shuttle. Five segments is one more segment than a Shuttle SRB. This will burn the standard Shuttle-shaped charge solid propellant called polybutadiene acrylonitride. The first stage will burn for 2.5 minutes, raising the CLV to an altitude of 59,000 m and a velocity of Mach 6.1. When the propellant is consumed the SRB will shut down and will be jettisoned. It will make a controlled fall into the ocean under a single parachute and will be recovered, and returned to the manufacturer for breakdown, cleaning, re-fueling, and re-use, in a similar manner to the present Shuttle SRBs. The Ares-I first stage will be manufactured by Alliant Techsystems, who produce the current Shuttle SRBs. The first two tests of the new pilot parachute were made in August 2007. A test subject was dropped from beneath a US Army Chinook helicopter at Yuma Proving Ground. The first in a series of drop tests for the new main recovery parachute for the longer SRB took place in October 2007. The parachutes used to recover the solid rocket boosters on both Ares launch vehicles and the Orion spacecraft will be refurbished in the existing Parachute Refurbishment Facility.

A new inter-stage adapter will mate the top of the first stage to the bottom of a new second stage. The adapter will carry separation rockets to ensure positive separation from the first stage when it is jettisoned. After completing that task, the interstage ring will also be jettisoned and will fall into the Atlantic Ocean. It will not be recovered. The interstage will be manufactured from composite materials, by Boeing, as part of their contract to produce the Ares-I second stage.

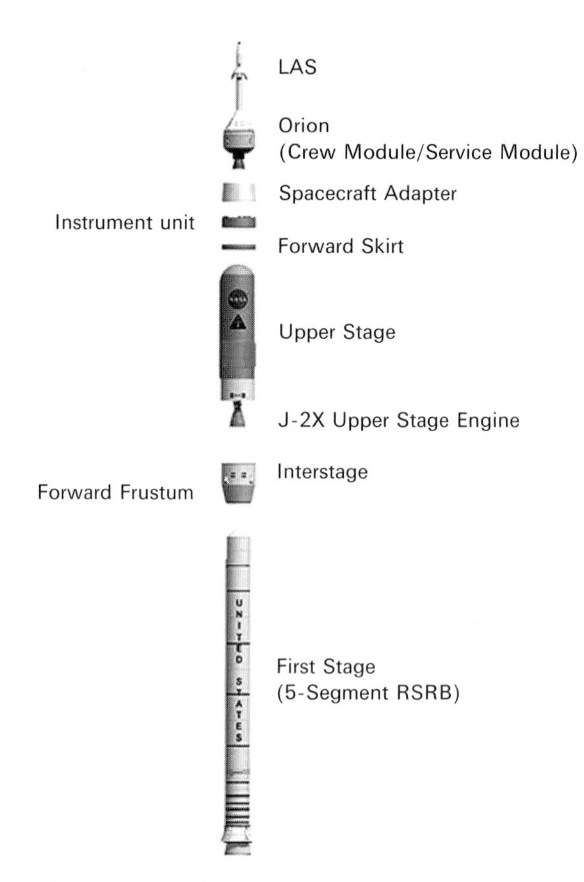

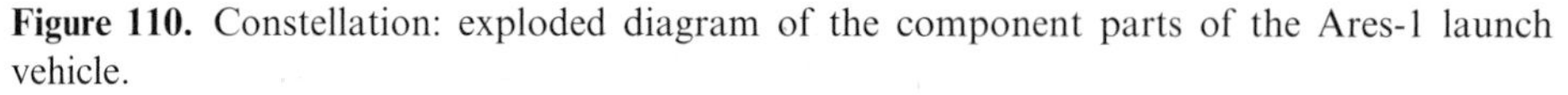

Figure 110. Constellation: exploded diagram of the component parts of the Ares-1 launch vehicle.

Second stage

The new second stage will be developed specifically for the Ares-I launch vehicle and will be powered by a single J-2X rocket motor burning liquid oxygen and liquid hydrogen to produce 1,300 kN of thrust. The second stage will lift its payload to an altitude of 116 km before shutting down and being jettisoned. The Orion spacecraft's service module propulsion system will complete the climb into a circular orbit at an altitude of 340 km. Boeing has been awarded a $515 million contract to support the NASA-led development of the Ares-I second stage and then produce the stage once definition is complete. Under the contract, Boeing will produce a single Ground Test Article, three Flight Fest Articles, and six production stages.

The second stage will be manufactured in the standard pattern, with the propellant tanks positioned one on top of the other and joined by an aluminium skirt. The main stir-welded aluminium stage structure and propellant tanks will be covered by insulating foam similar to that currently employed on the Shuttle's External Tank. The Shuttle's ever-present risk of damage from insulation foam falling off of the External Tank is negated by the fact that at launch the Orion spacecraft will sit at the top of the Ares-I launch vehicle. As a result any insulation foam shed from the exterior of the second stage during launch will already be beneath the spacecraft and should therefore be carried away by the launch vehicle's slipstream and not impact on the Orion spacecraft.

The J-2X is derived from the original J-2 re-startable rocket motor used on the Saturn-V's S-IVB third stage. The original motor could not be used because the aluminium alloy employed in its construction is no longer available, and some components used in the J-2 motor have since been banned for environmental reasons. The J-2X was originally part of a dual study with the J-2XD. The J-2X was to have served on the Ares-V launch vehicle and the J-2XD on the second stage of the Ares-I. The decision to use the J-2X on both vehicles was made in July 2007. On July 17, 2007, NASA awarded Pratt & Whitney Rocketdyne (P&WR) a $1.2 billion design and development contract to develop, produce, and test the first eight J-2X rocket motors. Only one will be a flight engine, with another serving as the motor for the Ares-I second-stage Propulsion Test Article, and six ground test articles. The contract runs through the end of December 2012.

In June 2007, NASA announced a contractor competition for the Ares-1 guidance avionics package, with bids to be submitted by July 30, and the contract expected to be awarded in November. The guidance package would prove inflight guidance to both Ares-I stages during the power flight phase. The package would be mounted on the second stage at Boeing's Michoud facility, where that stage will be manufactured.

"ORION" CREW EXPLORATION VEHICLE

After a review process in which contractor alliances, led by Lockheed-Martin and Boeing, produced designs for the CEV, Lockheed-Martin was named prime contractor for the new spacecraft on August 31, 2006. The Lockheed spacecraft was promptly named "Orion". Development of the new spacecraft was spread over three schedules.

- Schedule-1: September 2006–September 2013, a $3.9 billion contract to support the design, development, test, and evaluation of the Orion spacecraft.
- Schedule-2: September 2009–September 2019, a $3.5 billion contract to support post-development orders for the spacecraft.
- Schedule-3: September 2009–September 2019, a $750 million contract to support additional spacecraft engineering services.

Orion will be the replacement for the Shuttle when that vehicle is retired in 2010. The new spacecraft is being designed to carry four astronauts to ISS, or to the Moon, and six to Mars. It will consist of a conical crew module and cylindrical service module, which function as one spacecraft until just before re-entry, when the unprotected service module will be jettisoned. Only the crew module is protected to allow it to survive re-entry. Orion will have an overall appearance that is superficially similar to the Apollo Command and Service Module. The crew module will be a cone with a 5-metre diameter, a mass of 25 tonnes and 3 times the volume of the Apollo Command Module. The access hatch and windows will all be on one side of the vehicle, with a docking system and a transfer tunnel, surrounded by recovery parachutes, in the apex. Manoeuvring thrusters will be located around the base and in the apex of the module. The rounded base of the crew module will be covered by a circular heatshield, the backshell, for which Lockheed currently intends to use the Phenolic Impregnated Carbon Thermal Protection System developed by NASA Ames Research Centre. Meanwhile, JSC has also purchased Shuttle thermal protection material (blankets) for use on the sides of the cone where the heating regime is less severe. This purchase will also ensure that a Thermal Protection System is available for Orion's early flights to ISS, if problems delay the Thermal Protection System required for the more severe re-entry from a lunar or deep-space flight. The primary recovery zone will be on land, in the open spaces of North America. Final descent will be supported by parachutes, and landing impact loads will be negated by use of retrograde rockets, or inflatable airbags. The use of airbags with their heavy deployment and inflation systems would require the spacecraft's backshell to be jettisoned prior to their deployment, while the lighter solid propellant retrograde rockets could be mounted on the parachute harness and the backshell retained in place. Work continues to decide which of these systems will be used. A launch abort, or failed ascent to orbit would require a water landing in the Atlantic Ocean. The water-landing option will also be available as a back-up in the event that land landing is not possible at the end of a completed flight. The crew module will be reusable for up to ten flights.

The crew module will be constructed from aluminium lithium employing stir-welding technology and will provide $10.2\,m^3$ of habitable volume. Life support systems, providing a two-gas oxygen–nitrogen environment, will be manufactured by Hamilton Standard and the flight control avionics by Honeywell, both members of

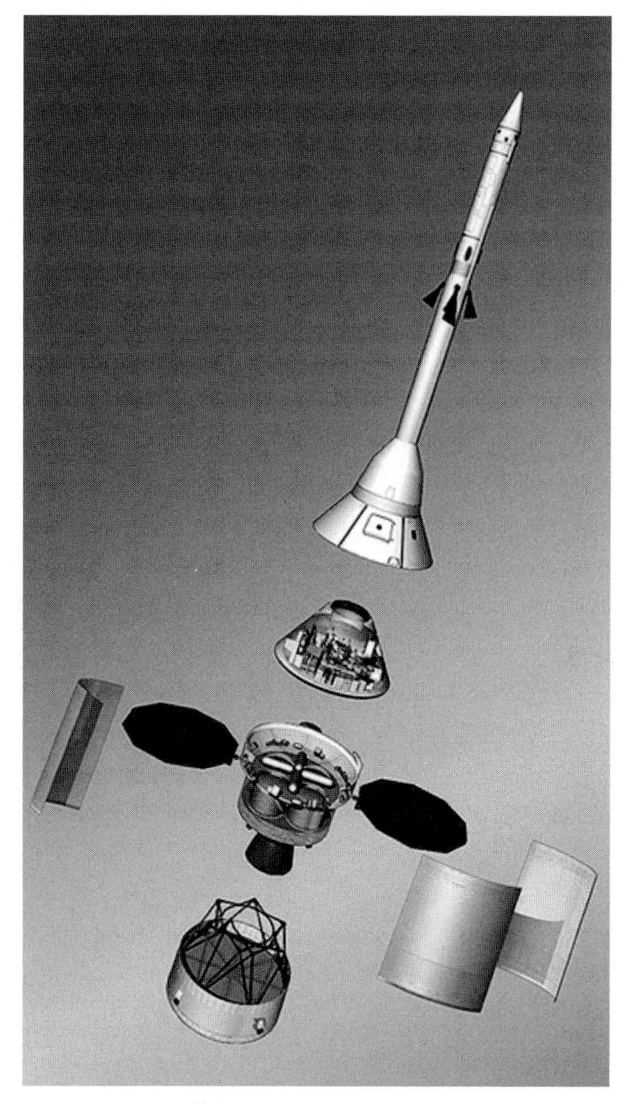

Figure 111. Constellation (early concept): the component parts of the new Orion spacecraft are from top to bottom: Launch Abort System with Boost Protection Cover, Crew Compartment, Service Module with circular Solar Array Wings, and launch vehicle adapter.

Lockheed-Martin's original CEV alliance. The flight avionics will be based on the automated systems employed on the Boeing-787 commercial jet liner. The spacecraft's main instrument panel will be a "glass cockpit", employing four large computer screens to display relevant information to the crew. The screens will be accessed via computer keypads and mechanical switches will be kept to a minimum. This is in keeping with modern military jet fighter technology, with which the present and future groups of pilot astronauts will be increasingly familiar. Rather than having repeated controls on both sides of the console, or having need for crew members to swap seats in order to perform particular functions, astronauts will merely call up the relevant data on the computer screen nearest to them. The computers will display easy-to-read visual representations of major systems, rather than the endless strings of numbers so common during the Apollo Moon flights. Almost the entire flight will be automated, with manual override available at critical points. Three primary computers would provide redundancy to the point that two could fail completely and the third could still return Orion to Earth from any point on a lunar flight. A fourth emergency computer will also be installed and will be completely independent of the three primary computers, to the point that it will employ different hardware and a different electrical supply. Rendezvous and docking will employ automated systems, probably based on NASA's Demonstration of Autonomous Rendezvous Technology (DART) technology, which was tested, not altogether successfully, in 2005. The Orion spacecraft would carry NASA's new Low Impact Docking System (LIDS).

The Orion service module will serve a similar role to its Apollo predecessor, namely the mounting for the spacecraft's main propulsion system and reaction control system, and storage of propellants, water, and avionics. The module will be a cylinder 5 metres in diameter and will contain the service module propulsion system, the principal propulsion system in the spacecraft. This engine, which will burn liquid methane in liquid oxygen, has yet to be developed. Manoeuvring thrusters will be mounted in clusters at 90° around the exterior of the module. The spacecraft's electrical power will come from two large circular photovoltaic arrays mounted at the rear of the service module. These will be folded at launch and deployed once the spacecraft is in orbit. The photovoltaic arrays will provide electricity to storage batteries in the spacecraft. A completely independent battery set will provide emergency power in the event of a major power system failure.

In December 2007, NASA selected Boeing to develop the guidance system for the Orion spacecraft. Jeffrey Hanley, Constellation programme manager for NASA, told a press conference:

> "Finally, with the last team in place, we can move on with the development of this new system . . . This last contract was a key piece that will now allow us to go to the preliminary design phase with a full team in the new year and begin to build and test this new system."

The new spacecraft is intended to return humans to the Moon, and then progress to Mars, but in the first instance it will be used for crew rotation on ISS. The first crewed flight to ISS is planned for no later than 2015. In comparison with the Shuttle, this new combination is estimated to be 10 times safer due to the in-line design of the stack, with the crewed spacecraft placed above the launch vehicle, rather than alongside it. The positioning of the Orion spacecraft, on top of a vertical launch vehicle, means that NASA can return to the use of the rocket-propelled Launch Abort System (LAS) to pull the crew module clear of a catastrophic launch vehicle failure during or just prior to launch. The Orion LAS design is similar to that used for Apollo, with the Crew Module sitting beneath a Boost Protection Cover. In the event of a launch vehicle emergency just prior to launch, or during the Ares-I first-stage boost phase, the solid propellant rocket motor in the LAS would be fired to lift just the Crew Module to an altitude from which it could make a safe parachute recovery. At that point the LAS would be jettisoned and the crew module would descend under its own parachutes. In the event that the LAS was not required it would be jettisoned at the same time that the Ares-I first stage was jettisoned. In both scenarios the LAS will fall into the Atlantic Ocean and will not be recovered. In August 2007, wind tunnel testing of various LAS configurations showed that a Sear-Haack design provided considerable aerodynamic advantages over earlier designs. The LAS will be developed by Orbital Sciences, while the solid propellant rocket motor for the escape rocket itself will be developed by Alliant Techsystems. In November 2007, the first boilerplate Orion spacecraft were being manufactured, for use during flight-tests of the LAS, which will be subjected to a series of pad abort and launch abort tests at the White Sands Missile Range, New Mexico, starting in November 2008.

Orion was originally to be developed in three configurations.

- Block-1A: for low-Earth orbital flights.
- Block-1B: an uncrewed cargo vehicle (subsequently cancelled).
- Block 2: for crewed lunar flights.

As the Shuttle programme works towards its final flight, NASA has begun planning the changes to the Launch Complex 39 facilities that it uses at Cape Canaveral.

The Ares-I/Orion spacecraft combination will be stacked on the mobile launch platform inside the Vertical Assembly Building (VAB), just like Apollo and the Shuttle before it; that procedure will take place in High Bay I. Ares-I/Orion is 150 feet taller than the present Shuttle and will be accessed by folding work platforms similar to those used to access the Saturn/Apollo vehicles.

Apollo's 3 Mobile Launch Platforms (MLPs) were re-configured to carry the Shuttle and will be reconfigured once more for the Ares launch vehicles. NASA is considering building a fourth MLP, so that two Ares-I and two Ares-V launch vehicles can be prepared at the same time. The new MLP design will include a Launch Umbilical Tower. The contract to design the Ares-1 MLP has been awarded to Reynolds, Smith & Hills Incorporated, a local company based in Merritt Island, Florida.

The two Apollo era Crawler Transporters that currently carry the Shuttle to the launchpad will continue in that role for the Ares launch vehicles. Although no additional work is required before the Crawler Transporter can carry the Ares-I to the launchpad, additional strengthening will be required in order to carry the Ares-V.

Launch Pad 39B will be converted to launch the Ares-I/Orion combination. Launch Pad 39A will be converted to launch the Ares-V. Work to convert Launch Complex 39, Pad B to take the Ares-I is planned to start in the spring of 2008. It will include removing the fixed and rotating service structures. A new launch tower will be constructed on the mobile launch platform, in a similar manner to that used for the Saturn/Apollo launch vehicle. Elevators and swing arms will give access to all areas of the Ares-I launch vehicle and the Orion spacecraft. The current emergency escape system, baskets sliding down a wire, will be replaced with a system that resembles a roller coaster, consisting of a car riding down a rail.

In the LaunchControl Centre next to the VAB, Firing Room 1 will be re-fitted to handle Ares-1 launches from Pad 39B. The room is presently vacant, but will be fitted with the ground support equipment required to support the preparation and launch of the Ares-I/Orion combination. The new vehicle is much simpler than the Shuttle, and therefore NASA is looking to use a smaller launch control team than 200 people required to launch a Shuttle.

The original Orion/Ares-I launch schedule was announced as

- April 2009: un-crewed Ares-1 four-segment SRB (inert fifth segment), dummy second stage, and ballast replacing the Orion spacecraft.
- October 2009: option to repeat of first flight if original fails.
- July 2012: un-crewed Ares-1 with five-segment SRB and live second stage.
- Late 2012: un-crewed flight of Ares-1 and Orion spacecraft.
- Fall 2014: un-crewed flight of Ares-1 and Orion Spacecraft.
- September 2014: first crewed flight of Ares-1 and Orion spacecraft.

When Project Constellation was announced, President Bush promised NASA a small annual budget increase to support the programme. Congress has failed to support that budget increase almost every year since then, with the exception of FY2007, when the budget request was actually increased by Congress. In late 2007, NASA announced that the first crewed Orion/Ares-1 launch had slipped into 2015 and all subsequent launches had been delayed accordingly.

As 2008 began, NASA announced that the first two Orion spacecraft to fly to ISS would each deliver a docking adapter, one of which would be left on PMA-2 on Harmony's ram and the other on PMA-3 at Node-3's nadir. One end of the adapter would provide for docking with the Russian-designed Androgynous Peripheral Attach System (APAS) currently used to dock the Shuttle to ISS, while the other end would support the American-designed LIDS to be used by the Orion spacecraft. During docking with ISS the LIDS docking system would be used to mount the docking adapter at the apex of the conical Orion Crew Module. The exposed APAS docking system would then be used to dock to the relevant PMA. At undocking, the Orion spacecraft's LIDS would be released, leaving the docking adapter mounted on the end of the PMA with its LIDS docking system exposed. Future Orion spacecraft would use their own LIDS docking systems to dock to the new adapters. The docking adapter has been named the APAS-To-LIDS Adapter System (ATLAS).

In May 2006, NASA officials and representatives of 13 other countries agreed 6 reasons that justified the Project Constellation effort to return human astronauts to the Moon. These are listed in the next section.

THE MOON'S PLACE IN A US-LED INTERNATIONAL INITIATIVE TO EXPLORE THE SOLAR SYSTEM

1. A training ground for human and robotic exploration of Mars and more distant destinations.
2. Scientific studies that answer fundamental questions about the early history of the solar system and provide sites for astronomical observatories.
3. A place to acquire the technical skills to sustain a human presence on another world.
4. A growth point for the global economy.
5. A means to forge new global partnerships and strengthen old ones.
6. A source of inspiration.

NASA's Deputy Administrator Shana Dale told the media:

> "These are huge endeavours we are embarking upon. We have seen the benefits of collaboration on the International Space Station. As we move forward, we want to make sure we are working very collaboratively with both the international and the commercial sector ... In the long run it makes for a much more sustainable program. This is definitely not just the United States doing this on its own."

America will continue to use ISS to concentrate its research on the reactions experienced by the human body to the unique environment of long-duration spaceflight. At the same time America will seek partners for Project Constellation. The new spacecraft will initially serve as the principal American crew delivery and recovery vehicle for ISS. In time it would carry humans back to the Moon. Although Orion would remain in lunar orbit while a number of Altair Lunar Surface Access Modules were used to develop a permanent base on the lunar surface, the ISS experience would not be forgotten. No doubt new hardware will be tested on ISS, but ISS will also give back to Constellation. When the Moon base is established the astronauts working on the surface would be rotated on 6-month increments, just as ISS crews are today. Their wellbeing will be supported by the medical data that today's astronauts are collecting on ISS. Their missions will be controlled using techniques developed by NASA and the Russians over decades of human spaceflight.

7

Postscript

When David Harland wrote the Postscript for *Creating the International Space Station* in 2001, the Shuttle-supported ISS was the only American human spaceflight programme funded by Congress. As I complete this Postscript in 2007, that is no longer true.

Following the loss of STS-107 in February 2003, the American human space programme was given a new set of priorities by President George W. Bush. The Shuttle would be used to complete ISS and then be retired in 2010. NASA would develop two new spacecraft and two new launch vehicles in order to return astronauts to the surface of the Moon, establish a permanent base there, and ultimately send a crew to set foot on Mars. President Bush has invited other nations to join America in Project Constellation, but so far none has signed up. Russia has recently announced its national space budget for the period up to 2015. Although ISS features prominently in that budget, there is no mention of Project Constellation. The Russians have stated that after 2015 they may reconsider their position regarding their participation in Project Constellation, but have made no promises. ESA and JAXA are only just beginning their major participation in the ISS programme, following the delivery of their laboratory modules. The delays to the station's construction (a two-year delay in launching Zvezda and a further three years following the loss of STS-107) have meant that those nations will not receive the length of use from their laboratories that they had originally planned. ESA and JAXA have so far not committed themselves to Project Constellation. China is the third nation to launch astronauts into orbit, and they have recently expressed an interest in becoming part of the ISS programme. In 2007, China sent a successful probe into lunar orbit, returning stunning photographs of Earth from that location. China has expressed an intention to place one of their astronauts on the lunar surface before the 13th American astronaut gets there.

NASA had originally planned to participate in the ISS programme until 2016, with the Shuttle operating throughout that period, and possibly until the mid-2020s.

The loss of STS-107 led NASA and the American government to admit the shortcomings of the Shuttle system, which is basically 1970s' technology, although some of the orbiters have undergone major upgrades. The President's public announcement that the Shuttle would be retired in 2010 meant that it would not be available to support ISS until 2016. The new Crew Exploration Vehicle, Orion, would not be available to fly crews into Earth orbit until 2015, and only then if NASA received sufficient funding to meet its optimistic early schedules for the new programme. Those funds have not been made available and Orion's schedule is already slipping. With no Orion spacecraft available, NASA will have no choice but to purchase seats on Russian Soyuz spacecraft in order to continue to have access to their hardware on ISS. This will be the second time that access to the station has only remained available to human crews because of Russian spacecraft and launch vehicles, and yet there are still many individual Americans who argue that the Russians should not have been invited to participate in the ISS programme and insist that Russian space hardware is antiquated, unreliable, and dangerous—for no other reason than because it is not American space hardware. They forget that Soyuz has been carrying cosmonauts into orbit since 1967 and that ISS is only permanently manned today because Zvezda is based on the Mir base block, which was itself based on the Russian experience in operating seven Salyut space stations in Earth orbit, starting in 1971. If it is not cancelled following the 2008 election, the Orion spacecraft and its Ares-I launch vehicle will be developed and its early crews will fly to ISS, where it may well replace Soyuz as the principal Crew Transfer Vehicle and CRV, holding four astronauts at a time, a task that would require two Soyuz spacecraft. American operators and any International Partners that do sign up to Project Constellation will gain confidence in the new spacecraft and its launch vehicle during those flights to ISS, while Constellation's "Altair" Lunar Surface Access Module and its Ares-V launch vehicle is developed and flight-tested. That confidence in and flight experience with the Orion/Ares-1 combination will make it easier for programme managers to decide when Project Constellation is ready to return humans to the lunar surface.

In 2007, even as Harmony was being installed on Unity in advance of its relocation on to Destiny's ram, the first boilerplate Orion spacecraft were under manufacture. They would be used to test the Launch Abort System and the Crew Module's parachute systems, as well as the landing system. These tests would be uncrewed. The first Solid Rocket Booster, for use on the first Ares-1 flight-test was also under development. Two successful drop tests of the parachute that would be used to recover that first stage had already taken place, using mass simulators, over the American desert.

The ISS has proved that nations with large cultural differences can work together in space in the name of science. Many of the ISS partner nations are politically opposed to the American-led invasions of Afghanistan and Iraq, but ISS is not a machine of war. It is a place of peaceful scientific research, and therefore individual nations continue to support it without contradicting their opposition to the two wars in question. The ISS is also a great showcase for national technological achievements. As I have stated, ISS is only crewed today because of Russia's Zvezda module and has only remained crewed throughout the past 7 years thanks to the Russian Soyuz and

Progress spacecraft and many of the routines established flying to the Russian Salyut and Mir space stations. The Pirs docking module provides EVA capability using the Russian Orlan-M pressure suit. The Russians have much to be proud of in the ISS programme.

The cargo-carrying capability of the American Shuttle has been indispensable. It has delivered the American Destiny laboratory module, the Z-1 truss with its attitude control system, and the P-6 ITS which provided temporary electrical power and ammonia cooling systems while the ITS was constructed. The large sections of the ITS have bolted together perfectly and the reconfigured electrical power and ammonia cooling systems have allowed the station to reach the point where it is now ready to support the European and Japanese laboratory modules. Shuttle can also deliver large quantities of supplies and take away equally large amounts of rubbish and unwanted items that might otherwise fill up the station making it difficult for the crew to perform their tasks. Shuttle also produces large amounts of water, just by running its fuel cells, which it does throughout each flight; that water can be bagged and left on the station each time the Shuttle visits, thus preventing the heavy liquid having to be transported to the station separately.

Canada's RMS on the Shuttles and the SSRMS on the station, along with the MBS, have also proved indispensable. Without them the station would not have been constructed so smoothly. Europe and Japan have both produced their crewed spacecraft modules to the highest standards demanded of such vehicles, and their robotic transfer vehicles will relieve the Shuttle of its cargo delivery and rubbish removal role when it is retired. All of these nations, 16 in total, have had to overcome mistrust and misgivings about working with different political, social, and even engineering and scientific cultures to make the ISS programme what it is today. The close-knit teamwork and the personal friendships that have developed across geographical and political borders are one of the greatest achievements of this multi-faceted programme.

Engineering demands during the ISS programme have included constructing the station itself, a mammoth task, and then just keeping the spacecraft systems functioning as they should. This had demanded regular maintenance and frequent repairs of equipment both inside and outside the station. On many occasions those repairs have only been possible after replacement parts have been delivered to the station on the next Progress, or the next Shuttle flight. If Project Constellation establishes a permanent base on the lunar surface that capability to supply spare parts in real time might still exist, but it will not be possible on the initial human flights to Mars. The lessons learnt from each failure on ISS must be applied in full to future spacecraft in an attempt to increase reliability to the point that the Mars crew's lives are not put in danger because of a minor mechanical failure.

Science on ISS has proved the Russian experience on Salyut and Mir, suggesting that two crew members are required just to perform maintenance and a third to perform science, is not necessarily correct. The automation of many of the experiments on ISS, allowing them to run in the background, without any regular input from the crew, along with the capability to have ground controllers activate and de-activate experiments on the station has relieved the crew of many tedious tasks.

Those experiments that do require a human input are managed as part of the flight plan and performed alongside the maintenance tasks. Many of the experiments performed on ISS have possible applications in future spacecraft. These include engineering experiments, materials exposure to see which materials perform best in the space environment, and plant growth experiments which might one day provide fresh food and even a natural oxygen production system as part of a hybrid Life Support System.

Following the announcement of Project Constellation in 2004, NASA concentrated its experiment programme on subjects directly applicable to long-duration spaceflight. This included experiments to show how the human body behaves in space and what adaptations are made naturally during long-term exposure to that environment. In short, NASA does not want to send astronauts to the surface of Mars and find that they are unable to function when they get there. Experiments on ISS might show that that will not happen or, more importantly, they might show how to prevent that from happening.

It remains to be seen if Project Constellation will be a national project by the richest and most technologically advanced nation on the Earth, or if the experience of the ISS programme will encourage other nations to join the quest. Will the human race leave Earth and explore the solar system, or will it be left to America to return to the Moon and press on to Mars alone?

America explores space because of what the space programme gives back to America. It encourages students to study the sciences and engineering, subjects that they will need if they want to be a part of the space programme. The demand for these subjects ensures that colleges and universities across the nation will teach them to the highest standards. Better educated students can only be a good thing for the future of America. In developing hardware and computer software for spaceflight, America's private companies expand their experience and their knowledge. The demands of human spaceflight regularly lead to new manufacturing techniques and new materials. New management techniques and new skills on the shop floor, as well as technological spin-off from the original products developed for spaceflight all add to America's manufacturing base, and that in its turn helps to improve their national economy. Large space programmes employ huge numbers of people across a vast range of skills. Those workers are paid by their companies, and they spend that money inside America, again helping to improve the national economy. Achievements accomplished inside a highly visible space programme add to America's national prestige around the world. Surely, America cannot be the only nation to see the upside of this arrangement.

So, what of ISS now that NASA is planning how to return to the Moon?

No longer is the International Space Station an end in itself, it has more meaning now than it ever did in the past. The ISS has become the Space Station that Werner von Braun foresaw in the famous articles that he wrote for *Colliers Magazine* in the early 1950s. The International Space Station has finally become an important stepping stone on the long road that is the age-old human desire to leave Earth and explore the solar system.

Appendix A

List of abbreviations

The following is a generic list of ISS acronyms. Not all of these acronyms are used in this volume:

RAM	**Forward, in the direction of travel**
WAKE	**Aft, opposite to ram**
ZENITH	**Directly above, space facing**
NADIR	**Directly below, Earth facing**
PORT	**Left side, opposite to starboard**
STARBOARD	**Right side, opposite to port**

2B	ISS SAW designation
4B	ISS SAW designation
AASS	Alternative Access to Space Station (US launch vehicle development programme)
AC	Assembly Complete
ACES	Atomic Clock Ensemble in Space
ACRV	Assured Crew Return Vehicle
ACU	Arm Control Unit
ADUM	Advanced Diagnostic Ultrasound in Microgravity experiment
ADVASC	ADVanced AStroCulture experiment
AEB	*Agência Espacial Brasiliera*
AFB	Air Force Base
ALTEA	Anomalous Long-Term Effects in Astronauts' central nervous system
AM	Airlock Module
ANDE	Atmospheric Neutral Density Experiment mirco-satellite
APAS	Androgynous Peripheral Attach System

APM	American Propulsion Module
APU	Auxiliary Power Unit
ARC	Ames Research Centre
ARIS	Active Rack Isolation System
ARIS-ICE	Active Rack Isolation System ISS Characterisation Experiment
ASC	*L'Agence Spatiale Canadienne*
ATCS	Active Thermal Control System
ATD	Advance Technology Demonstrator
ATLAS	APAS-To-LIDS Adapter System
atm	Atmospheres (a measure of pressure)
ATO	Abort To Orbit (Shuttle)
ATV	Autonomous Transfer Vehicle (European Space Agency)
ATV-CC	Automated Transfer Vehicle Control Centre
BCA	Battery Control Assembly
BCAT	Binary Colloidal Alloy Test
BCDU	Battery Charge/Discharge Unit
BGA	Beta Gimbal Assembly
BPS-PESTO	Biomass Production System-Photosynthesis Experiment and System Testing Operation
BSA	Battery Stowage Assembly
BSTC	Biotechnology Specimen Temperature Controller
BTR	BioTechnology Refrigerator
C&C	Command & Control
CAIB	Columbia Accident Investigation Board
CALET	CALorimetric Electron Telescope
CAM	Centrifuge Accommodation Module
Capcom	Capsule communicator
CAPL	CApillary Pumped Loop experiment
CBM	Common Berthing Mechanism
CBOSS	Cellular Biotechnology Operation Support System
CC	Control Centre
CCAA	Common Cabin Air Assembly
CCC	Contaminants Control Cartridge
CCTV	Crew & Cargo Transfer Vehicle
CDRA	Carbon Dioxide Removal Assembly (Destiny)
CEO	Crew Earth Observation experiment
CETA	Crew and Equipment Translation Aid/Assembly
CEV	Crew Exploration Vehicle
CEVIS	Cycle Ergometer with Vibration Isolation System
CGBA	Commercial Generic Bio-processing Apparatus
CHeCS	Crew Health Care System
CLV	Crew Launch Vehicle (Ares-I)
CM	Command Module (Apollo)
CM	Crew Module (Orion)

CME	Coronal Mass Ejection (solar flare)
CMG	Control Moment Gyro
CMRS	Crew Medical Restraint System
CMS	CounterMeasures System
CNES	*Centre National d'Etudes Spatiales* (French Space Agency)
COF	Columbus Orbital Facility
COL-CC	Columbus Control Centre
COSPAR	Committee On SPAce Research
COTS	Commercial Orbital Transport System
CPCG-H	Commercial Protein Crystal Growth-High-density experiment
CPDS	Charged Particle Directional Spectrometer
CRPCM	Canadian Remote Power Controller Module
CRV	Crew Return Vehicle
CSA	Canadian Space Agency
CSA-CP	Compound Specific Analyser-Combustion Products
CSLM	Coarsening of Solid–Liquid Mixtures experiment
CTB	Cargo Transfer Bag
CTV	Crew Transfer Vehicle
CWC	Contingency Water Container
DAFT	Dust and Aerosol-measuring Feasibility Test
DARPA	Defence Advanced Research Projects Agency
DART	Demonstration of Autonomous Rendezvous Technology
DC	Direct Current
DC-X	Delta Clipper-Experimental
DCM	Display and Control Module
DCM	Docking Compartment Module
DCPCG	Dynamically Controlled Protein Crystal Growth experiment
DCSU	Direct Current Switching Unit
DDCU	DC–DC Converter Unit
DELTA	Dutch Expedition for Life Science, Technology and Atmospheric research
DFRC	Hugh L. Dryden Flight Research Centre
DLA	Drive Lock Assembly
DLR	German Aerospace Centre
DMS	Data Management System
DoD	Department of Defence
DOS (Russian)	Long Duration Orbital Station
DOSMAP	DOSimetric MAPping experiment
DOSTELS	DOSimetric TELescopeS
DOUG	Dynamic Onboard Ubiquitous Graphics programme
DPS	De-orbit Propulsion Stage
EarthKam	Earth Knowledge acquired by middle school experiment
EAS	Early Ammonia Servicer
ECLS	Environmental Control and Life Support System

ECS	Exercise Countermeasures System
ECU	Electronic Control Unit
EDO	Extended Duration Orbiter
EDR	European Drawer Rack
EDV (Russian)	Water Storage Container
EELV	Extended Expendable Launch Vehicle
EF	Exposed Facility
EHS	Environmental Health System
ELC	EXPRESS Logistics Carrier
ELDO	European Launch Development Organisation
ELM	Experiment Logistics Module
ELV	Expendable Launch Vehicle
EMCS	European Modular Cultivation System
EMU	Extravehicular Mobility Unit
EP	EXPRESS Pallet
EPM	European Physiology Module
EPO	Education Payload Operation
EPS	Electrical Power System
ERA	European Robotic Arm
ES-ATV	*Evolution Storable,* Automated Transfer Vehicle (Ariane-V)
ESA	European Space Agency
ESP	External Stowage Platform
ESPAD	External Stowage Platform Attachment Device
ESR	European Stowage Rack
ESRO	European Space Research Organisation
EST	Eastern Standard Time (US Eastern)
ESTEC	European Space Research and TEchnology Centre
ET	External Tank (Shuttle)
ETC	European Transport Carrier
ETR	Eastern Test Range
EuTEF	European Technology Exposure Facility
EVA	Extravehicular Activity
EVARM	Extravehicular Activity Radiation Monitoring
EWA	Emittance Wash Application
ExPCA	EXPRESS Carrier Avionics
EXPPCS	EXPeriment on Physics of Colloids in Space
EXPRESS	EXpedite and PRocessing of Experiments to Space Station
F	Fahrenheit
FDI	Fluid Dynamics Investigation experiment
FGB	Functional Cargo Block (Zarya)
FMS	Force Movement Sensor
FMT	Financial Management Team
FOOT	FOOT/Ground Reaction Forces experiment
FPP	Floating Potential Probe
FRAM	Flight Releasable Attachment Mechanism

FRGF	Flight Releasable Grapple Fixture
FSA	Roscosmos, Russian Federal Space Agency
FSL	Fluid Science Laboratory
FY	Fiscal Year (budgetary year)
GAO	Government Accounting Office
GASMAP	Gas Analyser System Metabolic Analysis Physiology experiment
GB	Gigabyte
GCM	Gas Calibration Module
GCTC	Yuri Gagarin Cosmonaut Training Centre
GET	Ground Elapsed Time
GLONASS	Russian GLObal NAvigation Satellite System
GMT	Greenwich Mean Time
GN&C	Guidance Navigation and Control computer
GPS	Global Positioning System
GRC	John H. Glenn Research Centre
GSC	Guiana Space Centre
HCOR	High-rate Communications Outage Recorder
HDTV	High Definition TeleVision
HEDS	Human Exploration and Development of Space
HMS	Health Maintenance System
HOPE	H-II Orbital Plane Experimental
HPGA	High Pressure Gas Assembly
HRD	Human Resource Development
HRF	Human Research Facility
HTCI	High Technology and Commercialization Initiative
HTV	H-II Transfer Vehicle (Japan)
HUT	Hard Upper Torso
IAE	Institute of Aeronautics and Space
IAF	International Astronautical Federation
IAU	Interface Umbilical Assembly
IBMP	Institute for BioMedical Problems
ICC	Integrated Cargo Carrier (ATV)
ICM	Interim Control Module
ICSU	International Council of Scientific Unions
IEA	Integrated Equipment Assembly
IGO	Inspector General's Office
ILEWG	International Lunar Exploration Working Group
IMCE	ISS Management & Cost Evaluation Task Force (Young Committee)
IMU	Inertial Measurement Unit
IMV	Inter-Module Ventilation Assembly
InSPACE	Investigating the Structure of Paramagnetic Aggregates from Colloidal Emulsions
INV	INVentory Management System

IRU	In-flight Refill Unit
ISAS	Institute of Space and Aeronautical Science
ISPR	International Standard Payload Rack
ISS	International Space Station
ISSI	In Space Soldering Investigation
ISTP	Integrated Space Transportation System
ISU	International Space University (Strasbourg, France)
ITA	Integrated Truss Assembly
ITS	Integrated Truss Structure
IUA	Interface Umbilical Assembly
IV-CPDS	Intra-Vehicular Charged Particle Directional Spectrometer
JAXA	Japan Aerospace eXploration Agency
JELC	Japanese Experiment Logistics Carrier
JEM	Japanese Experiment Module
JEM-CAM	Japanese Experiment Module-Centrifuge Accommodation Module
JEM-ELM	Japanese Experiment Module-Experiment Logistics Module
JEM-ELM-EF	Japanese Experiment Module-Experiment Logistics Module-Exposed Facility
JEM-ELM-ES	Japanese Experiment Module-Experiment Logistics Module-Exposed Section
JEM-ELM-PS	Japanese Experiment Module-Experiment Logistics Module-Pressurised Section
JEM-PM	Japanese Experiment Module-Pressurised Module
JEM-RMS	Japanese Experiment Module-Remote Manipulator System
JPL	Jet Propulson Laboratory
JSC	Lyndon B. Johnson Space Centre
KhSC	Khrunichev State Research and Production Space Centre
KSC	John F. Kennedy Space Centre
KURS (Russian)	Automated docking system for Soyuz and Progress
LAS	Launch Abort System (Orion)
LC	Launch Complex
LCVG	Liquid Cooling and Ventilation Garment
LEE	Latch End Effector
LEO	Low-Earth Orbit
LES	Launch Escape System
LF	Logistics Flight
LH2	Liquid Hydrogen
LIDS	Low Impact Docking System
LiOH	Lithium hydroxide
LMC	Lightweight Multi-Purpose Experiment Support Structure Carrier
LOCAD-PTS	Lab-On-a-Chip Application Development-Portable Test System
Loop-A	Low-temperature Loop

Loop-B	Moderate-temperature cooling Loop
LOX	Liquid OXygen
LSAM	Lunar Surface Access Module
LSS	Life Support System
M	Modified (Progress designation)
MACE	Mid-deck Active Control Experiment
MACH	Multiple Application Customised Hitchhiker (numbered)
MAMS	Microgravity Acceleration Measurement System
MAXI	Monitor of All-sky X-ray Image
MBS	Mobile Base System
MBSU	Main Bus Switching Unit
MCA	Major Constituents Analyser
MCC	Mission Control Centre
MCOR	Medium-rate Communication Outage Recorder
MCS	Motion Control System
MDM	Multiplexer–DeMultiplexer
MELFI	Minus Eight-degree Laboratory Freezer for ISS
MEM (Russian)	Scientific Energy Module (original)
MEP (Russian)	Scientific Energy Platform (updated)
MEPE	Micro-encapsulation Electrostatic Processing Experiment
MEPSI	Micro-Electromechanical System Based PICOSAT Inspector
METOX	Metal Oxide
MGBX	Microgravity Science GloveBoX
MIRTS	Russian acronym for Zarya's battery charge controllers
MISSE	Materials on International Space Station Experiment
MIT	Massachusetts Institute of Technology
MLE	Mid-deck Locker Equivalent
MLM	Multipurpose Laboratory Module
MLP	Mobile Launch Platform
MMOD	MicroMeteoroid/Orbital Debris
MMT	Mission Management Team
MMU	Mass Memory Unit
MOC	Mobile Servicing System Operations Complex
MPAC/SEEDs	Micro-PArticle Capture and Space Environment Exposure Devices
MPESS-ND	Multi-Purpose Experiment Support Structure, Non-Deployable
MPLM	Multi-Purpose Logistics Module
MSFC	George C. Marshall Space Flight Centre
MSG	Microgravity Science Glovebox
MSS	Mobile Servicing System
MT	Mobile Transporter
MTCL	Moderate Temperature Cooling Loop
MTL	Moderate Temperature Control Loop
NAC	NASA Advisory Council

NAL	National Aerospace Laboratory
NASA	National Aeronautics and Space Administration
NASDA	NAtional Space Development Agency of Japan
NAVSTAR	NAVigation Signal Timing and Ranging (US satellite)
NEEMO	NASA Extreme Environments Mission Operation project
NEP (Russian)	Science and Power Platform
NOAX	Non-Oxide Adhesive eXperimental
NPO (Russian)	Production Enterprise
NRC	National Research Council
NRL	United States Naval Research Laboratory
NSBRI	National Space Biomedical Research Institute
NTM	Non-Terrestrial Material
NTO	Nitrogen tetroxide
NTSC	National Television Standards Committee
OAMS	Orbital Attitude Manoeuvring System
OBPR	Office of Biological & Physical Research
OBSS	Orbiter Boom Sensor System
OCA	Orbiter Communication Adapter
ODS	Orbiter Docking System
OGS	Oxygen Generation System
OKB	Special Design Bureau-1
OMB	NASA Office of Management and Budget
OMS	Orbital Manoeuvring System
OPF	Orbiter Processing Facility
OPS	Early Almaz Space Station Module
ORCS	Orbiter Reaction Control System
ORU	Orbital Replacement Unit
OSF	Office of Space Flight
OSP	Orbital Space Plane
OSVS	Orbiter Space Vision System
OTD	Orbital Transfer Device
OVC	Oxygen Ventilation Circuit
P	Port
P&WR	Pratt & Whitney Rocketdyne
P-6	Port-6
PA	Pressurised Adapter (Zarya)
PAD	Pad Abort Demonstration/Demonstrator
PAS	Payload Attach Structure
PAYCOM	PAYload COMmunication Manager
PC-3	Plasma Crystal-3 experiment
PCAS	Passive Common Attach System
PCBA	Portable Clinic Blood Analyser
PCE	Proximity Communications Equipment
PCG-EGN	Protein Crystal Growth-Enhanced Gaseous Nitrogen Dewar
PCG-STES	Protein Crystal Growth-Single Thermal Enclosure System

PCU	Plasma Contractor Unit
PDA	Payload Disconnect Assembly
PDGF	Power and Data Grapple Fixture
PDU	Power Drive Unit
PFMI	Pore Formation and Mobility Investigation
PGBA	Plant Generic Bio-processing Apparatus
PHALCON	Power, Heating, Articulation, Lighting and CONtrol Officer
PIC	Pyrotechnic Initiator Controller
PICOSAT	Inspector
PK-3 (Russian/German)	Crystal-3 experiment
PLSS	Portable Life Support System
PM	Pressurised Module
PMA	Pressurised Mating Adapter
POA	Payload Orbital Replacement Unit Accommodation
POC	Payload Operation Centre (Huntsville)
POC	Primary Oxygen Circuit
PPA	Pump Package Assembly
PROX OPS	PROXimity OPerationS
PSA	Power Supply Assembly
PSC	Physiological Signal Conditioner
PSRD	Prototype Synchrotron Radiation Detector
PTAB (Russian)	Storage Battery Current Regulator System
PTCS	Passive Thermal Control System
PuFF	Pulmonary Function Facility
PVGF	Power Video Grapple Fixture
PVR	PhotoVoltaic Radiator
RACU	Russian–American Conversion Unit
RAFT	RAdar Fence Transponder satellites
RCC	Reinforced Carbon–Carbon
RCS	Reaction Control System
RED	Restive Exercise Device
RGA	Rate Gyro Assembly
RJMC	Rotary Joint Motor Controller
RLV	Reusable Launch Vehicle
RM	Research Module
RMS	Remote Manipulator System
ROEU-PDA	Remotely Operated Electrical Umbilical-Power Distribution Assembly
ROI	Return On Investment
ROKVISS	Robotics Component Verification on ISS
ROOBA	Recharge Oxygen Orifice Bypass Assembly
RP-1	Kerosene (rocket fuel)
RPC	Remote Power Controller
RPCM	Remote Power Controller Module
RPM	Revolutions Per Minute

RRM	Russian Research Module
RSA	Russian Space Agency
RSC	Rocket and Space Corporation
RSP	Re-supply Stowage Platform
RTFTG	Return To Flight Task Group
RTLS	Return To Launch Site (Shuttle abort mode)
RV	Re-entry Vehicle
S	Starboard
S&M	Services and Mechanisms
S-0	Starboard-0
SAFAR	Simplified Aid for Extravehicular Activity Rescue
SAMS	Space Acceleration Measurement System
SARJ	Solar Alpha Rotation Joint
SASA	S-band Antenna Structural Assembly
SAW	Solar Array Wing
SCC	Station Control Computers (six laptops)
SEMS	Space Experiment Module Satchel
SFOG	Solid Fuel Oxygen Generator (oxygen candle)
SFP	Space Flight Participant (commercial crew member/not a professional astronaut)
SGANT	Space to Ground ANTenna
SIGI	Space Integrated Global Positioning Satellite/Inertial Navigation System
SLI	Space Launch Initiative
SLP-D1	SpaceLab Pallet, Deployable 1
SLP-D2	SpaceLab Pallet, Deployable 2
SM	Service Module (Zvezda)
SMILES	Superconducting Sub-MIllimeter-wave Limb Emission Sounder -26-
SMMOD	Service Module (Zvezda) Micrometeoriod and Orbital Debris Shield
SNFM	Serial Network Flow Monitor
SOC	Shuttle Operations Co-ordinator
SOLAR	SOLAR Monitoring Observatory
SPD	Spool Positioning Device
SPDM	Special Purpose Dexterous Manipulator
SPHERES	Synchronised Position Hold, Engage, Reorient Experiment Satellites
SRB	Solid Rocket Booster
SRR	Strategic Resources Review
SS	Space Shuttle
SSIPC	Space Station Integration and Promotion Centre
SSM	Russian acronym for Russian Docking and Stowage Module
SSME	Space Shuttle Main Engine

SSPF	Space Station Processing Facility
SSPTS	Station-to-Shuttle Power Transfer System
SSRMS	Space Station Remote Manipulator System
SSTO	Single Stage To Orbit
SSU	Sequential Shunt Unit
STC	Scientific and Technological Center (Moscow)
STS	Space Transportation System (Space Shuttle)
STSPTS	Station To Shuttle Power Transfer System
SUBSA	Solidification Using a Baffle in Sealed Ampules
SUITSAT	Suit Satellite
SVS	Space Vision System
T	Time of launch
$T+$	Time after launch (ground-elapsed time)
$T-$	Time before launch (countdown to launch)
TCS	Thermal Control System
TDRS	Tracking and Data Relay Satellite
TEPC	Tissue Equivalent Proportional Counter
TESS	TEmporary Sleep Station (Destiny)
TI	Terminal (phase) Initiation (Shuttle rendezvous manoeuvre)
TKS (Russian)	Transport Supply Ship
TKSC	Tsukuba Space Centre
TLD	Titan Launch Dispenser
TM	Transport Modified (Soyuz spacecraft)
TMA	Transport Modified-Anthropometric (Soyuz spacecraft)
TMG	Thermal Micrometeoroid Garment
TNSC	TaNegashima Space Centre
TORU (Russian)	Remote control (manual) docking system for Soyuz and Progress
TPS	Thermal Protection System
TRAC	Test of Reaction and Adaptation Capabilities experiment
TRRJ	Thermal Radiator Rotary Joint
TSC	Tele-science Support Centre
TsNIM	Central Research and Development Institute of Machine Building
TSS	Temporary Sleep Station
TSTO	Two-Stage To Orbit
TsUP	Mission Control Centre (Korolev, Moscow)
TUS	Trailing Umbilical System
TVIS	Treadmill with Vibration Isolation System
UABSA	Solidification Using A Baffle in Sealed Ampules
UCCAAS	Unpressurised Cargo Carrier Assembly Attachment System
UDM	Universal Docking Module
UDMH	Unsymmetrical DiMethyl Hydrazine
UF	Utility Flight
ULC	Unpressurised Logistics Carrier

ULF	Utilisation and Logistics Flight
UMA	Umbilical Mating Assembly
URM-D (Russian)	Universal Work Platform
USAF	United States Air Force
USHM	United States Habitat Module
USOC	User Support and Operations Centre
USOS	United States Operating Segment
UTMB	University of Texas Medical Branch
VAB	Vertical Assembly Building
VDC	Voltage, Direct Current
VDU	Video Distribution Unit
VIS	Vibration Isolation System (part of astronauts' treadmill)
VOA	Volatile Organic Analyser
WETA	Wireless Video System External Transceiver Antenna
WRS	Water Recovery System
X	eXperimental aircraft (American designation)
XPOP	*x*-axis Perpendicular to Orbital Plane
YVV	*y*-axis in Velocity Vector
Z	Zenith
ZCG	Zeolite Crystal Growth experiment

Appendix B

International Space Station Flight Log

Note: Number in brackets after Progress designations is number used in NASA press releases.

STS-108			
Endeavour			
Launched	17:19	05 December 2001	From Launch Complex 39, Pad B
Docked	15:03	07 December 2001	At PMA-2 on Destiny's ram
Undocked	12:28	15 December 2001	
Recovered	12:55	17 December 2001	At KSC
Objective	The first "utilisation" flight for the Destiny laboratory. Delivered Expedition-4 crew to ISS and returned the Expedition-3 crew to Earth. MPLM Raffaello was docked to Unity's nadir CBM at 12:55, December 8 and undocked at 14:20, December 14, 2001		

Progress M1-8 (7)			
Launched	15:13	21 March 2002	By Soyuz launch vehicle
Docked (hard)	15:58	24 March 2002	At Zvezda's wake
Undocked	04:23	25 June 2002	
De-orbited		25 June 2002	
Objective	Deliver logistics and propellant and take away trash		

STS-110			
Atlantis			
Launched	16:44	08 April 2002	From Launch Complex 39, Pad B
Docked	11:05	10 April 2002	At PMA-2 on Destiny's ram
Undocked	14:31	17 April 2002	
Recovered	12:27	19 April 2002	At KSC
Objective	Following the STS-109 solo flight to service the Hubble Space Telescope, STS-110 installed the S-0 segment on Destiny at 09:45, 11 April. This started the assembly of the Integrated Truss Structure		

Soyuz TM-34			
Launched	02:26	25 April 2002	
Docked	03:56	27 April 2002	At Zarya's nadir
Undocked	15:44	09 November 2002	
Recovered	06:04	10 November 2002	
Objective	ACRV replacement. Spaceflight participant Mark Shuttleworth visited the station with two Russian cosmonauts. They returned in Soyuz TMA-33		
STS-111			
Launched	17:23	05 June 2002	
Docked (soft)	12:25	07 June 2002	
Undocked	10:32	15 June 2002	
Recovered	13:58	19 June 2002	
Objective	Deliver Expedition-5 crew and return Expedition-4 crew to Earth. Deliver logistics and propellant and take away trash. Deliver Mobile Base system for SSRMS. MPLM Leonardo docked to Destiny's starboard CBM at 17:30, June 8 and removed at 16:11, June 14. Three EVAs performed, two to install MBS and one to repair wrist roll joint on SSRMS		
Progress M-46 (8)			
Launched	13:37	26 June 2002	By Soyuz launch vehicle
Docked (hard)	03:01	29 June 2002	At Zvezda's wake
Undocked	09:58	24 September 2002	
De-orbited		14 October 2002	
Objective	Deliver logistics and propellant and take away trash		
Progress M1-9 (9)			
Launched	13:37	25 September 2002	By Soyuz launch vehicle
Docked (hard)	13:07	29 September 2002	At Zvezda's wake
Undocked	10:59	01 February 2003	
Objective	Deliver logistics and propellant and take away trash		
STS-112			
Launched	15:46	07 October 2002	
Docked (soft)	11:17	09 October 2002	
Undocked	09:13	16 October 2002	
Recovered	11:44	18 October 2002	
Objective	Delivered S-1 truss and fixed it to the starboard side of S-0 Rruss at 09:36, October 10. Three EVAs completed to connect the two units together		
Soyuz TMA-1			
Launched	22:11	29 October 2002	
Docked	00:01	01 November 2002	At Pirs Docking Module
Undocked	18:40	03 May 2003	
Recovered	22:07	03 May 2003	
Objective	ACRV replacement. Returned Expedition-6 crew to Earth.		

STS-113			
Launched	19:50	23 November 2002	
Docked (soft)	16:59	25 November 2002	
Undocked	15:05	02 December 2002	
Recovered	14:37	07 December 2002	
Objective	Delivered P-1 Truss and fixed it to the port side of S-0 Truss. Three EVAs completed to connect the two units together		
Progress M-47 (10)			
Launched	07:59	02 February 2003	By Soyuz launch vehicle
Docked (hard)	09:49	04 February 2003	At Zvezda's wake
Undocked	18:48	27 August 2003	
De-orbited		27 August 2003	
Objective	Deliver logistics and propellant and take away trash		
Soyuz TMA-2			
Launched	23:54	26 April 2003	
Docked	01:56	28 April 2003	At Zarya's nadir
Undocked	18:17	27 October 2003	
Recovered	21:41	27 October 2003	
Objective	Two-man Expedition-7 crew up and down		
Progress M1-10 (11)			
Launched	06:34	08 June 2003	By Soyuz launch vehicle
Docked (hard)	07:15	11 June 2003	At Pirs
Undocked	15:42	04 September 2003	
De-orbited		03 October 2003	
Objective	Deliver logistics and propellant and take away trash		
Progress M-48 (12)			
Launched	21:48	29 August 2003	By Soyuz launch vehicle
Docked (hard)	23:40	30 August 2003	At Zvezda's wake
Undocked	03:36	30 January 2004	
De-orbited		30 January 2004	
Objective	Deliver logistics and propellant and take away trash		
Soyuz TMA-3			
Launched	01:38	18 October 2003	
Docked	03:16	20 October 2003	At Pirs
Undocked	16:52	28 April 2004	
Recovered	20:12	28 April 2004	
Objective	Two-man Expedition-8 crew up and down. ESA astronaut launched on Soyuz TMA-4 returned to Earth on Soyuz TMA-3		
Progress M1-11 (13)			
Launched	06:58	29 January2004	By Soyuz launch vehicle
Docked (hard)	08:13	31 January 2004	At Zvezda's wake
Undocked	05:19	24 May 2004	
De-orbited		03 May 2004	
Objective	Deliver logistics and propellant and take away trash. Progress used as test-bed to monitor vibrations for future Progress microgravity experiment platforms.		

Soyuz TMA-4			
Launched	23:19	18 April 2004	
Docked	12:01	21 April 2004	At Zarya's nadir
Undocked	18:08	23 October 2004	
Recovered	20:36	23 October 2004	
Objective	Two-man Expedition-9 crew up and down. ESA astronaut up returned on Soyuz TMA-3		
Progress M-49 (14)			
Launched	08:34	25 May 2004	
Docked	09:55	27 May 2004	At Zarya's nadir
Undocked	02:05	30 July 2004	
Objective	Deliver logistics and propellant and take away trash		
Progress M-50 (15)			
Launched	01:03	11 August 2004	By Soyuz launch vehicle
Docked	01:01	14 August 2004	At Zvezda's wake
Undocked	14:37	22 December 2004	
Objective	Deliver logistics and propellant and take away trash		
Soyuz TMA-5			
Launched	23:06	13 October 2004	
Docked	00:16	16 October 2004	At Pirs
Undocked	14:41	24 April 2005	
Recovered	18:08	24 April 2005	
Objective	Two-man Expedition-10 crew up and down. ESA astronaut from Soyuz TMA-6 crew down. Undocked from Pirs at 04:29, November 29, 2004 and docked to Zarya's nadir CBM at 04:53, November 29, 2004. Manoeuvre carried out to clear Pirs airlock for EVA		
Progress M-51 (16)			
Launched	17:19	23 December 2004	By Soyuz launch vehicle
Docked (hard)	18:58	25 December 2004	At Zvezda's wake
Undocked	11:06	27 February 2005	
Objective	Deliver logistics and propellant and take away trash		
Progress M-52 (17)			
Launched	14:09	28 February 2005	
Docked	15:10	02 March 2005	At Zarya's nadir
Undocked	16:16	15 June 2005	
Objective	Deliver logistics and propellant and take away trash		
Soyuz TMA-6			
Launched	20:46	14 April 2005	
Docked	23:19	16 April 2005	At Pirs
Undocked	17:42	10 October 2005	
Recovered	21:09	10 October 2005	
Objective	Two-man Expedition-11 crew up and down. ESA astronaut down in Soyuz TMA-5. Undocked from Pirs at 06:38, July 19, 2005 and docked to Zarya's nadir CBM at 07:08, the same day. Manoeuvre carried out to clear Pirs airlock for EVA. Spaceflight participant from Soyuz TMA-7 crew returned to Earth in this spacecraft.		

Progress M-53 (18)			
Launched	19:09	16 June 2005	By Soyuz launch vehicle
Docked (hard)	20:42	18 June 2005	At Zvezda's wake
Undocked	06:26	07 September 2005	
Objective	Deliver logistics and propellant and take away trash		
STS-114			
Launched	10:39	26 July 2005	
Docked (soft)	07:18	28 July 2005	
Undocked	02:24	06 August 2005	
Recovered	08:11	09 August 2005	
Objective	Return-to-flight mission following loss of STS-107 in 2003. Deliver logistics and propellant and take away trash. MPLM Leonardo docked to Destiny's starboard CBM July 30 and removed August 5, 2005. Three EVAs performed.		
Progress M-54 (19)			
Launched	09:08	08 September 2005	By Soyuz launch vehicle
Docked (hard)	10:42	10 September 2005	At Zvezda's wake
Undocked	06:06	03 March 2006	
Objective	Deliver logistics and propellant and take away trash.		
Soyuz TMA-7			
Launched	23:55	30 September 2005	
Docked	01:27	03 October 2005	On Pirs
Undocked	17:28	08 April 2006	
Recovered	20:48	08 April 2006	
Objective	ACRV replacement. This Expedition crew returned to Earth in Soyuz TMA-7. Spaceflight participant launched on this flight returned to Earth on Soyuz TMA-6. Undocked from Pirs at 03:46, November 18, 2005 and docked to Zarya's nadir port at 14:05 the same day, to clear Pirs for the crew's second EVA. Undocked from Zarya's nadir port at 02:49, March 20, 2006 and docked to Zvezda's wake port at 03:11, the same day, to make way for Soyuz TMA-8. Returned Brazilian astronaut from Soyuz TMA-8 crew to Earth.		
Progress M-55 (20)			
Launched	13:38	20 December 2005	By Soyuz launch vehicle
Docked (hard)	14:46	23 December 2005	At Pirs
Undocked	10:06	19 June 2006	
Objective	Deliver logistics and propellant and take away trash.		
Soyuz TMA-8			
Launched	21:30	29 March 2006	
Docked	22:19	01 April 2006	At Zarya's nadir
Undocked	17:53	28 September 2006	
Recovered	21:13	28 September 2006	
Objective	ACRV replacement. Launched first two members of Expedition-13 crew. This Expedition crew returned to Earth in Soyuz TMA-8. The third crew member, a Brazilian astronaut, returned to Earth on Soyuz TMA-7		

Progress M-56 (21)			
Launched	12:03	24 April 2006	By Soyuz launch vehicle
Docked (hard)	13:41	26 April 2006	At Zvezda's wake
Undocked	20:30	19 September 2006	
Objective	Deliver logistics and propellant and take away trash		
Progress M-57 (22)			
Launched	11:08	24 June 2006	By Soyuz launch vehicle
Docked (hard)	12:25	26 June 2006	At Pirs
Undocked	18:28	16 January 2007	
Objective	Deliver logistics and propellant and take away trash		
STS-121			
Launched	14:38	04 July 2006	
Docked (soft)	10:52	06 July 2006	
Undocked	06:08	15 July 2006	
Recovered	09:15	16 July 2006	
Objective	Carry MPLM Leonardo to ISS to restock stores. Leonardo docked to Destiny at 08:15, July 7, and undocked at 09:32, July 14, 2006. Three EVAs performed. Delivered third member of Expedition-13/14 crews.		
STS-115			
Launched	11:15	09 September 2006	
Docked (soft)	06:48	11 September 2006	
Undocked	08:50	17 September 2006	
Recovered	06:21	21 September 2006	
Objective	Deliver and install P-3/P-4 Truss, Solar Alpha Rotary Joint, and radiator.		
Soyuz TMA-9			
Launched	00:09	18 September 2006	
Docked	01:21	20 September 2006	At Zarya's nadir
Undocked	05:11	27 April 2007	
Recovered	08:31	27 April 2007	
Objective	ACRV replacement. Launched two members of Expedition-14 crew. Third member was already in place having been launched on STS-121. This Expedition crew returned to Earth in Soyuz TMA-9. The third crew member at launch, an American spaceflight participant, returned to Earth on Soyuz TMA-8. Undocked from Zvezda's wake port at 15:14, October 10, 2006 and docked to Zarya's nadir port at 15:34, the same day, to make way for Progress M-59. It was undocked from Zarya's nadir at 18:30 March 29 and docked to Zvezda's wake at 18:54, the same day.		
Progress M-58 (23)			
Launched	09:41	23 October 2006	By Soyuz launch vehicle
Docked (soft)	10:29	26 October 2006	At Zvezda's wake
Docked (hard)	13:00	26 October 2006	
Undocked	14:11	27 March 2007	
Deorbited		27 March 2007	
Objective	Deliver logistics and propellant and take away trash.		

STS-116			
Launched	20:47	09 December 2006	
Docked (soft)	17:12	11 December 2006	
Undocked	17:10	19 December 2006	
Recovered	05:32	22 December 2006	
Objective	Deliver P-5 Spacer Truss. Re-wire station's electrical system and bring P-4 Truss (delivered by STS-115) on-line		

Progress M-59 (24)			
Launched	21:12	17 January 2007	By Soyuz launch vehicle
Docked (hard)	21:59	19 January 2007	At Pirs
Undocked	18:07	01 August 2007	
Deorbited		01 August 2007	
Objective	Deliver logistics and propellant and take away trash		

Soyuz TMA-10			
Launched	13:31	07 April 2007	
Docked	15:10	09 April 2007	At Zarya's nadir
Undocked	03:14	21 October 2007	
Recovered	06:37	21 October 2007	
Objective	Deliver Expedition-15 crew/CRV replacement. This crew returned to Earth in Soyuz TMA-10. The spaceflight participant carried at launch returned to Earth on Soyuz TMA-9. Undocked from Zarya's nadir at 14:20, September 27, 2007 and relocated to Zvezda's wake, docking at 14:47, September 27, 2007		

Progress M-60 (25)			
Launched	23:25	11 May 2007	By Soyuz launch vehicle
Docked (hard)	01:10	15 May 2007	At Zvezda's wake
Undocked	19:37	18 September 2007	
De-orbited		25 September 2007	
Objective	Deliver logistics and propellant and take away trash		

STS-117			
Launched	19:38	08 June 2007	
Docked (soft)	15:36	10 June 2007	At Destiny's ram
Undocked	10:42	20 June 2007	
Recovered	16:49	22 June 2007	
Objective	Deliver and install S-3/S-4 Truss, Solar Alpha Rotary Joint, and radiator. Deliver Clayton Anderson to Expedition-15 crew. Recover Sunita Williams.		

Progress M-61 (26)			
Launched	13:33	02 August 2007	By Soyuz launch vehicle
Docked (hard)	14:40	05 August 2007	At Pirs' nadir
Undocked		2007	
De-orbited		2007	
Objective	Deliver logistics and propellant and take away trash		

STS-118			
Launched	18:36	08 August 2007	
Docked (soft)	14:02	10 August 2007	At Destiny's ram.
Undocked	07:56	19 August 2007	
Recovered	12:33	21 August 2007	
Objective	Deliver and install S-5 Truss. Test Station-to-Shuttle Power transfer System. Deliver SpaceHab single module of logistics		
Soyuz TMA-11			
Launched	09:22	10 October 2007	
Docked	10:50	12 October 2007	At Zarya's nadir
Undocked		2007	
Recovered		2007	
Objective	Deliver Expedition-16 crew/CRV replacement. This crew returned to Earth in Soyuz TMA-11. The spaceflight participant carried at launch returned to Earth on Soyuz TMA-10		
STS-120			
Launched	11:38	23 October 2007	
Docked (soft)	08:35	25 October 2007	
Undocked	05:32	05 November 2007	
Recovered	13:01	07 November 2007	
Objective	Deliver Node-2, Harmony, and temporarily install it on unity. Relocate P-6 ITS and deploy SAWs.		

All times US Eastern.

Appendix C

Extravehicular activity

Flight and astronauts	*Start time*	*Date*	*End time*	*Date*	*Duration*
STS-108					
Godwin and Tani	12:52	09:12:01	15:04	09:12:01	4h 12 min
Expedition-4					
Onufrienko and Walz (Pirs)	15:59	14:01:02	22:02	14:01:02	6h 03 min
Onufrienko and Bursch (Pirs)	10:19	25:01:02	16:18	25:01:02	5h 59 min
Walz and Bursch (Quest)	06:38	20:02:02	12:25	20:02:02	5h 47 min
STS-110 (Quest)					
Smith and Walheim	10:36	11:04:02	18:24	11:04:02	7h 48 min
Ross and Morin	10:09	13:04:02	17:39	13:04:02	7h 30 min
Smith and Walheim	09:48	14:04:02	16:15	14:04:02	6h 27 min
Ross and Morin	10:29	16:04:02	17:06	16:04:02	6h 37 min
STS-111 (Quest)					
Chang-Diaz and Perrin	11:27	09:06:02	18:41	09:06:02	7h 14 min
Chang-Diaz and Perrin	11:20	11:06:02	16:20	11:06:02	5h 00 min
Chang-Diaz and Perrin	11:16	13:06:02	18:33	13:06:02	7h 17 min
Expedition-5 (Pirs)					
Korzun and Whitson	05:23	16:08:02	09:48	16:08:02	4 h 25 min
Korzun and Treschev	01:27	22:08:02	06:48	22:08:02	5 h 21 min
STS-112 (Quest)					
Wolf and Sellers	11:21	10:10:02	18:22	10:10:02	7 h 01 min
Wolf and Sellers	10:31	12:10:02	16:35	12:10:02	6 h 04 min
Wolf and Sellers	10:11	14:10:02	16:47	14:10:02	6 h 36 min
STS-113 (Quest)					
López-Alegría and Herrington	14:49	26:11:02	21:35	26:11:02	6 h 45 min
López-Alegría and Herrington	13:36	28:11:02	19:46	28:11:02	6 h 10 min
López-Alegría and Herrington	14:25	30:11:02	21:25	30:11:02	7 h 00 min

(*continued*)

Flight and astronauts	*Start time*	*Date*	*End time*	*Date*	*Duration*
Expedition-6 (Quest)					
Bowersox and Pettit	07:50	15:01:03	14:41	15:01:03	6 h 51 min
Bowersox and Pettit	08:40	08:04:03	15:06	08:04:03	6 h 26 min
Expedition-8 (Pirs)					
Foale and Kaleri	16:17	26:02:04	20:12	26:02:04	3 h 55 min
Expedition-9 (Pirs)					
Padalka and Fincke	17:56	24:06:04	18:10	24:06:04	0 h 14 min
Padalka and Fincke	17:19	30:06:04	22:59	30:06:04	5 h 40 min
Padalka and Fincke	02:58	03:08:04	07:28	03:08:04	4 h 30 min
Padalka and Fincke	12:43	03:09:04	18:04	03:09:04	5 h 21 min
Expedition-10 (Pirs)					
Chiao and Sharipov	02:43	26:02:05	08:11	26:02:05	5 h 28 min
Chiao and Sharipov	01:25	28:03:05	05:55	28:03:05	4 h 30 min
STS-114 (Discovery)					
Noguchi and Robinson	05:46	30:07:05	12:35	30:07:05	6 h 50 min
Noguchi and Robinson	04:42	01:08:05	23:56	01:08:05	7 h 14 min
Noguchi and Robinson	04:48	03:08:05	10:49	03:08:05	6 h 01 min
Expedition-11 (Pirs)					
Krikalev and Phillips	03:02	18:08:05	20:00	18:08:05	4 h 58 min
Expedition-12 (Quest)					
McArthur and Tokarev	10:32	07:11:05	15:54	07:11:05	5 h 22 min
McArthur and Tokarev	18:44	03:02:06	00:27	04:02:06	5 h 43 min
Expedition-13					
Vinogradov and Williams (Pirs)	18:48	01:07:06	01:19	02:06:06	6 h 31 min
Williams and Reiter (Quest)	10:04	03:08:06	16:58	03:08:06	5 h 54 min
STS-121 (Quest)					
Sellers and Fossum	09:17	08:07:06	16:48	08:07:06	7 h 31 min
Sellers and Fossum	08:14	10:07:06	15:01	10:07:06	6 h 47 min
Sellers and Fossum	06:20	12:07:06	14:31	12:07:06	7 h 11 min
STS-115 (Quest)					
Stafanyshyn-Piper and Tanner	05:17	12:09:06	11:43	12:09:06	6 h 26 min
Burbank and MacLean	05:05	13:09:06	12:16	13:09:06	7 h 11 min
Stafanyshyn-Piper and Tanner	06:00	15:09:06	12:42	15:09:06	6 h 42 min
Expedition-14					
López-Alegría and Tyurin (Pirs)	19:17	22:11:06	01:53	23:11:06	5 h 38 min
López-Alegría and Williams (Quest)	11:14	31:01:07	18:09	31:01:07	7 h 55 min
López-Alegría and Williams (Quest)	08:38	04:02:07	15:49	04:02:07	7 h 11 min
López-Alegría and Williams (Quest)	08:26	08:02:07	15:06	08:02:07	6 h 40 min
López-Alegría and Tyurin (Pirs)	05:27	22:02:07	11:45	22:02:07	6 h 18 min

Flight and astronauts	*Start time*	*Date*	*End time*	*Date*	*Duration*
STS-116 (Quest)					
Curbeam and Fuglesang	15:31	12:12:06	22:07	12:12:06	6 h 36 min
Curbeam and Fuglesang	14:41	14:12:06	19:41	14:12:06	5 h 00 min
Curbeam and Williams	14:25	16:12:06	21:56	16:12:06	7 h 31 min
Curbeam and Fuglesang (unplanned)	14:12	18:12:06	00:44	19:12:06	6 h 38 min
Expedition-15					
Yurchikhin and Kotov (Pirs)	15:05	30:05:07	20:30	30:05:07	5 h 25 min
Yurchikhin and Kotov (Pirs)	10:23	06:06:07	16:00	06:06:07	5 h 37 min
Yurchikhin and Anderson (Quest)	07:25	23:07:07	15:06	23:07:07	7 h 41 min
STS-117					
Reilly and Olivas	16:02	11:06:07	22:17	11:06:07	6 h 15 min
Forrester and Swanson	14:03	13:06:07	20:33	13:06:07	7 h 16 min
Reilly and Olivas	13:38	15:06:07	20:40	15:06:07	7 h 58 min
Forrester and Swanson	12:25	17:06:07	18:54	17:06:07	6 h 29 min
STS-118 (Quest)					
Mastracchio and Williams	12:28	11:08:07	18:45	11:08:07	6 h 17 min
Mastracchio and Williams	11:32	13:08:07	18:00	13:08:07	6 h 28 min
Mastracchio and Anderson	10:38	15:08:07	16:05	13:08:07	5 h 28 min
Williams and Anderson	09:17	18:08:07	14:19	18:08:07	5 h 02 min
STS-120 (Quest)					
Parazynski and Wheelock	06:02	26:10:07	12:16	26:10:07	6 h 14 min
Parazynski and Tani	05:32	28:10:07	12:05	28:10:07	6 h 33 min
Parazynski and Wheelock	05:45	30:10:07	12:53	30:10:07	7 h 08 min
Parazynski and Wheelock	06:03	03:11:07	13:43	03:10:07	7 h 19 min
Expedition-16 (Quest)					
Whitson and Malenchenko	03:54	09:11:07	10:49	09:11:07	6 h 55 min
Whitson and Tani	05:10	20:11:07	12:26	20:11:07	7 h 16 min
Whitson and Tani	04:50	24:11:07	11:54	24:11:07	6 h 04 min

All times shown are US Eastern.

Index

Printing: Mercedes-Druck, Berlin
Binding: Stein + Lehmann, Berlin